99

Advances in Biochemical Engineering/Biotechnology

Series Editor: T. Scheper

Advances in Biochemical Engineering/Biotechnology

Series Editor: T. Scheper

Recently Published and Forthcoming Volumes

Biotechnology for the Future
Volume Editor: Nielsen, J.
Vol. 100, 2005

Gene Therapy and Gene Delivery Systems
Volume Editors: Schaffer, D. V., Zhou, W.
Vol. 99, 2005

Sterile Filtration
Volume Editor: Jornitz, M. W.
Vol. 98, 2006

Marine Biotechnology II
Volume Editors: Le Gal, Y., Ulber, R.
Vol. 97, 2005

Marine Biotechnology I
Volume Editors: Le Gal, Y., Ulber, R.
Vol. 96, 2005

Microscopy Techniques
Volume Editor: Rietdorf, J.
Vol. 95, 2005

Regenerative Medicine II
Clinical and Preclinical Applications
Volume Editor: Yannas, I. V.
Vol. 94, 2005

Regenerative Medicine I
Theories, Models and Methods
Volume Editor: Yannas, I. V.
Vol. 93, 2005

Technology Transfer in Biotechnology
Volume Editor: Kragl, U.
Vol. 92, 2005

Recent Progress of Biochemical and Biomedical Engineering in Japan II
Volume Editor: Kobayashi, T.
Vol. 91, 2004

Recent Progress of Biochemical and Biomedical Engineering in Japan I
Volume Editor: Kobayashi, T.
Vol. 90, 2004

Physiological Stress Responses in Bioprocesses
Volume Editor: Enfors, S.-O.
Vol. 89, 2004

Molecular Biotechnology of Fungal β-Lactam Antibiotics and Related Peptide Synthetases
Volume Editor: Brakhage, A.
Vol. 88, 2004

Biomanufacturing
Volume Editor: Zhong, J.-J.
Vol. 87, 2004

New Trends and Developments in Biochemical Engineering
Vol. 86, 2004

Biotechnology in India II
Volume Editors: Ghose, T. K., Ghosh, P.
Vol. 85, 2003

Biotechnology in India I
Volume Editors: Ghose, T. K., Ghosh, P.
Vol. 84, 2003

Proteomics of Microorganisms
Volume Editors: Hecker, M., Müllner, S.
Vol. 83, 2003

Biomethanation II
Volume Editor: Ahring, B. K.
Vol. 82, 2003

Biomethanation I
Volume Editor: Ahring, B. K.
Vol. 81, 2003

Gene Therapy and Gene Delivery Systems

Volume Editors: David V. Schaffer, Weichang Zhou

With contributions by
N. E. Altaras · J. G. Aunins · M. E. Davis · R. K. Evans · M. Ferrari
A. Geall · B. Goff · J. C. Grieger · J. Heidel · G. Hermanson · P. Hobart
A. Kamen · J. O. Konz · R. Langer · S. R. Little · N. Loewen
M. Manthorpe · S. Mishra · E. M. Poeschla · A. Rolland
R. J. Samulski · D. V. Schaffer · J. J. Wolf · J. H. Yu · W. Zhou

Springer

Advances in Biochemical Engineering/Biotechnology reviews actual trends in modern biotechnology. Its aim is to cover all aspects of this interdisciplinary technology where knowledge, methods and expertise are required for chemistry, biochemistry, micro-biology, genetics, chemical engineering and computer science. Special volumes are dedicated to selected topics which focus on new biotechnological products and new processes for their synthesis and purification. They give the state-of-the-art of a topic in a comprehensive way thus being a valuable source for the next 3–5 years. It also discusses new discoveries and applications. Special volumes are edited by well known guest editors who invite reputed authors for the review articles in their volumes.
In references *Advances in Biochemical Engineering/Biotechnology* is abbeviated *Adv Biochem Engin/Biotechnol* and is cited as a journal.

Springer WWW home page: http://www.springeronline.com
Visit the ABE content at http://www.springerlink.com/

Library of Congress Control Number: 2005933540

ISSN 0724-6145
ISBN-10 3-540-28404-4 Springer Berlin Heidelberg New York
ISBN-13 978-3-540-28404-8 Springer Berlin Heidelberg New York
DOI 10.1007/11542766

Springer is a part of Springer Science+Business Media

springer.com

Printed in Germany

Cover design: *Design & Production* GmbH, Heidelberg
Typesetting and Production: LE-TEX Jelonek, Schmidt & Vöckler GbR, Leipzig

Printed on acid-free paper 02/3141 YL – 5 4 3 2 1 0

Series Editor

Prof. Dr. T. Scheper

Institute of Technical Chemistry
University of Hannover
Callinstraße 3
30167 Hannover,Germany
scheper@iftc.uni-hannover.de

Volume Editors

Dr. David V. Schaffer

Department of Chemical Engineering
and Helen Wills Neuroscience Institute
University of California
201 Gilman Hall
Berkeley, CA 94720-1462, USA
Schaffer@berkeley.edu

Dr. Weichang Zhou

Process Sciences and Engineering
Protein Design Labs, Inc.
34801 Campus Drive
Fremont, CA 94555, USA
wzhou@pdl.com

Editorial Board

Advances in Biochemical Engineering/Biotechnology Also Available Electronically

For all customers who have a standing order to Advances in Biochemical Engineering/Biotechnology, we offer the electronic version via SpringerLink free of charge. Please contact your librarian who can receive a password or free access to the full articles by registering at:

springerlink.com

If you do not have a subscription, you can still view the tables of contents of the volumes and the abstract of each article by going to the SpringerLink Homepage, clicking on "Browse by Online Libraries", then "Chemical Sciences", and finally choose Advances in Biochemical Engineering/Biotechnology.

You will find information about the

- Editorial Board
- Aims and Scope
- Instructions for Authors
- Sample Contribution

at springeronline.com using the search function.

Attention all Users of the "Springer Handbook of Enzymes"

Information on this handbook can be found on the internet at springeronline.com

A complete list of all enzyme entries either as an alphabetical Name Index or as the EC-Number Index is available at the above mentioned URL. You can download and print them free of charge.

A complete list of all synonyms (more than 25,000 entries) used for the enzymes is available in print form (ISBN 3-540-41830-X).

Contents

Adv Biochem Engin/Biotechnol (2005) 99: 1–5
DOI 10.1007/10_001

Published online: 8 November 2005

Gene Therapy and Gene Delivery Systems as Future Human Therapeutics

David V. Schaffer[1] (✉) · Weichang Zhou[2]

[1]Department of Chemical Engineering and Helen Wills Neuroscience Institute, University of California, 201 Gilman Hall, Berkeley, CA 94720-1462, USA
Schaffer@cchem.berkeley.edu

[2]Process Sciences and Engineering Protein Design Labs, Inc., 34801 Campus Drive, Fremont, CA 94555, USA
WZHOU@PDL.COM

Although its underlying concepts date back to the 1960s, gene therapy as a modern molecular medicine is a relatively young field that was born in 1989 with the transfer of a drug resistance gene marker to a patient's lymphocytes in the first gene transfer clinical trial [1]. This and other early trials, which utilized early generation retroviral vectors, provided evidence of safe *in vivo* gene transfer for potential therapeutic benefit but also highlighted the need for progress on many fronts, including deeper molecular knowledge of disease target pathologies, the development of enhanced therapeutic cargoes, further insights into gene delivery mechanisms, engineering of enhanced vector systems, and the improvement of vector production processes. This special volume of *Advances in Biochemical Engineering and Biotechnology* provides a broad, state of the art view of the modern field of gene therapy from academic and industrial viewpoints. It demonstrates that major progress has been made in many aspects including large scale GMP manufacturing technologies in the past decade and a half, which enable not only clinical trials, but also potentially commercialization of these vectors for widespread applications in humans.

Gene therapy, the delivery of therapeutic genes to a patient for therapeutic benefit, offers the potential for permanent cures of diseases. Since its inception, however, the biggest challenge in the field has revolved around the word "delivery". Nucleic acids are labile macromolecules that are challenging to deliver to the insides of cells, and a broad spectrum of vectors has accordingly been engineered to facilitate delivery. On one end are in a sense the most synthetic vectors, naked or unformulated plasmid DNA, which are under development for various targets including DNA vaccines as well as cancer and cardiovascular applications. Manthorpe et al. (pp.) discuss how, due to its relative simplicity, plasmid DNA systems have rapidly advanced, with major progress in the optimization of genetic elements for transgene expression, large scale GMP production, and vector administration. These plasmid DNA systems are now in clinical trials for numerous applications.

To expand the applicability of plasmid DNA to other tissues and disease targets, the DNA is complexed with synthetic components such as lipids and polymers, conceptually similar to the packaging of nucleic acids into a viral particle. Heidel et al. (pp.) discuss the basic mechanisms of DNA transport through cells and describe how polymeric systems can be designed to surmount barriers in cellular targeting and intracellular delivery. Furthermore, they raise the very important concept that as different modules or activities are added to a vector, from targeting to endosomal escape to nuclear transport, the vector must be assembled as a functionally integrated system that presents the correct capabilities at the correct junctures as it progresses inside a cell. In a sense, this approach will bring synthetic systems functionally closer to their viral counterparts. Furthermore, Little and Langer (pp.) discuss the application of synthetic systems to a very important class of applications, cancer vaccines. This application builds upon fundamental knowledge of immunology and antigen presentation to develop vaccine systems that harness cellular immunity, allowing the body to recognize a cancer as foreign and then eliminate it. Furthermore, this application perhaps best harnesses the current advantages of synthetic systems: high-level, safe, and transient gene expression.

At the other end of the vector spectrum, viral vectors have made significant progress in safety and efficiency. Adenoviral vectors, developed in the early 1990s, have progressed for a broad number of applications described by Altaras et al. (pp.). Highly advanced manufacturing processes using modern mammalian cell and virus propagation and sophisticated purification technologies, stable liquid formulation, and vector product release and characterization methods have been developed for first generation adenoviral vectors, serving as models for the development of large scale GMP production systems for other viral vectors. As a result, the extremely efficient delivery capabilities of these vectors can be harnessed towards potential commercial use for a number of widespread applications, including genetic vaccines and cancer therapeutics.

In parallel, viral vector systems have developed, becoming more synthetic in nature. In addition to adenoviral vectors "gutted" of all cargo, which Altaras et al discuss, new lentiviral and adeno-associated viral vector systems lacking all viral genes in the vector have advanced significantly during the late 1990s and early 2000s. As Loewen and Poeschla discuss, lentiviral vectors have emerged as an important vector system that exhibits the capability of very high efficiency for gene delivery to a broad range of cell types and tissues resulting in sustained gene expression. Furthermore, a number of safer vectors derived from nonhuman lentiviruses have been developed, and the resulting gene expression from integrated genomes does not appear to suffer from the transcriptional silencing that affected simple retroviral vectors.

These vectors have enjoyed success in a number of animal models and have entered into clinical trials.

Likewise, as discussed by Grieger and Samulski (pp.), since their inception in the late 1980s, adeno-associated viral vectors have enjoyed rapid development into very high efficiency vectors in numerous tissues and several clinical trials. The development of enhanced production and purification technologies for these vectors has facilitated its clinical development. The recent discovery of a large number of alternate adeno-associated virus (AAV) serotypes with different properties promises to yield a collection of vectors with different capabilities for application to a number of disease targets. Finally, since it is composed of DNA surrounded by a relatively simple protein shell, AAV is perhaps the mammalian virus conceptually most related to a synthetic particle.

As discussed by both Grieger and Samulski as well as Yu and Schaffer (pp.), it is not necessary to always "settle" for what nature has provided during viral vector development. To optimize viruses for human therapeutic applications, the vectors often require re-engineering. For example, is the development of rational approaches for targeted gene delivery, an effort that has benefited from improved basic knowledge of viral structure-function relationships. Furthermore, the development of library and directed evolution approaches mimics the natural process of viral evolution to create vectors with novel and attractive properties. Collectively, by engineering or manipulating the properties of viral systems at the molecular level, these approaches bring viruses closer to the level of molecular control over vector properties enjoyed by synthetic systems, making the spectrum between synthetic and viral systems continuous.

There are a number of challenges that remain in vector development and commercial manufacturing. Targeted gene delivery promises to minimize side effects and enhance therapeutic efficacy. Furthermore, to extend the expression duration of both synthetic and viral vector systems, vector integration or stable episomal maintenance is required, and progress in targeted integration and episome engineering will enhance the safety of this process. Finally, further enhancements in vector production systems will improve the safety and economics of gene medicines.

In parallel to vector development, there has been significant progress in the development of enhanced cargos. With the completion of the human genome sequence, gene therapy can serve as a direct conduit to translate basic knowledge of the molecular pathology underlying disease into direct therapeutic benefit. For example, as alluded to by Heidel et al, the delivery of nucleic acids to induce RNA interference promises to expand the therapeutic potential of clinical genetic medicines even further. To date, nearly 1000 gene therapy clinical trials have been conducted (http://www.wiley.co.uk/wileychi/genmed/clinical/). These have established that the vast majority of vectors and gene products are extremely safe for human use. In addition, there has been preliminary success in several trials, including trials for hemophilia, cancer and cardiovascular disease. Furthermore, the first gene therapy product, adenoviral vector delivery of the tumor suppressor p53 for cancer therapy, has been clinically approved in China.

Finally, fundamental advances in vector and cargo development promise to yield further successes as they are translated to the clinic. Consequently, considerable product, process, analytical, and formulation development challenges remain towards large scale GMP manufacturing of these vectors for clinical trials and potential commercial applications. Even though many critical development issues for manufacturing of first generation Adenovirus type 5 vectors and DNA plasmids have been successfully solved, as described in this special volume, efforts need to be focused on many next generation adenoviral and other viral vectors in development to enable large scale GMP production. In addition, more sophisticated analytical techniques need to be developed to enable the characterization of these viral vector products at the same level as well-characterized biologics such as monoclonal antibodies. These technologies would ensure a better fundamental understanding of these viral vector product characteristics and aid their production with consistent quality attributes.

There has sometimes been speculation that the field of gene therapy could be progressing faster. This is a healthy question to pose, and one for which deeper insights can be gained by examining the progression of other fields in biotechnology and molecular medicine. Since they were first derived thirty years ago in 1975, monoclonal antibodies (mAbs) have established themselves as a major new generation of human therapeutics, with 18 approved antibodies (one of which was withdrawn from the market in February, 2005) and two antibody fusion proteins to date in the United States by the FDA and a projected 2005 sales of more than 10 billion US dollars (http://www.fda.gov/cder/biologics/biologics_table.htm). However, hindsight reveals that the road to their development was not smooth. A recent review article discusses the fact that "By the end of the 1980s enthusiasm for therapeutic mAbs was waning. It was further eroded by the pharmaceutical problems of mAbs as they were expensive to produce, needed specialist expertise to administer, and were often associated with considerable toxicity." (Glennie and Johnson 2000) Fundamental advances in molecular engineering (e.g. humanization) and bioprocess development were successfully achieved to overcome many of these problems and transform early promises into commercial human therapeutics, which have been shown to be highly clinically beneficial in the treatment of various diseases.

Gene therapy product candidates, with their many differences from protein-based therapeutics, will be a new class of human therapeutics for clinical needs that are currently not being met. They promise a great future and in the meantime offer new and distinct challenges ranging from delivery to production. The creative and comprehensive work discussed in this special volume shows strong potential to provide the fundamental biology and engineering advances needed to overcome these challenges and deliver on the promise of clinical gene therapy.

References

1. Rosenberg SA, Aebersold P, Cornetta K, Kasid A, Morgan RA, Moen R, Karson EM, Lotze MT, Yang JC, Topalian SL et al (1990) Gene transfer into humans – immunotherapy of patients with advanced melanoma, using tumor-infiltrating lymphocytes modified by retroviral gene transduction. N Engl J Med 323(9):570–578
2. Glennie MJ, Johnson PW (2000) Clinical trials of antibody therapy. Immunol Today 21(8):403–410

Adv Biochem Engin/Biotechnol (2005) 99: 7–39
DOI 10.1007/10_002

Published online: 1 November 2005

Molecular Conjugates

Jeremy Heidel · Swaroop Mishra · Mark E. Davis (✉)

Chemical Engineering, 210-41, California Institute of Technology, Pasadena, CA 91125, USA
mdavis@cheme.caltech.edu

Abstract Molecular conjugates are nanometer-sized entities consisting of synthetic materials (lipids, polycations, targeting agents, and so on) and nucleic acids. These composites are delivery vehicles that function to provide the transport of nucleic acids to sites of action. Recently, great progress has been made in the construction of these nonviral delivery vehicles and the understanding of how they function in cells and animals. Here, we review some of the important issues in assembling molecular conjugates and understanding their behavior in biological fluids, cells, and animals. One of the largest challenges in the field of molecular conjugates is how to integrate the components into a workable system that exploits the combined attributes of the components without suffering losses due to the assembly of the system. We discuss some of the difficulties involved in the assembly of a functioning delivery system for in vivo use.

Keywords Nonviral · Polycations · Liposomes · Targeting · Nuclear localization sequence (NLS)

Abbreviations

ASGPR	asialoglycoprotein receptor
bp	base pairs
CD	cyclodextrin
EGFR	epidermal growth factor receptor
FR	folate receptor
h	hour(s)
IFN-γ	interferon-γ
IL-6	interleukin-6
IL-12	interleukin-12
i.v.	intravenous
min	minute(s)
NAD(P)H	dihydronicotinamide adenine dinucleotide (phosphate)
NF-κB	nuclear factor-κB
NLS	nuclear localization signal
nm	nanometer(s)
PBS	phosphate-buffered saline
pDNA	plasmid DNA
PEG	polyethylene glycol
PEI	polyethylenimine
shRNA	small hairpin RNA
siRNA	small interfering RNA
Tf	transferrin
TfR	transferrin receptor

1 Introduction

Nonviral gene delivery vectors must exhibit a variety of properties to achieve their function. First, they must condense nucleic acids into small particles approximately 100 nm or less in diameter, and confer protection from degrading factors that exist in serum and in cells. The gene delivery particles must be taken up by cells that have been targeted, direct the nucleic acids to an appropriate intracellular destination (cytoplasm for small interfering RNA (siRNA) and nucleus for plasmid DNA (pDNA)), release the nucleic acids to allow their action, and exhibit minimal toxicity (Fig. 1). Initially, many investigations addressed these objectives in a serial fashion. This series approach to the development of non-viral vectors has produced an array of different materials that combine with nucleic acids to form small particles that are capable of entering cells. Measured in terms of transfection efficiency, however, these materials have yet to rival viruses as gene delivery agents. Although the limiting intracellular barrier(s) to nonviral gene delivery remain(s) poorly defined, many recent approaches have sought to better understand and enhance: i) the escape from the endocytic pathway, ii) vector unpackaging, iii) cytoplasmic persistence of nucleic acids, and iv) nuclear delivery. A potential intracellular barrier that has not been extensively addressed is the poor cyto-

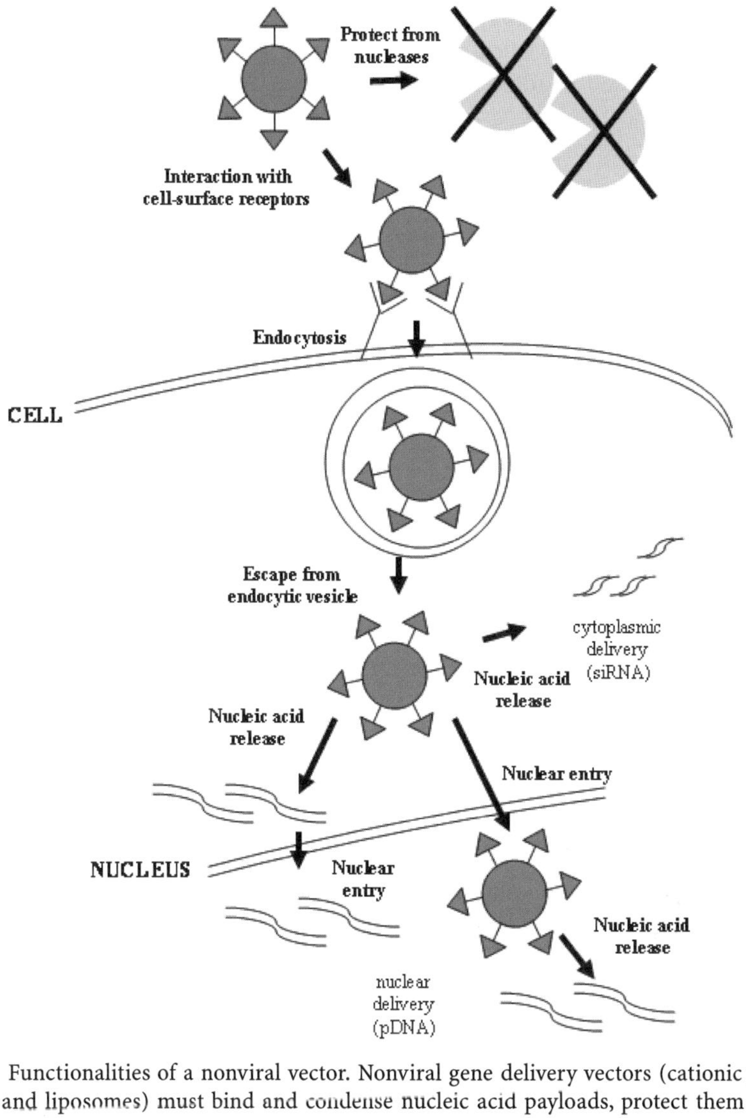

Fig. 1 Functionalities of a nonviral vector. Nonviral gene delivery vectors (cationic polymers and liposomes) must bind and condense nucleic acid payloads, protect them from nuclease degradation, allow for cellular uptake, permit escape from endocytic vesicles, and release the nucleic acids to permit their function. It remains unclear whether pDNA must be released from the vector prior to nuclear entry or should be directed to the cell nucleus within intact vector-pDNA complexes

plasmic mobility of nucleic acids above 2000 bp in length, and we discuss this issue in more detail below.

The synthetic nature of nonviral systems allows for facile, well-defined modification. Although readily-available materials have not yet provided an ideal vector, identification of their inadequacies has prompted rationally-designed modifications as well as the development of new materials. These changes have generally been made in a modular fashion, with new components incorporated to address specific barriers. These combinations of functional components point to the eventual construction of an engineered system for nonviral gene delivery. The serial approach is giving way to a systems approach where the various barriers to delivery are simultaneously considered in the materials design. For example, the initial emphasis on identifying materials that bind and condense nucleic acids may have underappreciated the importance of their subsequent intracellular release [1]. Attention has now turned to vectors with a weaker binding strength [2] or whose binding of nucleic acids is disrupted following cellular uptake [3–6].

Due to shortcomings such as high toxicity, instability to physiological salt and serum and poor transfection efficiency, early nonviral systems showed little in vivo applicability. Modifications of existing systems to address these problems have brightened the prospects for effective in vivo gene delivery. However, these alterations, often involving changes to the chemical structure of the delivery vectors or nucleic acids, have also affected gene delivery performance in unanticipated ways. For example, Mishra et al. demonstrated that conferring salt stability to polyethylenimine-DNA complexes with a poly(ethylene glycol) coating dramatically changes the morphology of the endocytosed entities and significantly reduces the resulting gene expression in vitro [7]. The need for a systems approach extends to in vitro investigations, as modifications intended for in vivo applicability can significantly affect both in vitro and in vivo performance.

To achieve in vivo applicability, non-viral vectors must be developed that can be prepared in a reproducible manner with defined composition and properties. For almost all nonviral vectors, there are no well-established formulations that give homogeneous vehicles, and the product characteristics are not defined in a consistent and quantitative manner. For example, it is likely that polycation-nucleic acid complexes formulated at a high charge ratio (ratio of positive charge centers on the polycation to negative charge centers to the nucleic acids) do not contain all the polycation in the complexed state [8]. Uncomplexed polycation may well contribute to the in vitro transfection efficiency of nonviral vectors, but cannot be expected to do so in vivo, where it would not necessarily be distributed in association with the complexes.

Human medical applications are likely possible with nonviral gene delivery vectors, provided that improvements can be made to give appropriate therapeutic indices. The advantages of these systems include an ability to avoid the DNA size limitations and immunogenicity that are possible with some viral vectors, and also straightforward formulation with easily-manufactured, relatively low-cost materials that will assist in providing large-scale therapeutics at acceptable costs.

Nonviral gene delivery remains an area of steady progress. Here, we discuss issues that need to be addressed in order to reach the ultimate goal of molecular conjugates that are true human therapeutics. We discuss barriers in formulation, extracellular transport, and intracellular transport. We close with a discussion that emphasizes a systems approach to the creation of these complex nanometer-sized assemblies. While molecular conjugates can involve lipids and polycations, our discussion primarily concerns polycations.

2 Formulation barriers

As with viruses, scale-up of the production of nonviral gene delivery particles to the commercial scale faces significant hurdles. Viral gene delivery particles can be prepared in large quantities using "natural" cellular production systems. For virus production, plasmids containing genes for viral assembly and the desired DNA payload are typically cotransfected into "packaging cells" in vitro, wherein the viruses are assembled and secreted in large quantity. This preparation scheme allows for scale-up, but has the potential to be complicated by the introduction of contamination and/or genetic mutation. However, a key issue is that the synthesis and assembly processes are all performed by biological entities that naturally perform these complex steps. Nonviral vectors must be prepared using synthetic materials by relying upon traditional reaction chemistry and purification (such as chromatography, filtration) schemes to separate the desired product from other components that may be present within the reaction mixture; similar issues are faced for production (such as extrusion with liposomes) and purification of the final assembled product. These processes create the potential for greater batch-to-batch variability, and variations in properties such as polycation molecular weight distribution and complex size have been shown to affect gene delivery efficiency [9] and toxicity [10].

The mechanism of viral production within packaging cells ensures that each virion contains precisely the same number of copies of the genetic payload. Because nonviral complex formation relies upon self-assembly, there is a potential for greater heterogeneity within the resulting formulations. Factors such as order-of-addition, concentration of components, and charge ratio all affect the mean and polydispersity of the complex diameter. Further, as will be discussed later in this chapter, many nonviral formulations contain multiple components in addition to the vector and nucleic acid, such as stabilizing polymers and cell-targeting ligands. Additional interformulation heterogeneity may be introduced if the method of assembling these constituents does not provide sufficient control.

Non-uniformity within a single nonviral formulation presents a separate concern. An increasing body of evidence suggests that significant quantities of vector within a formulation may remain unbound to the nucleic acid [8]. Although a positive (greater than 1 : 1) charge ratio is required for complete condensation of the nucleic acid, it likely results in the presence of unbound material. While researchers have shown that it is possible to quantify this free material and remove it from a formulation (for example by filtration (Fig. 2)) prior to administration, the overwhelming majority of published experiments fail to consider or account for this material and its role in toxicity and transfection efficiency. Uncomplexed vector within nonviral formulations further complicates interpretation of results and the validity of comparisons between in vitro and in vivo experiments. Very recently, Wagner and coworkers addressed this issue in cultured cells and mice using complexes of polyethylenimine (PEI) and DNA with vs. without prepurification by size exclusion chromatography to remove free polymer [11]. While polyplex size and surface charge (zeta potential) did not change upon purification, high doses of the purified complexes in cultured cells showed enhanced transfection efficiency that was linked to lower toxicity. Removal of free material also led to reduced symptoms of toxicity in mice and significantly lowered

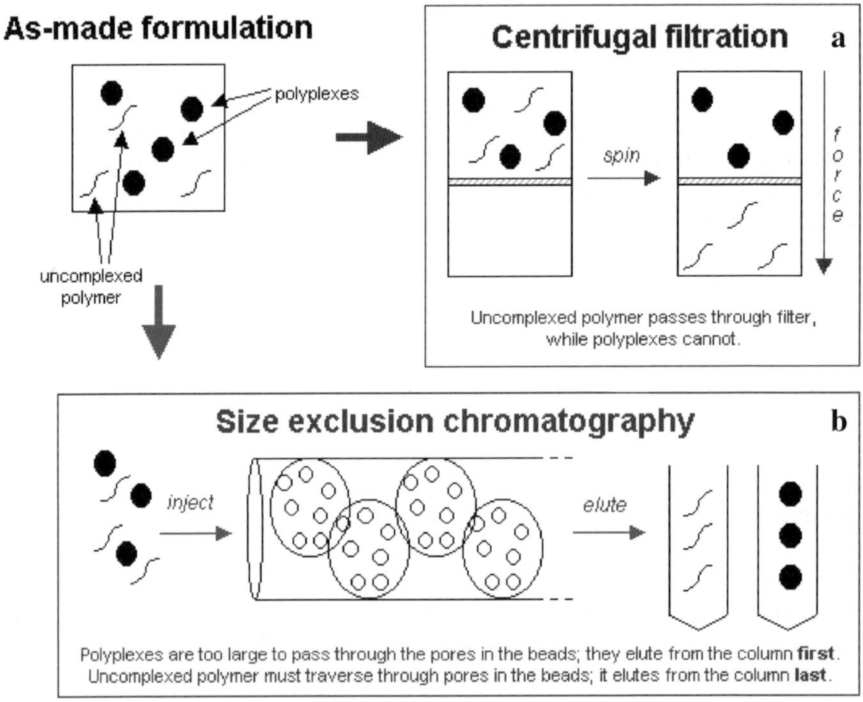

Fig. 2 Separation of uncomplexed polymer from a formulation. Nonviral formulations are typically made at a positive (greater than 1 : 1) charge ratio to ensure that all of the nucleic acid is condensed; this results in an excess of polymer that remains uncomplexed. Methods to remove this material from the polyplexes include **a** centrifugal filtration or **b** size exclusion chromatography

lung-specific gene expression. This investigation clearly revealed the impact of free polycation in vitro and in vivo.

Therapeutic doses of nonviral gene delivery systems necessitate formulations and administration of sufficiently high concentrations of the complexes to make the issues of immunogenicity, toxicity, and the fate of the decomposition products from the molecular conjugates relevant. Cationic polymers, both alone and in formulations with DNA, have been shown to bind serum proteins, activate the complement system [12] and to induce erythrocyte aggregation in vitro, although reduction of the formulation net charge and stabilization by PEGylation (necessary for in vivo application, as discussed below) reduces these effects strongly [13, 14]. Similar effects, including induction of several inflammatory cytokines (such as IFN-γ, IL-6, and IL-12) and potent activation of the complement cascade have been observed with lipoplexes upon intravenous injection at therapeutic doses in mice [15]. It has been further observed that the sequence of DNA being delivered by nonviral vectors – in particular the presence of "CpG motifs" – can contribute greatly to complex immunogenicity [16].

Nonviral complexes can also exhibit significant toxicity at therapeutic doses. Injection of liposomes in mice can induce leukopenia, thrombocytopenia, and elevation of serum transaminases indicative of hepatocyte necrosis [15]. Recent work with charged gold nanoparticles suggests that their cationic nature may be responsible for toxicity; in cultured mammalian fibroblasts, human erythrocytes, and bacteria, cationic particles were moderately toxic, whereas anionic particles were essentially nontoxic [17]. While understanding the precise determinants of immunogenicity and the toxicity of nonviral formulations continues to grow, a comprehensive knowledge of the corresponding molecular mechanisms remains elusive. This is an area of research that needs greater emphasis for molecular conjugates, and minimization of these barriers is going to be necessary for practical application of molecular conjugates in humans.

3
Extracellular barriers

Well-designed, nonviral gene therapeutics of sizes 25–100 nm in diameter will be suitable for systemic administration and will be able to target (a) particular cell type(s) within the body. A number of significant obstacles exist for these particles from the point of injection to contact of cell surface receptors in mammals that stand to prevent their entry into the cells they are designed to affect. The particles must remain discrete (non-aggregated) colloids within the complex milieu of the blood, staying small enough to be endocytosed. They should also possess a targeting moiety that promotes their preferential uptake by the desired target cells. Both of these objectives need to be accomplished without eliciting a response from the host immune system that is designed to identify and destroy such "foreign" material. This section will review selected recent advances in understanding and overcoming each of the major potential obstacles, including strategies for salt- and serum-stabilization and the incorporation of specific targeting ligands.

3.1 Stability and immunogenicity

Stability within the blood is a fundamental necessity for a successful formulation for systemic administration. The electrostatic nature of the self-assembly between cationic vector (polycations, liposomes, ...) and anionic nucleic acid (pDNA, siRNAs, oligonucleotides, ...) has the potential to be unstable in an environment with high ionic strength, such as blood. Indeed, most simple polycation/DNA complexes exhibit rapid aggregation upon exposure to physiological levels of salt: the complexes interact with each other. Further, these charged colloids may fail to reach their target cells intact if they possess excess positive or negative charge, owing to possible interactions with extracellular matrix proteins and cell surface phospholipids and glycosaminoglycans or anion-scavenging cell surface proteins, respectively. Thus, formulations are typically prepared with a minimum of net charge and are modified with hydrophilic, neutral polymers (most notably polyethylene glycol (PEG)) that help to shield the charge.

Stabilization via introduction of PEG ("PEGylation") to nonviral complexes can be performed in several different ways. Addition of a monofunctionalized, amine-reactive PEG derivative to polycations (such as PEI and poly-L-lysine) or liposomes leads to a covalent attachment of PEG; the extent of PEGylation can be altered by modifying the reaction conditions. Such covalent PEGylation can typically be performed either before [18–20] or after [13] condensation of the nucleic acid by the polycation, depending on the polycation/DNA ratio used. Alternatively, Pun and coworkers have employed a noncovalent PEGylation scheme that involves tethering PEG to a hydrophobic small molecule (adamantane) that forms strong, noncovalent inclusion compounds with β-cyclodextrin (β-CD) moieties within the polycation [21]. This noncovalent PEGylation strategy has been used with CD-modified PEIs in mice to dramatically lower toxicity and to redirect biodistribution from the lung (typical site of deposition for nonstabilized PEI aggregates in vivo) to the liver [22]. This strategy provides physical separation between sites of PEGylation and those of interaction with DNA. Both covalent and noncovalent PEGylation have been utilized to give complexes that retain their size in PBS, cell culture medium (with or without serum), and 100% serum.

While it is used most commonly, PEGylation is not the only means to provide stabilization of nonviral gene therapy complexes. One alternative comes from the laboratory of Seymour, where it has been demonstrated that modification of polyplexes with a multivalent polymer, *N*-(2-hydroxypropyl)methacrylamine (PHPMA), can provide lateral stabilization (see steric stabilization via monovalent PEG) and extend polyplex half-life in the mouse bloodstream from less than 5 min to over 90 min [23]. This polymer has also been employed to coat adenoviral vectors and similarly extended their circulation time [24]. Perouzel and coworkers have developed carbohydrate-lipid conjugates ("neoglycolipids") that have enhanced the stability (reduced aggregation in high ionic strength medium) and transfection efficiencies of ternary liposome/peptide/DNA formulations [25]. Finally, Wu, Nantz and coworkers have developed poly(cationic lipid)s (PCLs) that have shown stability in serum for up to 24 h [26] and reduced binding to serum proteins compared to common liposomal formulations [27].

Beyond being salt- and serum-stabilized, nonviral gene delivery complexes must evade interactions with nontarget cells and the immune system of the host animal. While a variety of mechanisms (such as complement activation) can be invoked, the specific chemical nature of the vector and the amount of "free" (uncomplexed) material are thought to be critical in determining the extent of immunogenicity. Finsinger and coworkers [28] have shown that incorporation of PEG and removal of free material from copolymer formulations sharply reduce complement activation in vitro. Further, complexes have been shown to interact with red blood cells in a molecular weight-dependent manner [29], causing a reduction in the amount of administered dose available to reach the desired target cells. Continued efforts to purify and characterize stabilized complexes will lead to reduced nonspecific interactions and heightened efficacy [11].

3.2
Targeted systems for receptor-mediated delivery

A successful strategy for targeting a particular target cell type for delivery must identify and exploit a distinguishing feature of the cell surface. The most common strategy is to direct delivery to a unique cell surface receptor that is overexpressed by the desired target cells. The cognate ligand (can be a small molecule, protein or antibody) is incorporated within the complexes and, upon binding to the receptor on the target cell surface, induces cellular uptake (receptor-mediated endocytosis). Some typical examples of ligands and receptors that have received significant attention in the literature on molecular conjugates are listed in Table 1. Many common types of cancer are distinguished by overexpression of a particular cell surface molecule (such as transferrin receptor, folate receptor, HER2/neu, and so on), as are particular cell types within the body (including asialoglycoprotein receptor in hepatocytes and mannose receptor in macrophages). Successful strategies to target each of these receptors are enumerated here using representative large and small molecule-targeting ligands.

Table 1 Receptor-ligand combinations exploited by targeted formulations. A variety of cell surface molecules that are highly expressed in particular cell types (both normal and cancerous) are targeted with cognate ligands (small molecules or antibodies) incorporated within nonviral gene delivery formulations

Ligand	Target cell type(s)	Ligand/Delivery System [Reference]
Transferrin Receptor (TfR)	Many cancers	**Transferrin** Covalent attachment to PEI [33] Tf-PEG noncovalent attachment to CDP [34], [35] **Anti-TfR Single Chain Antibodies** Physical mixing with cationic liposomes [31] **Anti-TfR Antibodies** PEG-Ab incorporation in liposomes [32]
Folate Receptor (FR)	Many cancers	**Folate** Covalent attachment to lipids via PEG [37, 39] Protamine/DNA polyplexes within lipid-PEG-folate liposomes [40]
HER2/neu	Many cancers, including breast and ovarian	**Anti-HER2 Antibodies** Conjugation of Ab fragments to PEGylated liposomes [122]
Asialoglycoprotein Receptor (ASGPR)	Hepatocytes	**Galactose** Conjugation to lipids within PEGylated liposomes [41] Direct conjugation to PEI [43] Triantennary ligand conjugated to PEG-lipid [44]
Mannose Receptor (MR)	Macrophages	**Mannose** Mannose-conjugated lipids with PEI [46] Mannosylated polylysine [47]

3.2.1 Transferrin receptor (TfR)

Transferrin receptor (TfR) is expressed in all nucleated cells in the body and is up-regulated in cancerous cells, presumably because their rapid growth and division increases their requirement for iron [30]. TfR binding of iron-bound (diferric) transferrin (Tf), the natural ligand for TfR, at the cell surface induces endocytosis through clathrin-coated pits. Consequently, TfR is frequently selected as a target for preferential receptor-mediated uptake of nonviral gene delivery complexes by tumors that overexpress it (reviewed by [30]). Various strategies to incorporate either Tf or anti-TfR antibodies (or antibody fragments) into such complexes have been employed with varying degrees of success.

Several research groups have successfully incorporated anti-TfR antibodies into liposomal gene delivery formulations (denoted lipoplexes). Yu et al. combined anti-TfR single chain antibody fragments with cationic liposomes by physical mixing [31]. pDNA encoding GFP (green fluorescent protein) was then added, and the resulting complexes injected i.v. into tumor-bearing mice. The inclusion of the antibody fragments was shown to increase ($\sim$ 3 fold) the GFP expression within the tumors (as measured by Western blot).

Anti-TfR antibodies have also been covalently tethered to lipoplexes suitable for in vivo administration. Zhang and coworkers attached each of two antibodies to polyethylene glycol (PEG) within PEGylated liposomes in a controlled and quantifiable manner [32]; these antibodies targeted murine TfR (mTfR) and human insulin receptor (hIR). Plasmids encoding anti-EGFR (epidermal growth factor receptor) shRNAs were complexed with this formulation and injected i.v. in mice that had previously received an injection of human glioma cells into the brain. The anti-mTfR antibodies allowed liposome transport across the blood-brain barrier, and the anti-hIR antibodies conveyed targeting to the human glioma cells. The formulations were quite successful in decreasing tumor EGFR expression and vascular density, and weekly administration of these targeted liposomes gave a strong (88%) increase in survival rates of adult mice.

As with liposomes, polycations have been conjugated with Tf to endow the resulting complexes (denoted polyplexes) with tumor targeting. Hildebrandt and coworkers covalently attached Tf to polyethylenimine (PEI) and used the resulting polymer to target delivery of a luciferase-expressing plasmid to N2A (murine neuroblastoma) tumors in mice [33]. Using in vivo whole-animal bioluminescent imaging, the authors demonstrated robust tumor targeting of luciferase expression. Tf has also been noncovalently incorporated into polyplexes by means of inclusion compound formation. Pun et al. synthesized PEG/Tf conjugates containing a terminal adamantane moiety that associates with the hydrophobic cavity of β-cyclodextrin (β-CD) subunits within the polycation backbone [34]. When injected i.v., this system delivered fluorescently labeled DNAzymes (catalytic, short ssDNAs) preferentially to subcutaneous tumors in tumor-bearing mice [35]. In the absence of the transferrin, the complexes did not enter tumor cells.

As described above, both liposomal and polymeric nonviral gene delivery formulations have been effectively modified to target TfR for delivery to tumors. Given the wealth of knowledge about this target and its possible ligands, the breadth of success seen thus far, and the vast number of possible gene targets for TfR-expressing tumors, research in this area is likely to remain strong.

3.2.2 Folate receptor (FR)

As with transferrin receptor (TfR), cell surface receptors for folate (folate receptors, FRs) are overexpressed in a variety of cancers in humans. Folate, which binds FR with a $\sim$ 30-fold higher affinity than the natural ligand for FR (6S-*N*-5-methyltetrahydrofolate), is often incorporated into nonviral gene delivery systems to target these types of tumors (reviewed by [36]). Polymer- and/or lipid-based systems engineered to target FR have been prepared to deliver plasmids or oligonucleotides with some success.

Hofland and coworkers [37] synthesized modified lipids that contained pendant folate moieties conjugated via a polyethylene glycol (PEG) linker. These lipid-PEG-folate liposomes were used to deliver pDNA containing the chloramphenicol acetyltransferase (CAT) gene to BALB/c mice that possessed a lung carcinoma (M109, introduced by subcutaneous injection). Intravenous administration of these targeted liposomes led to significant CAT expression within the tumor and a sharp (up to hundredfold) reduction in lung expression (the usual site of maximal expression for PEG-lipids). Similarly, Rait and coworkers prepared folate-containing liposomes to deliver phosphorothioate oligonucleotides to human breast cancer tumors in mice [38]. A combination therapy including these folate-targeting liposomes and docetaxel strongly inhibited xenograft growth compared to untreated mice or those receiving either treatment alone. Leamon and coworkers modified oligonucleotide-complexed liposomes with a variety of folate-PEG-lipid conjugates and found strong, receptor-mediated uptake in cultured FR-expressing cells but failed to observe a corresponding improvement in delivery to FR-bearing tumors in mice [39].

An additional level of complexity was introduced by Bruckheimer and coworkers, who precomplexed pDNA (containing a thymidine kinase gene) with protamine prior to lipid-PEG-folate liposome formulation [40]. Effective targeting was suggested by increased uptake of fluorescently-labeled complexes (1.6-fold enhancement of uptake) in cultured murine breast adenocarcinoma cells versus an untargeted (non-folate-containing) formulation. Dual administration of targeted formulations with gancyclovir in immunocompetent mice expressing breast tumors resulted in a significant reduction in tumor volume compared to relevant control groups.

3.2.3 Lectins

The targeting of transferrin receptor and folate receptor described above is applicable for tumors that overexpress these receptors. A variety of cell surface molecules provide accessible targets for delivery to noncancerous cells as well, based upon their naturally nonuniform distribution across the various cells types of the body. Here, we discuss one such group of target molecules – lectins – which are carbohydrate-binding cell surface proteins expressed by many normal (noncancerous) cell types. One such lectin, the asialoglycoprotein receptor (ASGPR), recognizes certain galactose-containing structures and is highly expressed on liver parenchymal cells. To target these hepatocytes, numerous strategies have been employed to incorporate galactose or lactose (a galactose-containing disaccharide) in nonviral vectors.

PEGylated liposomes containing a galactosylated cholesterol derivative have exhibited preferential uptake by parenchymal cells of the liver (7.7-fold greater uptake than nonparenchymal cells) upon intravenous administration in mice [41]. Hosseinkhani and coworkers reported results from intravenous injection of mice with PEGylated polyplexes made with spermine (an oligoamine) that was covalently modified with dextran (a polysaccharide) [42]. They observed preferential transfection of parenchymal liver cells in a manner that could be abrogated by preinjection of arabinogalactan or galactosylated bovine albumin (but not mannosylated bovine albumin), implicating ASGPR in polyplex uptake. Similarly, galactose has been included in polyplex and lipoplex formulations to target hepatocytes via ASGPR. Galactosylated PEI (Gal-PEI) was shown to give galactose-inhibitable gene expression in cultured murine hepatocytes and it achieved superior delivery to hepatocytes (vs. unmodified PEI) after injection in mice [43]. Frisch et al. attached a triantennary galactosyl ligand to a PEG-modified lipid delivery system and demonstrated targeted transfection of cultured human hepatocytes [44]. These results are surprising and somewhat confusing, given that the complexes examined had a diameter of around 150 nm, more than twice as large as the reported 70 nm upper size limit for endocytosis via interaction with the ASGPR [45].

Another lectin, mannose receptor (MR), is highly expressed by a number of subtypes of macrophages, cells that are targets for gene therapy to treat Gaucher's disease and HIV-1 infection. Incorporation of mannose residues into nonviral gene delivery vectors has been shown to be effective in targeting macrophages in vitro and in vivo. For example, Sato and coworkers incorporated mannose-conjugated lipids and polyethylenimine (PEI) into a delivery system for plasmid DNA to target murine macrophages [46]. The resulting system gave improved uptake (vs. non-mannose-containing formulations) and transfection efficiency in mouse peritoneal macrophages and could be inhibited significantly by addition of a soluble competitor (mannan) to the transfection solution, indicating that uptake is receptor-mediated, as expected. Targeting of mannosylated complexes to liver macrophages has been demonstrated in vivo as well. Injection of pDNA encoding chloramphenicol acetyltranferase (CAT) complexed to mannosylated polylysine in mice led to liver-specific uptake (> 80%) and gene expression (detectable only in liver) [47].

Taken in total, these results suggest that lectins can serve as accessible and appropriate target molecules for effective targeting of nonviral gene delivery particles to cell types that express them at high levels.

3.3 Future directions: Control and characterization of complexes

While numerous early successes have been achieved in targeting nonviral gene delivery complexes to particular cell types in vivo, a tremendous opportunity to improve upon vector design and complex efficacy through better quantitative characterizations of these ligand-containing systems remains. This will likely be necessary before the practical application of these assemblies as true therapeutics can be achieved. Three of the most important parameters to be considered are the control and quantification of: i) complex size and uniformity, ii) uncomplexed material and iii) ligand density. Selected recent results of investigations involving these issues are discussed below.

3.3.1 Complex size

The average hydrodynamic diameter of nonviral complexes is most commonly measured by dynamic light scattering (DLS) and electron microscopy (EM). Diameters of less than 200 nm are frequently observed and generally considered to be small enough for clathrin-dependent endocytosis [48]. It is worth emphasizing that such measurements are most relevant when they are made in an environment that mimics physiological conditions as much as possible (salt concentrations, pH, serum, and so on). If not sufficiently stabilized, complexes may appear small and stable when measured in water but demonstrate rapid aggregation in the presence of physiological levels of salt and/or serum. Also, with regard to targeted systems, it has been demonstrated that particular cell-surface receptors have unique size restrictions for receptor-mediated endocytosis. For example, Rensen and coworkers used tritiated, galactose-containing liposomes to demonstrate that the asialoglycoprotein receptor (ASGPR) requires a complex diameter of 70 nm or less for recognition and processing [45]. Knowledge of this critical size threshold for each cell surface receptor of interest is essential to guide formulation development that is most likely to achieve effective targeting and endocytosis.

3.3.2 Uncomplexed material

When nucleic acids are condensed by cationic polymers or liposomes, they are often formulated at an excess of positive charge (that is, a + / – charge ratio greater than 1), as is required for full nucleic acid condensation. Consequently, and as was discussed earlier, the potential exists for this "excess" vector to remain in the formulation as free/uncomplexed material, creating heterogeneity in the mixture that is administered to cultured cells and in vivo. One study estimated that 40–60% of polyethylenimine (PEI) added to pDNA (at a 8/1 N/P ratio) remained uncomplexed [28], while a more quantitative investigation estimated free PEI to be $\sim$ 86% of the total [8]. Thus, the majority of the polycations in such formulations are not associated with the DNA, and the role of this free polymer in transfection (uptake, intracellular trafficking, DNA release) remains unclear. It is reasonable to hypothesize that the performance of such a heterogeneous system during in vitro transfections (where the mixture is static within a well of a plate or a Petri dish) may differ widely from that in vivo, where additional transport processes within flowing blood figure to separate complexes and free material from each other. Removal of uncomplexed material prior to administration of complexes will result in a more well-defined, homogeneous therapeutic, but examples of such treatment within the literature are scarce [11].

3.3.3 Ligand density

While there are numerous demonstrations of effective targeting via ligand-containing complexes, few include any quantitative assessment of ligand density or comparison of conjugation methods. Zhang et al. incorporated antibodies within PEGylated liposomes using covalent conjugation to PEG and quantified the loading (for example 43–87 antibody molecules per liposome) through inclusion of trace amounts of radiolabeled antibodies [32]. Kunath et al. directly added galactose to PEI through reductive amination at a variety of degrees of substitution in a manner that was readily quantified [49]. Such quantification is uncommon in targeted delivery systems and is particularly meaningful because of the conjugation strategy employed for these liposomes. The incorporation of ligands simply through physical mixing or other noncovalent attachment (for example inclusion compounds, see [21]), may, in contrast, be more difficult to quantify owing to greater potential for differences in formulation stoichiometries between measurement and therapeutic conditions. Since complexes will contain numerous ligand molecules, there is the possibility that the targeted particles can have binding affinities for the surface of cells exceeding that of the ligand alone because of the cooperativity of multiple binding events on the particles. This effect has been observed with synthetic gene delivery particles but not quantified in sufficient detail. Thus, the incorporation of low-affinity ligands has the ability to show high affinity through multiple binding events. This feature merits further investigation, especially in vivo, since it may provide avenues for targeting without the use of expensive antibodies.

The importance of ligand conjugation and flexibility on receptor binding has been examined, and it has been shown that multiple bonds between ligand and particle reduce the targeting rate significantly (by up to 50%) compared to constructs where ligands are attached via a unique aliphatic chain [50]. Continuing efforts to develop effective targeted nonviral gene delivery systems should place emphasis on the establishment of an optimal ligand conjugation scheme that is reproducible and quantifiable and allows sufficient ligand accessibility and flexibility for maximal receptor interaction and uptake.

4 Intracellular barriers

The intracellular barriers to efficient, non-viral gene delivery include aspects of trafficking within and escape from the endocytic pathway, unpackaging of nucleic acids from the delivery vector, mobility and persistence within cytoplasm and delivery to the cell nucleus. By exploiting the synthetic nature of molecular conjugates, many chemical modifications have been made in an effort to address these perceived barriers. Examples include enhancement of endosomal escape with buffering components, of vector unpackaging with selectively labile vectors, of nuclease resistance with modified nucleic acids, and of nuclear delivery using targeting sequences. One area that has not been effectively addressed is the poor cytoplasmic mobility of cationic polymer-nucleic acid complexes (polyplexes) and of larger unpackaged nucleic acids. Below, we review the recent literature on the intracellular delivery of nucleic acids by nonviral systems.

4.1 Trafficking within and escape from endocytic vesicles

A net positive surface charge facilitates facile entry of polycation-nucleic acid particles to cultured cells. Electron microscopy has been used to visualize the apparent endocytosis of pDNA complexed with cationic lipids [51] or with cationic polymers [7]. Both reports show images of gene delivery particles contained within invaginations of the cell membrane. Using polylysine-pDNA complexes, Mislick and Baldeschwieler demonstrated that this uptake occurs through binding of cationic particles to anionic cell-surface proteoglycans [52]. The recent model of Kopatz et al. further implicates syndecans, a class of proteoglycans common to adherent cells [53]. Noting that pathogens are able to divert syndecan binding to enter cells, this model suggests that the clustering of syndecan molecules on the cell surface and their intracellular binding of actin can induce particle engulfment and cell entry. Goncalves and coworkers recently suggested that transfection efficiency is greatest when the means of uptake is clathrin-dependent endocytosis [54]; this mode of entry places limits on particle size to about 200 nm [48].

Active in all cells, nonspecific uptake through binding of proteoglycans is not useful for targeted delivery in mammals. Particles for in vivo gene delivery should incorporate ligands to target specific cell types and allow particles to enter cells through receptor-mediated endocytosis. Following uptake, the particles are delivered to the endocytic pathway. For many delivery vectors, cells transfected with fluorescently-labeled polycations and/or nucleic acids display punctate cytoplasmic staining indicative of sequestration in the endocytic pathway. Using transmission electron microscopy of cells transfected in vitro, Mishra et al. have observed vesicles containing intracellular polyplexes and their aggregates [7]. Vesicles of the endocytic pathway undergo active transport along intracellular microtubules. Consistent with the hypothesis of polyplex entrapment by these vesicles, Suh et al. have shown that the movement of polyethylenimine (PEI)-plasmid DNA (pDNA) complexes within cells includes an active mode that can be impaired by depolymerization of microtubules [55].

The vesicles comprising the endocytic pathway become progressively more acidic as endocytosed materials proceed towards lysosomes. There is strong evidence that endocytosed polyplexes experience this reduction in pH as they move through the endocytic pathway. Multiple researchers have used changes in fluorescence or used confocal or electron microscopy to show polyplex components in acidic and/or endocytic compartments [7, 54, 56–59]. Two groups have used bulk ratiometric fluorescence assays to quantify the acidification experienced by intracellular polyplexes. By conjugating both a pH-sensitive and a pH-insensitive label to pDNA [60] or to polycationic delivery vectors [61], these investigators were able to calculate the average pH environment of intracellular complexes made with PEI or polylysine and to show that these pH values are reduced over time.

These studies collectively show that many types of polyplexes proceed through the endocytic pathway and accumulate in the perinuclear region in lysosomes, where nucleases and other degradative enzymes present an undesirable environment for the delivered nucleic acids. It has been generally assumed that avoidance of lysosomes can give rise to improved gene delivery with polyplexes. This hypothesis has been supported by enhancements in reporter gene expression observed for transfections of many types of polyplexes in the presence of the endosomotropic agent chloroquine [62–66].

The influenza virus exploits the increasingly acidic pH of endocytic vesicles to trigger reorganization of its hemagglutinin protein and escape from the endocytic pathway. In an effort to engineer endosomal escape in nonviral gene delivery by a similar method, Kyriakides et al. developed pH-sensitive membrane-disrupting polymers [67]. Observing the strong in vitro hemolytic performance of poly(propylacrylic acid) at pH values less than 6.5, these investigators incorporated the polymer in nonviral formulations and measured increases in reporter gene expression both in vitro and in vivo.

PEI (polyethylenimine) is one delivery vector that gives high transfection efficiencies that are unaffected by the presence of chloroquine. This observation gave rise to the "proton sponge" hypothesis, wherein PEI is naturally endosomotropic due to its numerous amine residues with pK_a values in the physiological range [68]. In this hypothesis, the amine residues of PEI are able to buffer the acidification of the endosome, thereby impeding its progression on the endocytic pathway and/or facilitating its rupture. Kichler et al. demonstrated that acidification plays a role in PEI-mediated transfection in vitro; inhibition of endosome acidification with bafilomycin A1 led to a sharp drop in reporter gene expression [69]. The "proton sponge" hypothesis is supported by the recent results of Sonawane and coworkers, who observed that, relative to poorly-buffering polylysine-pDNA complexes, polyplexes prepared with strongly-buffering PEI or polyamidoamine gave reduced acidification, increased swelling, improved buffering capacity, and heightened osmotic fragility of endosomes [70].

The aforementioned bulk ratiometric fluorescence assays, however, did not fully support the "proton sponge" hypothesis for PEI. For the case of dual-labeled pDNA, PEI showed apparent buffering relative to polylysine [60]; for the case of dual-labeled delivery vectors, however, the apparent buffering with PEI was restricted to only one of the three cell lines examined [61]. It was also shown recently that acetylation of PEI improved its transfection efficiency despite a concomitant reduction in buffering capacity [71]. Mishra et al. have observed intact polyplexes in cytoplasm with both PEI and a nonbuffering β-cyclodextrin-containing polymer, demonstrating that a buffering delivery vector is not a requirement for endosomal escape of intact polyplexes [7]. Funhoff et al. questioned the general applicability of the "proton sponge" hypothesis upon observing poor transfection efficiency of a new polymer with a pK_a that should supply buffering capacity [72]. Erbacher et al. further showed that chloroquine can dissociate polycations from pDNA, suggesting that vector displacement is another means by which this buffering agent may enhance transfection efficiency in vitro [62].

Despite questions about the "proton sponge" hypothesis and the benefit of buffering elements, buffering moieties have been added to many vectors in an effort to mimic the purported endosomotropic activity of chloroquine and PEI. One such moiety is histidine, a natural amino acid whose side chain becomes protonated at pH ~ 6. Midoux et al. prepared a histidine-substituted polylysine that generated much greater in vitro reporter gene expression than unmodified-polylysine transfection and even unmodified-polylysine transfection in conjunction with chloroquine treatment [73]. Transfection efficiency in vitro was markedly improved with a low extent of histidine substitution and continued to improve slightly with additional histidine substitution. As with PEI-pDNA transfection, reporter gene expression in an in vitro histidylated polylysine-pDNA transfection was reduced when endosome acidification was inhibited with bafilomycin A1. While Midoux et al. used histidine substitution throughout the polylysine backbone, Davis et al. pursued an alternate strategy of histidine substitution only at the termini of a β-cyclodextrin containing polymer [74]. Davis et al. observed that histidine substitution produced increases during in vitro reporter gene expression for the same level of pDNA delivery to cells, supporting the hypothesis that buffering moieties somehow contribute to the defeat of intracellular barriers.

Rather than applying direct histidine conjugation, some researchers have pursued buffering through incorporation of polyhistidine. Pack et al. found that the transfection efficiency of ternary complexes of pDNA, polylysine, and glycosylated polyhistidine was not significantly improved relative to simple pDNA-polylysine complexes [75]. Noting that polyhistidine had been shown to have fusogenic properties at mildly acidic pH [76], Benns and coworkers prepared a polyhistidine-graft-polylysine copolymer as a gene delivery vector. The β-galactosidase expression resulting from transfection with the copolymer was superior to that of simple polylysine [77]. The purported fusogenic contribution was limited, however, as chloroquine treatment produced further increases in transfection efficiency for both the copolymer and unmodified polylysine. Although the importance of buffering thus remains an open question relative to factors such as vector displacement, it is likely that buffering components offer contributions to enhanced gene delivery.

The engineering of endosomal escape processes with molecular conjugates should account for both the nature and morphology of the intravesicular polyplexes. Mishra et al. used transmission electron microscopy to show that unmodified polycation-pDNA complexes enter cells as large aggregates with 3–5 times the as-formulated diameter [7]. Still larger aggregates were observed as aggregate-containing vesicles appeared closer to the nucleus. The membranes of these vesicles conformed to the contorted, bulbous shape of the aggregates therein. When PEI-pDNA complexes were coated with polyethylene glycol (PEG) to confer extracellular salt stability, the particles' intracellular morphology was also seen to be affected. Intracellular PEGylated particles were observed as discrete entities contained in smooth, circular vesicles, and the particles retained their as-formulated size (Fig. 3). The observation of unmodified polyplexes as large aggregates in intracellular vesicles may be related to the superior transfection efficiency seen with those polyplexes relative to their PEGylated variants. Only a slight swelling of the large aggregates could lead to vesicle rupture, while this is not the case with the PEGylated particles. Further studies will be needed to address these issues.

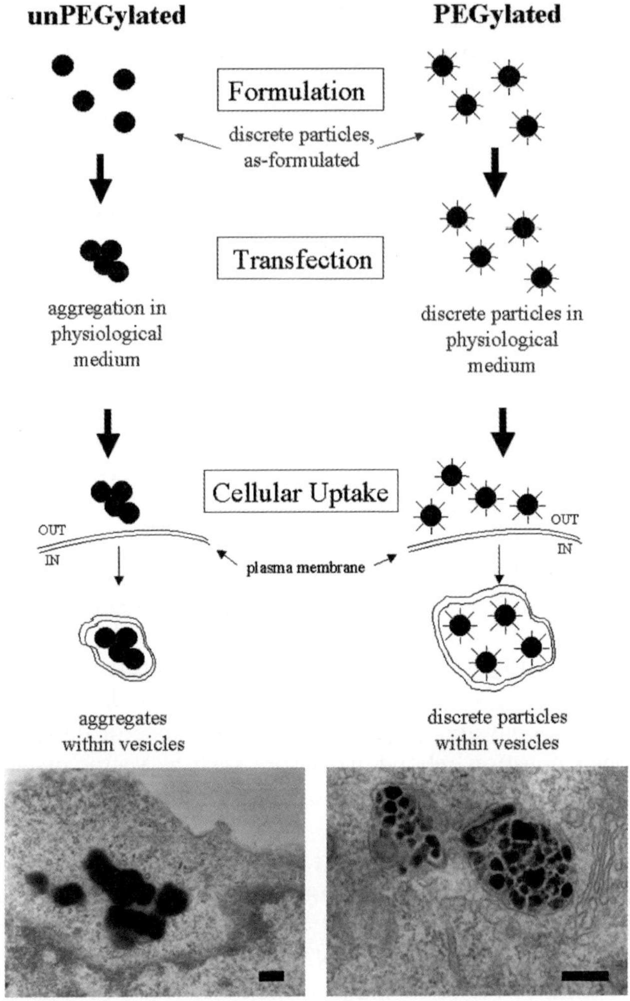

Fig. 3 Stabilization of polyplexes affects uptake and trafficking. Unstabilized complexes aggregate in the transfection medium; these enter cells and traffic through them as large clumps. In contrast, PEGylation of nonviral complexes allows for uptake of discrete entities that remain distinct throughout the endocytic pathway. The intracellular complexes can be observed in the transmission electron microscope. *Scale bars*, 200 nm

4.2
Vector unpackaging

The unpackaging of nucleic acids from delivery vectors is essential to the gene delivery process. Nucleic acids function through interactions with cellular components such as proteins and other nucleic acids, and these cellular components may remain inaccessible to nucleic acids that remain bound to delivery vectors. Reduced transcriptional activity has been observed for pDNA bound to cationic dendrimers [78] or to polylysine [1]. Prolonged binding of pDNA and more severe inhibition of its transcriptional activity are together correlated with an increase in polylysine length [1]. Koping-Hoggard et al. used chitosan to demonstrate a correlation between ease of polyplex dissociation and increases in transfection efficiency [2]. As described above, the results of Erbacher et al. indicate that the endosomotropic agent chloroquine produces enhancements in transfection efficiency in vitro by dissociating the polycations from pDNA in the delivered polyplexes [62].

A proposed mechanism for vector unpackaging is displacement of nucleic acids by intracellular polyanions. The kinetics and ease of this unpackaging are related to the chemical structure of the delivery vector. A series of β-cyclodextrin-containing polymers prepared by Hwang et al. demonstrated a clear relationship between the polymer-pDNA binding strength and the number of methylene units separating charge centers on the polymer [79]. Itaka et al. observed that linear and branched PEI are displaced from pDNA with similar kinetics, but showed using fluorescence resonance energy transfer (FRET) that, in cells, linear PEI is displaced from pDNA far more readily than the branched variant [58]. Based on these studies, it is unlikely that polycations and/or lipid delivery systems that bind sufficiently strong to nucleic acids to afford stability in vivo will spontaneously release the nucleic acid to any great extent once inside cells. Thus, these systems should be designed so that the transition from extracellular to intracellular compartments triggers a reduction in the binding strength of the delivery vector.

Degradable delivery vectors have been designed in an effort to facilitate vector unpackaging. Various researchers have sought to construct vectors that will fall apart within cells and render the polyplexes unstable. Unfortunately, the optimal kinetics and intracellular location of this degradation are unknown, and it is difficult to predict where and how quickly a new vector will exhibit degradation. For example, Forrest et al. prepared highly crosslinked mimics of 25 kDa branched PEI by reacting 800 Da PEI with short, linear diacrylates [4]. The resulting conjugates were sensitive to hydrolysis at both near-neutral and mildly acidic pH. An intercalating-dye exclusion assay, however, showed that polyplexes produced with these conjugates did not exhibit pDNA release in vitro even after 24 h in hydrolyzing conditions. The increasingly acidic pH of endocytic vesicles can also be exploited to induce vector unpackaging. Choi et al. demonstrated that polycation-pDNA particles could be prepared with an acid-labile lipid at pH 8.5 and would rapidly degrade at pH 5.3 [5]. The pH-sensitive lipid-pDNA particles showed sharp increases in transfection efficiency relative to similar, pH-insensitive particles.

Other designs have sought to exploit the reducing environment found within cells. In an effort to provide extracellular particle stabilization that would be discarded after cellular uptake, Carlisle et al. coated PEI-pDNA polyplexes with poly[*N*-(2-hydroxypropyl)methacrylamide] through a reducible disulfide linkage [6]. The coating could be removed by reduction with dithiothreitol, and transfections with coated particles showed that the resultant reporter gene expression could be modulated by the extent of degradable coating and the strength of the intracellular reducing environment. Rather than using disulfide bond reduction to remove a stabilizing coating, Saito and coworkers designed polyplexes to become endosomolytically active in a reducing environment [80]. The cationic DNA-binding peptide protamine and the endosomolytic protein listeriolysin O were conjugated through a disulfide bond, rendering listeiolysin O inactive. This protamine-lysteriolysin O copolymer was designed to degrade in acidic vesicles and release active listeriolysin O to facilitate the endosomal escape of the remaining polyplex. In fact, an optimized loading of listeriolysin O to the polyplexes produced pronounced enhancements in reporter gene expression.

The internal polyplex structure can also be rendered sensitive to a reducing environment. A modified polylysine vector prepared by Pichon et al. has its amine charge centers connected to the polymer backbone by disulfide bonds [3]. Polyplexes prepared with this modified polylysine can be disrupted by reduction with dithiothreitol or glutathione and they exhibit improvements in transfection efficiency relative to unmodified polylysine. Miyata et al. prepared polyplexes using a thiolated PEG-polylysine copolymer, then transferred the polyplexes to an oxidative solution to crosslink the copolymers within the polyplexes [81]. Their results indicated that the extent of intrapolyplex crosslinking and the density of the polymers' cationic charge both contribute to pDNA release upon reduction of the disulfide crosslinks.

Similarly, McKenzie and coworkers prepared a series of synthetic peptides as delivery vectors, and observed that inclusion of cysteine residues allowed for spontaneous disulfide bond formation upon pDNA condensation [82, 83]. The disulfide bond formation conferred a reduction in particle size and stability against shear stress, and the authors ascribed increases in transfection efficiency to enhanced pDNA release upon intracellular reduction of the disulfide bonds. For these materials, variation in the number of cysteine residues can be used to control the extent of disulfide crosslinking and the resulting stability.

The reversible polymerization of liposomes presents another opportunity for the preparation of degradable complexes, and polymerized lipoplexes have been designed with susceptibility to degradation by reducing conditions. Balakirev et al. used lipoic acid, a molecule containing an unstable dithiolane ring, to prepare reactive amphiphiles that polymerized under oxidative conditions to condense pDNA into small particles [84]. The researchers implicated intracellular reduction by glutathione and NAD(P)H in the improved transfection efficiency of the reducible lipoplexes.

4.3
Cytoplasmic persistence and mobility

For certain applications, it may be sufficient to ensure the nucleic acids escape from the endocytic pathway and are released from delivery vectors. RNA interference approaches, for example, are designed to function in the cytoplasm and often rely on small oligonucleotides that can readily diffuse through the cytoplasm [85]. Unfortunately, delivery to the cytoplasm is inadequate for complete genes or plasmids that must access the cell's transcriptional machinery; these materials must traverse the cytoplasm and reach the cell nucleus.

Barring direct association of polyplexes and intracellular microtubules, observations of active, microtubule-dependent intracellular transport of PEI-pDNA complexes [55] suggest that polyplex motion in cytoplasm occurs in the context of entrapment in endocytic vesicles. Intact polyplexes are typically dense and dozens or hundreds of nanometers in diameter, and those that escape endocytic vesicles cannot be assumed to exhibit significant movement in cytoplasm or to traffic to the cell nucleus. Diffusion in the cytoplasm is restricted by the structure of the cytoskeleton and the severe crowding presented by proteins, organelles, and other macromolecules [86, 87]. The apparent immobility of intact polyplexes in cytoplasm suggests that these particles can only access the cell nucleus during cell division, when there is cytoplasmic mixing and degradation of the nuclear membrane. Reports of intact polyplexes visualized in the cell nucleus are uncommon in the literature. One hypothesis describing the intracellular trafficking of polyplexes suggests that the cytoplasm presents a "dead end" for intact polyplexes due to their large size and lack of active transport mechanisms. It is reasonable to assume that current systems will undergo vector unpackaging and/or degradation before or upon deposition in the cytoplasm.

Although vector unpackaging is necessary for the eventual function of the delivered nucleic acids, its consequence in the endocytic pathway or in the cytoplasm is exposure of nucleic acids to a destructive environment. Lechardeur et al. found that cytoplasmic nucleases limit pDNA to a half-life in cytoplasm of only 50–90 min [88]. Pollard et al. confirmed these observations, finding a pDNA half-life in cytoplasm of 2 h, and suggested various means of confronting this hazard, including saturation or inhibition of cytoplasmic nucleases or the use of artificial nucleic acids that would not be susceptible to the nucleases [89]. Ribeiro et al. took a parallel approach to demonstrate that replacement of specific labile sites in pDNA can confer some resistance to nuclease degradation without sacrificing the in vitro expression levels obtained from that pDNA [90]. Similarly, Layzer and coworkers discovered chemical modifications to siRNAs that greatly enhance the half-life in plasma without compromised efficacy in vivo [91].

The size of the nucleic acid cargo affects its intracellular mobility. Polyplexes are typically used to deliver pDNA of 5000 or more base pairs (bp).

Unpackaged pDNA of this size is unlikely to escape degradation by traversing the cytoplasm. Lukacs et al. used fluorescence recovery after photobleaching (FRAP) to measure the size-dependent diffusion of nucleic acids in cells [85]. The diffusion coefficient of linear DNA fragments in cytoplasm dropped severely as the DNA length increased from 100 to 500 bp, and fragments longer than 2000 bp exhibited little to no diffusion. These results suggest that, for most plasmids, diffusion alone cannot facilitate transport of unpackaged pDNA from cytoplasm into the cell nucleus.

4.4 Nuclear delivery

While the precise mechanism of intracellular trafficking remains unknown, polyplexes are able to deliver some pDNA to the cell nucleus; however, the amount is likely only a small fraction of that delivered to cells [92]. Without incorporation of components to enhance nuclear delivery, significant pDNA delivery to the cell nucleus has rarely been visualized by microscopy. In their multiple particle tracking experiments, Suh and coworkers ascribed the absence of fluorescently-labeled PEI in cell nucleus to their use of nondividing cells [55]. Fluorescence in situ hybridization (FISH) by Langle-Rouault et al. showed unpackaged pDNA only in cytoplasm, not in the nucleus, even though a hundredfold increase in luciferase expression had indicated the authors had successfully designed a system for improved nuclear delivery [93].

Indirect confirmation of nuclear delivery from simple polycation-pDNA particles is provided by the numerous reports of non-zero expression levels from exogenous pDNA. The expression of reporter genes indicates that some pDNA has reached the cell nucleus and been transcribed. Despite efforts at direct visualization, it remains unclear whether this pDNA reaches the nucleus in the form of polyplexes, polyplex fragments, or unpackaged pDNA. Using in vitro transfection with PEI, Godbey et al. observed fluorescently labeled PEI and pDNA in the cell nucleus [9]. The observations were made 4 h post-transfection and the PEI and pDNA appeared to be associated.

Using transmission electron microscopy of cells transfected in vitro, Mishra et al. explored the intracellular fate of polyplexes prepared with PEI or with a β-cyclodextrin-containing polymer [7]. These efforts did not reveal intact polyplexes in the cell nucleus. However, by using immunolabeling of cells transfected with biotin-labeled pDNA, the investigators were able to visualize unpackaged pDNA in both cytoplasm and nucleus of transfected cells (Fig. 4). In contrast to the results of Godbey et al., these observations are consistent with a hypothesis of cytoplasmic vector unpackaging followed by nuclear import of unpackaged pDNA, either intact or as digested fragments. In experiments using flow cytometry of cells transfected with fluorescently labeled pDNA, James and Giorgio noted but discounted the possible impact of

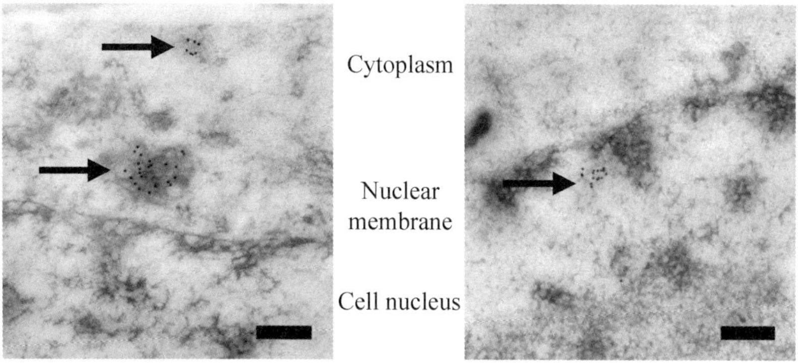

Fig. 4 Observation of unpackaged intracellular pDNA through immunolabeling. Unpackaged intracellular nucleic acids can be visualized in the transmission electron microscope [7]. Here, polyethylenimine was used to deliver biotin-labeled pDNA to BHK-21 cells in vitro. Cells were fixed and prepared in thin sections, and an antibiotin antibody was used to label intracellular pDNA with colloidal gold. Clusters of gold particles (*arrows*) in the cytoplasm (*left*) or cell nucleus (*right*) indicate the presence of pDNA outside of intact polyplexes. *Scale bars,* 200 nm

degraded pDNA [94]. It is unknown whether pDNA fragments can enter the nucleus more easily than intact pDNA (a possibility supported by the aforementioned results of Lukacs and coworkers [85]) or contribute measurably to observed transfection efficiencies.

Although the mechanism of nuclear delivery is not clear, cell division appears to play a significant role. Brunner et al. showed that the transfection efficiencies of branched PEI and polylysine exhibit a strong dependence on cell cycle [95]. Similarly, multiple investigators have seen that mitosis produces an increase in transfection efficiency with liposomes [96–98]. Using microinjection of pDNA to cytoplasm and examining a range of pDNA doses, Ludtke et al. showed that compared to nondividing cells, roughly twice as many dividing cells expressed a reporter gene [99]. A widely accepted explanation for the positive correlation between transfection efficiency and cell division is that cytoplasmic pDNA can access the nucleus more easily upon the breakdown of the nuclear envelope in mitosis. Given the poor diffusivity of pDNA in cytoplasm, pDNA introduction to the cell nucleus may also be facilitated by the reorganization of the intracellular milieu during cytokinesis.

In marked contrast to their work with other delivery agents, Brunner and coworkers observed that transfection with linear PEI was not affected by the cell cycle [95, 100]. Consistent with this result, confocal microscopy by Wightman et al. showed pDNA association with the nucleus for linear PEI but not branched PEI [101]. Linear PEI shares its unusual cell-cycle independence with viral vectors, which can exploit viral mechanisms for nuclear delivery, and electroporation. A molecular level explanation of how linear PEI and electroporation achieve cell cycle independence could provide valuable insights for improving nuclear delivery of nonviral vectors.

In the design of synthetic gene delivery systems, reliance on cell division for nuclear delivery should be avoided because of the desire to target nondividing cells in vivo. As transfection efficiency is not abolished entirely in nondividing cells, existing systems likely achieve some nuclear delivery through one or more alternative mechanisms. Although the numerous nuclear pore complexes allow small solutes and proteins to pass freely through the nuclear envelope, pDNA (but not complexes unless they are very small [102]) may be able to be transported into the cell nucleus through these pores via a passive or active transport mechanism. Too large to diffuse through the nuclear pore complexes by passive diffusion, karyophilic proteins are directed to the nucleus by a class of peptides known as nuclear localization signals (NLSs) [103–106]. NLS-containing proteins are carried to the cell nucleus through a nuclear pore complex in a stepwise, energy-dependent fashion. Efforts to incorporate a nuclear targeting component in a synthetic gene delivery system have focused on exploiting this natural pathway for nuclear delivery [107].

Conjugation to an NLS has been shown to enhance the nuclear delivery of carboplatin, a small molecule anticancer agent [108]. Analogously, some investigators have attempted to enhance nuclear delivery of delivered nucleic acids by direct labeling with NLSs. Sebestyen et al. utilized covalent attachment of an NLS to a linearized pDNA construct with a functional reporter gene [109]. When the construct was applied to digitonin-permeabilized cells, the NLS did not affect the resulting expression levels. These authors subsequently showed that an NLS markedly enhanced nuclear delivery of linear pDNA in microinjected cells only for DNA under 1500 bp [110]. Consistent with this result, a microinjected 900 bp linear expression cassette produced greater reporter gene expression when conjugated to the NLS. Moving beyond this size limit, Zanta et al. prepared a 3500 bp linear DNA fragment capped by a single terminal NLS [111]. When administered to cultured cells with a cationic lipid or with PEI, the NLS-labeled construct produced greater luciferase expression than an unlabeled construct or pDNA. However, the effect of the NLS was less pronounced in nondividing cells or with increasing DNA doses.

The aforementioned reports all incorporated the NLS of the SV40 large T antigen. Early identification of this NLS has left it well-characterized in the literature, but there is no guarantee this classical NLS is the most suitable option for nonviral delivery of nucleic acids. The wide variety of NLSs is accompanied by many independent import pathways. Although proteins directed for nuclear import are not necessarily restricted to a single pathway [104], the simultaneous function of independent pathways may indicate that certain cargoes are best suited for specific pathways [112]. Cellular nucleic acids enter the nucleus on a distinct pathway from many proteins, as evidenced by the nuclear import of RNA-containing small nuclear ribonucleoprotein particles [113].

Rather than using the SV40 large T antigen NLS, Subramanian et al. conjugated the M9 NLS to a DNA-binding peptide and used cationic lipids to deliver the conjugate-pDNA complexes to cultured cells [114]. These M9-containing lipoplexes exhibited significant improvements in transfection efficiency relative to simple pDNA-lipid transfection, but formulations that excluded the peptide conjugate in favor of one or both of its unconjugated components also produced increases in reporter gene expression. Further, the number of peptides necessary to elicit these effects was enormous. As such, it is difficult to assess the extent to which the M9 NLS contributed to nuclear delivery through the expected mechanism. Using peptide nucleic acid (PNA) clamps to bind NLS peptides to linearized 5000 bp pDNA, Bremner et al. observed little enhancement in transfection efficiency with any of four NLSs: the SV40 large T antigen NLS, an extended version of this NLS, the M9 NLS, and the NLS of the human T cell leukemia virus type 1 Rex protein [115]. Only the minor increase with the extended SV40 large T antigen NLS was statistically significant.

The variable performance of NLSs in nonviral delivery could stem from multiple factors, including inhibitory interactions between NLSs and nucleic acids. The NLSs studied by Bremner et al. displayed electrostatic interactions with nucleic acids, and conjugation to pDNA impeded NLS binding to transport receptors [115]. Further, it is not clear that NLSs, whose natural function is in protein transport, can contribute to the transport of large pDNA. Another concern is a dearth of knowledge on the optimal NLS loading number and density. In their experiments with microinjected or digitonin-permeabilized cells, Ludtke et al. observed a positive correlation between enhanced nuclear delivery and a greater number of NLSs per linear DNA construct [110]. An excessive NLS loading, however, may cause different regions of a single DNA construct to be directed simultaneously through multiple nuclear pore complexes, leaving the nucleic acid unable to enter the nucleus efficiently or entirely. Zanta et al. emphasized this point, designing their gene construct with a single, effectual NLS and estimating that the likelihood of simultaneous entry through multiple nuclear pore complexes was significantly enhanced for nucleic acids above 1000 bp [111].

Rather than incorporating an NLS, some investigators have sought to exploit the nuclear import machinery in an indirect manner. As shown by Kaneda et al., delivery of nuclear proteins concurrent with lipofection of pDNA enhances the pDNA nuclear delivery and gene expression [116]. These authors suspected that complexes of pDNA and nuclear proteins could protect pDNA from nucleases and facilitate pDNA transfer to the nucleus. Under the assumption that endogenous, newly-synthesized nuclear proteins can function in the same way, a number of investigators have proposed a "piggybacking" mechanism for pDNA delivery to the nucleus [112, 117]. As newly-synthesized nuclear proteins will be in the cytoplasm and many will contain both a DNA-binding element and an NLS, this hypothesis suggests that DNA-binding elements of the nuclear proteins bind to unpackaged pDNA in the cytoplasm and carry this pDNA into the nucleus as they are carried there by the nuclear import machinery.

Dean and coworkers lent credence to the piggybacking hypothesis by demonstrating that the sequence of pDNA could affect its nuclear import, and that this import, requiring certain cytoplasmic factors, could be inhibited by restricting transport through the nuclear pore complex [117, 118]. Langle-Rouault et al. demonstrated a sequence-dependent effect using a plasmid containing the Epstein-Barr origin of replication, oriP [93]. The researchers used lipofection or microinjection to transfect a cell line that stably expressed Epstein-Barr nuclear antigen 1. Consistent with a hypothesis of enhanced nuclear delivery through interaction beween oriP and Epstein-Barr nuclear antigen 1, the oriP-containing plasmid showed a hundredfold transfection efficiency enhancement in the stably-expressing cell line. Similarly, Vacik et al. showed that expression of pDNA containing portions of the smooth muscle gamma actin promoter was markedly enhanced in cultured cells that express transcription factors specific to smooth muscle cells [119]. The enhanced expression was observed both in smooth muscle cells (SMCs) and in epithelial cells stably transfected to express an SMC serum response factor. Mesika et al. used the transcription factor NF-κB to develop a controllable method of nuclear delivery through piggybacking [120]. NF-κB is found predominantly in the cytoplasm unless its nuclear localization is activated by a particular stimulus. These investigators prepared pDNA with a repeating sequence of NF-κB binding sites, used a cationic dendrimer to transfect cells in vitro, and monitored luciferase reporter gene expression as a function of NF-κB stimulation. In the four cell lines investigated, an order of magnitude enhancement in gene expression was observed for transfection with the modified pDNA in conjunction with activation of NF-κB. Confocal microscopy verified that NF-κB activation improved nuclear delivery of the modified pDNA, leading the authors to conclude that the modified pDNA was binding NF-κB in the cytoplasm and accompanying it to the nucleus upon NF-κB activation.

For all systems, the morphology of the nucleic acids may have an impact on nuclear delivery efficiency. Studies in nonviral gene delivery tend to focus on closed circular pDNA, which is resistant to exonucleases and gives higher transfection efficiency than its linearized variant [115]. However, the best results to date with NLSs have utilized linearized nucleic acids [110, 111, 115]. In part, the disparate performance of circular and linearized pDNA may be due to differences in cytoplasmic mobility.

Existing nonviral delivery systems do not traverse the nuclear membrane with ease. NLSs naturally act to deliver cellular proteins to the nucleus, but the NLS function may not be transferable to larger entities such as pDNA thousands of bp in length. Further, the methods currently under study for nuclear delivery, including NLSs and piggybacking, may be severely undermined by the relative immobility of pDNA in cytoplasm if this immobility impedes the delivered nucleic acids from encountering the relevant proteins. The development of an efficient nuclear delivery method remains an area of opportunity.

4.5
Future directions: Intracellular barriers

A variety of means are under development to address the intracellular barriers experienced by nucleic acids that are delivered by molecular conjugates. For example, the acidity of the endocytic pathway presents an opportunity for the engineering of endosomotropic mechanisms, and one approach to inducing the dissociation of polyplexes utilizes the reducing environment of the cell. Such modifications may become more effective with elucidation of the preferred time and location of endosomal escape and vector unpackaging. Efforts to engineer improved persistence and nuclear delivery are also underway.

To date, nonviral approaches have not produced a full rival to the superior transfection efficiency of viral vectors. A particular intracellular barrier that remains to be addressed is the poor cytoplasmic mobility of larger nucleic acids. pDNA that is unpackaged from delivery vectors in the cytoplasm is likely to be degraded before it can diffuse to its place of function or to bind proteins that can confer nuclease protection and nuclear delivery. A mechanism for promoting cytoplasmic mobility, perhaps in conjunction with nuclear delivery, likely presents an opportunity for further improvements in transfection efficiency. As nonviral systems are commonly characterized by transient expression, opportunities also remain in prolonging expression. A successful approach by Nakai et al. achieved integration of exogenous pDNA by linearization; integration was further enhanced with incorporation of virally-derived inverted terminal repeats [121].

Recent research demonstrates that the intracellular trafficking of nonviral formulations is affected by the particles' morphology, size, and cellular uptake mechanism [7, 54]. For example, Goncalves et al. recently demonstrated that the sizes of polylysine-pDNA complexes dictate whether they enter cells through clathrin-dependent or -independent pathways [54]. These authors used a variety of methods to present detailed characterizations of the materials' intracellular destination(s) and observed that clathrin-dependent internalization was more conducive to enhanced transfection. Because the physiochemical nature of the delivery formulation affects both its cellular uptake and its intracellular trafficking, progress in intracellular gene delivery requires this elucidation of mechanisms for physiologically-relevant systems that are salt- and serum-stabilized, appropriately sized, and targeted to specific cell types.

5
Overall summary: A systems approach

As illustrated in the sections above, the challenges of creating a nonviral delivery system are many. However, we believe that one of the most daunting ones is the integration of the components into a workable system that combines the attributes of the components without suffering losses because of their integration. Examples of the detrimental effects of integration already exist. For example, PEGylation of polyplexes can endow them with serum stability but also greatly inhibit gene expression. The key issue is to develop delivery systems that provide the appropriate spatio-temporal functions; the system must perform particular functions in the right places at the right times to provide for effective transport of the nucleic acid to the desired site of action in mammals. As mentioned above, we believe that this cannot be done without some type of adaptive system. Thus, systems that sense their surroundings and respond to spatio-temporal environmental changes are likely to have better success than systems that are not dynamic. Early examples of these systems do exist [74], but much further progress using this approach will be necessary to bring these entities from laboratory systems to practical "bedside" medicines.

References

1. Schaffer DV, Fidelman NA, Dan N, Lauffenburger DA (2000) Biotechnol Bioeng 67:598
2. Koping-Hoggard M, Varum KM, Issa M, Danielsen S, Christensen BE, Stokke BT, Artursson P (2004) Gene Ther 11:1441
3. Pichon C, LeCam E, Guerin B, Coulaud D, Delain E, Midoux P (2002) Bioconjug Chem 13:76
4. Forrest ML, Koerber JT, Pack DW (2003) Bioconjug Chem 14:934
5. Choi JD, MacKay JA, Szoka FC (2003) Bioconjug Chem 14:420
6. Carlisle RC, Etrych T, Briggs SS, Preece JA, Ulbrich K, Seymour LW (2004) J Gene Med 6:337
7. Mishra S, Webster P, Davis ME (2004) Eur J Cell Biol 83:97
8. Clamme JP, Azoulay J, Mely Y (2003) Biophys J 84:1960
9. Godbey WT, Wu KK, Mikos AG (1999) P Natl Acad Sci USA 96:5177
10. Kunath K, von Harpe A, Fischer D, Petersen H, Bickel U, Voigt K, Kissel T (2003) J Control Release 89:113
11. Boeckle S, von Gersdorff K, van der Piepen S, Culmsee C, Wagner E, Ogris M (2004) J Gene Med 6:1102
12. Plank C, Mechtler K, Szoka FC Jr, Wagner EW (1996) Hum Gene Ther 7:1437
13. Ogris M, Brunner S, Schuller S, Kircheis R, Wagner E (1999) Gene Ther 6:595
14. Brus C, Petersen H, Aigner A, Czubayko F, Kissel T (2004) Bioconjug Chem 15:677
15. Tousignant JD, Gates AL, Ingram LA, Johnson CL, Nietupski JB, Cheng SH, Eastman SJ, Scheule RK (2000) Hum Gene Ther 11:2493

16. Tousignant JD, Zhao H, Yew NS, Cheng SH, Eastman SJ, Scheule RK (2003) Hum Gene Ther 14:203
17. Goodman CM, McCusker CD, Yilmaz T, Rotello VM (2004) Bioconjug Chem 15:897
18. Kwok KY, McKenzie DL, Evers DL, Rice KG (1999) J Pharm Sci 88:996
19. Petersen H, Fechner PM, Martin AL, Kunath K, Stolnik S, Roberts CJ, Fischer D, Davies MC, Kissel T (2002) Bioconjug Chem 13:845
20. Kursa M, Walker GF, Roessler V, Ogris M, Roedl W, Kircheis R, Wagner E (2003) Bioconjug Chem 14:222
21. Pun SH, Davis ME (2002) Bioconjug Chem 13:630
22. Pun SH, Bellocq NC, Liu A, Jensen G, Machemer T, Quijano E, Scluep T, Wen S, Engler H, Heidel J, Davis ME (2004) Bioconjug Chem 15:831
23. Oupicky D, Ogris M, Howard KA, Dash PR, Ulbrich K, Seymour LW (2002) Mol Ther 5:463
24. Green NK, Herbert CW, Hale SJ, Hale AB, Mautner V, Harkins R, Hermiston T, Ulbrich K, Fisher KD, Seymour LW (2004) Gene Ther 11:1256
25. Perouzel E, Jorgensen MR, Keller M, Miller AD (2003) Bioconjug Chem 14:884
26. Wu J, Lizarzaburu ME, Kurth MJ, Liu L, Wege H, Zern MA, Nantz MH (2001) Bioconjug Chem 12:251
27. Liu L, Zern MA, Lizarzaburu ME, Nantz MH, Wu J (2003) Gene Ther 10:180
28. Finsinger D, Remy JS, Erbacher P, Koch C, Plank C (2000) Gene Ther 7:1183
29. Ward CM, Read ML, Seymour LW (2001) Blood 97:2221
30. Li H, Qian ZM (2002) Med Res Rev 22:225
31. Yu W, Pirollo KF, Yu B, Rait A, Xiang L, Huang W, Zhou Q, Ertem G, Chang EH (2004) Nucleic Acids Res 32:e48
32. Zhang Y, Zhang Y-f, Bryant J, Charles A, Boado RJ, Pardridge WM (2004) Clin Can Res 10:3667
33. Hildebrandt IJ, Iyer M, Wagner E, Gambhir SS (2003) Gene Ther 10:758
34. Bellocq NC, Pun SH, Jensen GS, Davis ME (2003) Bioconjug Chem 14:1122
35. Pun SH, Bellocq NC, Cheng J, Grubbs BH, Jensen GS, Davis ME, Tack F, Brewster M, Janicot M, Janssens B, Floren W, Bakker A (2004) Can Biol Ther 3:e31
36. Zhao XB, Lee RJ (2004) Adv Drug Deliv Rev 56:1193
37. Hofland HEJ, Masson C, Iginla S, Osetinsky I, Reddy JA, Leamon CP, Scherman D, Bessodes M, Wils P (2002) Mol Ther 5:739
38. Rait AS, Pirollo KF, Xiang L, Ulick D, Chang EH (2002) Mol Med 8:475
39. Leamon CP, Cooper SR, Hardee GE (2003) Bioconjug Chem 14:738
40. Bruckheimer E, Harvie P, Orthel J, Dutzar B, Furstoss K, Mebel E, Anklesaria P, Paul R (2004) Can Gene Ther 11:128
41. Managit C, Kawakami S, Nishikawa M, Yamashita F, Hashida M (2003) Int J Pharm 266:77
42. Hosseinkhani H, Azzam T, Tabata Y, Domb AJ (2004) Gene Ther 11:194
43. Morimoto K, Nishikawa M, Kawakami S, Nakano T, Hattori Y, Fumoto S, Yamashita F, Hashida M (2003) Mol Ther 7:254
44. Frisch B, Carriere M, Largeau C, Mathey F, Masson C, Schuber F, Scherman D, Escriou V (2004) Bioconjug Chem 15:754
45. Rensen PCN, Sliedregt LAJM, Ferns M, Kieviet E, van Rossenberg SMW, van Leeuwen SH, van Berkel TJC, Biessen EAL (2001) J Biol Chem 276:37577
46. Sato A, Kawakami S, Yamada M, Yamashita F, Hashida M (2001) J Drug Target 9:201
47. Nishikawa M, Takemura S, Yamashita F, Takakura Y, Meijer DKF, Hashida M, Swart PJ (2000) J Drug Target 8:29
48. Rejman J, Oberle V, Zuhorn IS, Hoekstra D (2004) Biochem J 377:159

49. Kunath K, von Harpe A, Fischer D, Kissel T (2003) J Control Release 88:159
50. Olivier V, Meisen I, Meckelein B, Hirst TR, Peter-Katalinic J, Schmidt MA, Frey A (2003) Bioconjug Chem 14:1203
51. Labat-Moleur F, Steffan AM, Brisson C, Perron H, Feugeas O, Furstenberger P, Oberling F, Brambilla E, Behr JP (1996) Gene Ther 3:1010
52. Mislick KA, Baldeschwieler JD (1996) P Natl Acad Sci USA 93:12349
53. Kopatz I, Remy JS, Behr JP (2004) J Gene Med 6:769
54. Goncalves C, Mennesson E, Fuchs R, Gorvel JP, Midoux P, Pichon C (2004) Mol Ther 10:373
55. Suh J, Wirtz D, Hanes J (2003) P Natl Acad Sci USA 100:3878
56. Fajac I, Allo JC, Souil E, Merten M, Pichon C, Figarella C, Monsigny M, Briand P, Midoux P (2000) J Gene Med 2:368
57. Grosse S, Tremeau-Bravard A, Aron Y, Briand P, Fajac I (2002) Gene Ther 9:1000
58. Itaka K, Harada A, Yamasaki Y, Nakamura K, Kawaguchi H, Kataoka K (2004) J Gene Med 6:76
59. Rosenkranz AA, Yachmenev SV, Jans DA, Serebryakova NV, Murav'ev VI, Peters R, Sobolev AS (1992) Exp Cell Res 199:323
60. Akinc A, Langer R (2002) Biotechnol Bioeng 78:503
61. Forrest ML, Pack DW (2002) Mol Ther 6:57
62. Erbacher P, Roche AC, Monsigny M, Midoux M (1996) Exp Cell Res 225:186
63. Gonzalez H, Hwang SJ, Davis ME (1999) Bioconjug Chem 10:1068
64. Midoux P, Mendes C, Legrand A, Raimond J, Mayer R, Monsigny M, Roche AC (1993) Nucleic Acids Res 21:871
65. Oupicky D, Carlisle RC, Seymour LW (2001) Gene Ther 8:713
66. Zhang X, Sawyer GJ, Dong X, Qiu Y, Collins L, Fabre JW (2003) J Gene Med 5:209
67. Kyriakides TR, Cheung CY, Murthy N, Bornstein P, Stayton PS, Hoffman AS (2002) J Control Release 78:295
68. Behr JP (1997) Chimia 51:34
69. Kichler A, Leborgne C, Coeytaux E, Danos O (2001) J Gene Med 3:135
70. Sonawane ND, Szoka FC, Verkman AS (2003) J Biol Chem 278:44826
71. Forrest ML, Meister GE, Koerber JT, Pack DW (2004) Pharmaceut Res 21:365
72. Funhoff AM, van Nostrum CF, Koning GA, Schuurmans-Nieuwenbroek NME, Crommelin DJA, Hennink WE (2004) Biomacromolecules 5:32
73. Midoux P, Monsigny M (1999) Bioconjug Chem 10:406
74. Davis ME, Pun SH, Bellocq NC, Reineke TM, Popielarski SR, Mishra S, Heidel JD (2004) Curr Med Chem 11:1241
75. Pack DW, Putnam D, Langer R (2000) Biotechnol Bioeng 67:217
76. Wang CY, Huang L (1984) Biochemistry 23:4409
77. Benns JM, Choi JS, Mahato RI, Park JS, Kim SW (2000) Bioconjug Chem 11:637
78. Bielinska AU, Kukowska-Latallo JF, Baker JR (1997) Biochim Biophys Acta 1353:180
79. Hwang SJ, Bellocq NC, Davis ME (2001) Bioconjug Chem 12:280
80. Saito G, Amidon GL, Lee K-D (2003) Gene Ther 10:72
81. Miyata K, Kakizawa Y, Nishiyama N, Harada A, Yamasaki Y, Koyama H, Kataoka K (2004) J Am Chem Soc 126:2355
82. McKenzie DL, Kwok KY, Rice KG (2000) J Biol Chem 275:9970
83. McKenzie DL, Smiley E, Kwok KY, Rice KG (2000) Bioconjug Chem 11:901
84. Balakirev M, Schoehn G, Chroboczek J (2000) Chem Biol 7:813
85. Lukacs GL, Haggie P, Seksek O, Lechardeur D, Freedman N, Verkman AS (2000) J Biol Chem 275:1625
86. Luby-Phelps K (2000) Int Rev Cytol 192:311

87. Lechardeur D, Lukacs GL (2002) Curr Gene Ther 2:183
88. Lechardeur D, Sohn KJ, Haardt M, Joshi PB, Monck M, Graham RW, Beatty B, Squire J, O'Brodovich H, Lukacs GL (1999) Gene Ther 6:482
89. Pollard H, Toumaniantz G, Amos JL, Avet-Loiseau H, Guihard G, Behr JP, Escande D (2001) J Gene Med 3:153
90. Ribeiro SC, Monterio GA, Prazeres DMF (2004) J Gene Med 6:565
91. Layzer JM, McCaffrey AP, Tanner AK, Huang Z, Kay MA, Sullenger BA (2004) RNA 10:766
92. Pollard H, Remy JS, Loussouarn G, Demolombe S, Behr JP, Escande D (1998) J Biol Chem 273:7507
93. Langle-Rouault F, Patzel V, Benavente A, Taillez M, Silvestre N, Bompard A, Sczakiel G, Jacobs E, Rittner K (1998) J Virol 72:6181
94. James MB, Giorgio TD (2000) Mol Ther 1:339
95. Brunner S, Sauer T, Carotta S, Cotten M, Saltik M, Wagner E (2000) Gene Ther 7:401
96. Tseng WC, Haselton FR, Giorgio TD (1999) Biochim Biophys Acta 1445:53
97. Mortimer I, Tam P, MacLachlan I, Graham RW, Saravolac EG, Joshi PB (1999) Gene Ther 6:403
98. Escriou V, Carriere M, Bussone F, Wils P, Scherman D (2001) J Gene Med 3:179
99. Ludtke JJ, Sebestyen MG, Wolff JA (2002) Mol Ther 5:579
100. Brunner S, Furtbauer E, Sauer T, Kursa M, Wagner E (2002) Mol Ther 5:80
101. Wightman L, Kircheis R, Rossler V, Carotta S, Ruzicka R, Kursa M, Wagner E (2001) J Gene Med 3:362
102. Liu G, Li D, Pasumarthy MK, Kowalczyk TH, Gedeon CR, Hyatt SL, Payne JM, Miller TJ, Brunovskis P, Fink TL, Muhammad O, Moen RC, Hanson RW, Cooper MJ (2003) J Biol Chem 278:32578
103. Gorlich D (1998) EMBO J 17:2721
104. Mattaj IW, Engelmeier L (1998) Annu Rev Biochem 67:265
105. Bednenko J, Cingolani G, Gerace L (2003) J Cell Biol 162:391
106. Weis K (2003) Cell 112:441
107. Munkonge FM, Dean DA, Hillery E, Griesenbach U, Alton EWFW (2003) Adv Drug Deliver Rev 55:749
108. Aronov O, Horowitz AT, Gabizon A, Fuertes MA, Perez JM, Gibson D (2004) Bioconjug Chem 15:814
109. Sebesteyen MG, Ludtke JJ, Bassik MC, Zhang G, Budker V, Lukhtanov EA, Hagstrom JE, Wolff JA (1998) Nat Biotech 16:80
110. Ludtke JJ, Zhang G, Sebestyen MG, Wolff JA (1999) J Cell Sci 112:2033
111. Zanta MA, Belguise-Valladier P, Behr JP (1999) P Natl Acad Sci USA 96:91
112. Boulikas T (1997) Intl J Oncol 10:301
113. Michaud N, Goldfarb D (1992) J Cell Biol 116:851
114. Subramanian A, Ranganathan P, Diamond SL (1999) Nat Biotech 17:873
115. Bremner KH, Seymour LW, Logan A, Read ML (2004) Bioconjug Chem 15:152
116. Kaneda Y, Iwai K, Uchida T (1989) Science 243:375
117. Dean D (1997) Exp Cell Res 230:293
118. Wilson GL, Dean BS, Wang G, Dean DA (1999) J Biol Chem 274:22025
119. Vacik J, Dean BS, Zimmer WE, Dean DA (1999) Gene Ther 6:1006
120. Mesika A, Grigoreva I, Zohar M, Reich Z (2001) Mol Ther 3:653
121. Nakai H, Montini E, Fuess S, Storm TA, Meuse L, Finegold M, Grompe M, Kay MA (2003) Mol Ther 7:101
122. Park JW, Kirpotin DB, Hong K, Shalaby R, Shao Y, Nielsen UB, Papahadjopoulos D, Marks JD, Benz CC (2001) J Control Release 74:95

Adv Biochem Engin/Biotechnol (2005) 99: 41–92
DOI 10.1007/10_003

Published online: 8 November 2005

Plasmid Vaccines and Therapeutics: From Design to Applications

Marston Manthorpe · Peter Hobart · Gary Hermanson · Marilyn Ferrari · Andrew Geall · Blake Goff · Alain Rolland (✉)

Vical Incorporated, 10390 Pacific Center Court, San Diego, CA 92121, USA
arolland@vical.com

Abstract In the late 1980s, Vical and collaborators discovered that the injection into tissues of unformulated plasmid encoding various proteins resulted in the uptake of the plasmid by cells and expression of the encoded proteins. After this discovery, a period of technological improvements in plasmid delivery and expression and in pharmaceutical and manufacturing development was quickly followed by a plethora of human clinical trials testing the ability of injected plasmid to provide therapeutic benefits. In this chapter, we summarize in detail the technologies used in the most recent company-sponsored clinical trials and discuss the potential for future improvements in plasmid design, manufacturing, delivery, formulation and administration. A generic path for the clinical development of plasmid-based products is outlined and then exemplified using a case study on the development of a plasmid vaccine from concept to clinical trial.

Keywords Anthrax · Anthrax vaccine · Biologics License Application · Clinical trial · Company-sponsored clinical trial · DNA vaccine · Formulation · Gene delivery · Gene expression systems · Gene therapy · Genetic immunization · Genetic vaccination · Investigational New Drug Application · Needle-free devices · Plasmid · Plasmid design · Plasmid DNA · Plasmid formulation · Plasmid manufacture · Vaccination · Vaccine

Abbreviations

AD	autodisable or autodestruct
Ad	adenovirus vector
AIDS	acquired immunodeficiency syndrome
APC	antigen presenting cell
arg	arginine auxotroph gene
AVA	Anthrax Vaccine Adsorbed
BAK	benzalkonium chloride
BD	Becton Dickinson and Company
BGH	bovine growth hormone
BLA	Biologics License Application
CAD	coronary artery disease
CCC	covalently closed circular
CDC	The Centers for Disease Control and Prevention
CF	cystic fibrosis
CFTR	cystic fibrosis transmembrane conductance regulator
cGMP	current good manufacturing practice
CITE	cap-independent translational entry
CMV	cytomegalovirus
CTAB	cetyltrimethylammonium bromide
CTL	cytotoxic T lymphocytes
DC-Chol	3-*N*-(*N*′,*N*′-dimethylaminoethane)-carbamoyl cholesterol
DDAB-PC	dodecylammonium bromide-phosphatidyl choline
DEL-1	development-regulated endothelial locus-1
DMRIE	(+/−)-*N*-(2-hydroxyethyl)-*N*,*N*-dimethyl-2,3-bis(tetradecyloxy)-1-propanaminium bromide
DODAC	*N*,*N*-Dioleyl-*N*,*N*-dimethylammonium chloride
DOPE	1,2-dioleoyl-sn-glycero-3-phosphoethanolamine
DOSPER	1,3-di-oleoyloxy-2-(6-carboxy-spermyl)-propylamide
DOTMA	*N*-[1-(2,3-dioleyloxy)propyl]-*N*,*N*,*N*-trimethylammonium chloride
DPyPE	1,2-diphytanoyl-sn-glycero-3-phosphoethanolamine
E. coli	*Escherichia coli*
EF	*Bacillus anthracis* edema factor
ELISA	enzyme-linked immunosorbant assay
FACS	fluorescence-activated cell sorting
FDA	U.S. Food and Drug Administration
FGF	fibroblast growth factor
FHIT	fragile histidine triad
Flu	influenza
GAP-DLRIE	aminopropyl-dimethyl-bis-dodecyloxy-propanaminium bromide
GAP-DMORIE	(+/−)-*N*-(3-aminopropyl)-*N*,*N*-dimethyl-2,3-bis(cis-9-tetradecenyloxy)-1-propanaminium bromide)
GLP	good laboratory practices

GM-CSF	granulocyte-macrophage colony stimulating factor
GMP	good manufacturing practices
HBsAg	hepatitis B surface antigen
HIV-1	human immunodeficiency virus-1
HPLC	high pressure liquid chromatography
HPV	human papillomavirus
HSP	heat shock protein
HSV-TK	Herpes simplex virus-thymidine kinase
ID	intradermal
IFN	interferon
IL	interleukin
IN	intranasal
IP	intraperitoneal
IRES	internal ribosomal entry site
IT	intratumoral
IV	intravenous
IVT	intraventricular
JE	Japanese encephalitis
KDa	kilodaltons
LAL	*Limulus amoebocyte* lysate assay
LF	*Bacillus anthracis* lethal factor
MHC	major histocompatibility complex
MLV	multilamellar vesicles
mRBG	modified-rabbit beta-globin
Naked DNA	unformulated plasmid (e.g., in saline or phosphate-buffered saline)
NIAID	National Institute of Allergy and Infectious Diseases
NLS	nuclear localization signal
NP	influenza nucleoprotein
ORF	open reading frame
ori	bacterial origin of replication
OTC	ornithine transcarbamylase
PA	*Bacillus anthracis* protective antigen
PAD	peripheral arterial disease
PEG	polyethylene glycol
PEG-DSPE	monoethoxyl polyethylene-glycol-distearoylphosphatidylethanolamine
PINC	protective, interactive, noncondensing
PLG	poly(lactide-co-glycolide)
POC	proof of concept
POE	polyoxyethylene
POP	polyoxypropylene
Pulm	pulmonary
PVP	poly(vinyl pyrrolidone)
RSV	respiratory syncytial virus
RT-PCR	reverse transcription polymerase chain reaction
SC	subcutaneous
SCID	severe combined immunodeficiency
SPLP	stabilized plasmid-lipid particles
SV40	Simian virus-40
TA	taurocholic acid
TAP	transporter associated with processing

TFA	trifluoroacetic acid
TLR	toll-like receptor
TM-TPS	CellFECTIN®; N, NI, NII, NIII-Tetramethyl-N, NI, NII, NIII-tetrapalmitylspermine
TPA	tissue plasminogen activator
UNICEF	United Nations Children's Fund
UTR	untranslated region
Vaxfectin™	GAP-DMORIE + DPyPE
VCL-6292	Vical plasmid 6292 encoding *Bacillus anthracis* protective antigen
VCL-62952	Vical plasmid 62952 encoding *Bacillus anthracis* lethal factor
VEGF	vascular endothelial growth factor
VTEU	Vaccine and Treatment Evaluation Unit
WHO	World Health Organization

1 Introduction

The full sequencing of the human genome combined with an increased understanding of the role of several disease-related genes has positioned somatic gene therapy and genetic vaccination amongst the most exciting and promising biotechnological advances. Gene therapy can be viewed as a way to produce proteins using the patient's own cells as mini-bioreactors. Gene therapy holds promise for replacing a missing or defective gene within the patient's cells and for supplementing the body with the production of proteins that can prevent or treat disease [1–3]. The development of new genetic vaccines that have the ability to generate specific anti-pathogen or anti-tumor immune responses is being explored to provide innovative prophylactic or therapeutic treatments for infectious disease and cancer [4–9].

Although numerous clinical trials have been conducted worldwide [10], to date the unequivocal clinical validation of gene therapy products (genetic vaccines and genetic therapies) still remains a critical milestone. If one compares the evolution of recombinant proteins and monoclonal antibodies with the evolution of gene therapy, one sees parallels among the various stages and timing that these biotechnology products had to pass before achieving wide acceptance and recognition by the medical, financial and public communities. Historically, the first phase of optimism and scientific excitement for recombinant proteins and monoclonal antibodies was followed by doubts. A period of about ten years elapsed between early preclinical/clinical success and the first commercial success. Based on such comparators, analysts have predicted that a first gene-based product may receive Biologics License Application (BLA) approval around 2005 in the United States and/or Europe. In fact, the world's first gene therapy product, Gendicine, received market approval in China in early 2004, the product being an adenoviral vector expressing a tumor suppressor gene (termed p53) for the treatment of patients with head and neck squamous cell carcinoma [11, 12].

Since the initiation of the first therapeutic gene therapy clinical trial 15 years ago [13], the field of human gene therapy has gone through several cycles of hope, hype, concern, disbelief and hope again. The unfortunate death in 1999 of a patient with ornithine transcarbamylase (OTC) deficiency following the intrahepatic infusion of an adenoviral vector, and recent cases of T-cell leukemia in X-linked severe combined immunodeficiency (SCID) infants following administration of a retroviral vector [14, 15], have left the public disillusioned and wondering about the future of gene therapy, in particular when using viral vectors. Fortunately, the curing of 15 children (to date) suffering from X-linked SCID in France and the United Kingdom, and the first market approval of the gene therapy product in China have spurred public interest and stimulated the imagination and impetus of the scientific and medical community.

To date, approximately a thousand clinical trials have been reported worldwide (approximately 70% in the USA, 25% in Europe and 5% in other countries) by companies and academic or governmental institutions. Most clinical trials are in Phase 1 (64%) or Phase 1–2 (20%) with only 13% in Phase 2 and less than 3% in Phase 3. The trials comprise treatments of acquired diseases such as cancer (66%), cardiovascular ($\sim$ 8%), neurological and infectious disease ($\sim$ 7%) and inflammatory diseases as well as genetic diseases ($\sim$ 10%) such as hemophilia, cystic fibrosis, SCID and muscular dystrophy. The majority of these clinical trials employ viral vectors such as modified adenoviruses, adeno-associated viruses, retroviruses, poxviruses and Herpes simplex viruses (see Table 1).

A small proportion of clinical trials involve the use of plasmid, either as unformulated plasmid (so-called "naked DNA" [16]) or as formulated plasmid. Many gene therapy clinical trials use anti-cancer approaches such as cytokines for immunotherapy, tumor antigens for vaccines, tumor-suppressor genes, and "suicide gene/prodrugs" for tumor cell inhibition. The treatment of cardiovascular disorders, such as peripheral artery disease (PAD), coronary artery disease (CAD) and restenosis, represents a fast-growing area for gene therapy. Several companies are attempting to reverse ischemic conditions by producing new blood vessels using genes encoding angiogenic factors such as VEGFs (vascular endothelial growth factors), FGFs (fibroblast growth factors), DEL-1 (development-regulated endothelial locus-1) and HGF (Hepatocyte Growth Factor). The first plasmid-based vaccine clinical trials were initiated in the mid-1990s and several trials in which an immune response to the encoded immunogen is induced are in progress against the pathogens causing AIDS, hepatitis B, cytomegalovirus (CMV)-associated diseases, Flu, Ebola, malaria and anthrax. Plasmid-based vaccines may provide several advantages over current live-attenuated and killed virus and recombinant subunit vaccines, such as: i) generation of more focused im-

Table 1 Prevalence of gene delivery systems in clinical trials [10]

Delivery System	# of trials	% of trials
Retrovirus	261	26.4
Adenovirus	256	25.9
Poxvirus	56	5.7
Vaccinia virus	32	3.2
Herpes simplex virus	30	3
Adeno-associated virus	25	2.5
Poxvirus + vaccinia virus	15	1.5
Lentivirus	2	0.2
Naked DNA	150	15.2
Cationic lipids	85	8.6
Gene gun	5	0.5
Others	70	7.3
Total	987	100

mune responses, emphasizing antibodies or MHC class I-restricted responses (cytotoxic T lymphocytes); ii) better potential for multivalent vaccination; iii) a generic manufacturing process, and; iv) use of standardized quality control and storage conditions that are independent of the encoding gene(s).

The field of gene therapy has constantly evolved since its inception, moving from *ex vivo* approaches (the first clinical trial in 1990 involved transduction of the adenosine deaminase gene with a viral vector in a SCID patient's blood cells *ex vivo* prior to autologous reinjection [13]) to gene-based products that can be directly administered *in vivo*. The field has also evolved from the treatment of genetic disorders to the prevention and treatment of acquired diseases. Furthermore, emphasis has been placed on improving the safety of viral vectors and the efficiency of nonviral systems, and more recently on various devices such as the use of the "gene gun" or other needle-free injectors, which have been used in clinical studies to facilitate the delivery of plasmids to target cells.

Effective gene therapy requires the control of both the location and function of therapeutic genes at specific target sites within the patient's body. Delivery and expression systems have to be rationally designed and assembled to address the key events that might be limiting plasmid distribution from the site of administration to entry into the nucleus of targeted cells [17]. Most gene administration methods are based on viral and synthetic delivery systems that are designed to safely and effectively transport the gene(s) of interest into target cells. Viruses (which are not covered in the present review) have evolved over millions of years to efficiently infect mammalian cells. Although various manipulations have been made to viruses to produce vectors that are not infectious, but are still able to efficiently transduce host cells, several major issues remain associated with their manufacturing, administration and antigenicity. Current viral vector families and an example of each that are being pursued by gene therapy companies include *Retroviridae* (lentiviruses), *Adenoviridae* (adenoviruses), *Parvoviridae* (adeno-associated viruses) and *Togaviridae* (alphaviruses).

Nonviral gene-based medicines and vaccines include naked DNA and formulated plasmids [18]. The percentage of current clinical trials using such systems is much smaller (only 20%) than for viral vectors and the number of companies focusing on nonviral products is relatively small. Nevertheless, the number of publications on synthetic gene delivery systems has increased exponentially for more than a decade and many new delivery systems are being explored in preclinical research and development.

Naked DNA in saline was first shown to be able to transfect skeletal muscle in 1990 [19]. Transfection efficiency with unformulated plasmid is typically low and thus requires high plasmid doses. Depending on the route of administration, expression can be transient (skeletal muscle being one of the exceptions) and therefore the major application of that technology to date has been vaccination. The use of a biolistic device, the "gene gun", which requires adsorption of naked DNA onto gold particles, has also been the focus of plasmid-based vaccine development for infectious diseases and cancers.

Plasmids have also been formulated with a variety of synthetic gene delivery systems but only a few have reached the clinical stage [1]. Amongst a large group of synthetic delivery elements and formulations that have been tested *in vitro* and *in vivo*, cationic lipid-based and noncondensing polymer systems are predominantly being tested in clinical trials. Of the synthetic gene delivery systems, cationic lipid-based delivery systems have been the most widely used for the delivery of plasmids, in indications such as cancer immunotherapy, cystic fibrosis, pulmonary disorders, restenosis, coronary and peripheral artery diseases, as well as cancer and infectious disease. Cationic lipid systems vary in their composition and physicochemical characteristics (for example the chemical structure of the cationic lipid, the addition of a colipid, excipient, the ratio of cationic lipid to colipid, the ratio of cationic lipid to DNA, the particle size). Protective, interactive, noncondensing polymers (PINC) have been used in humans for the direct administration of cytokine genes to accessible tumors as well as for the intramuscular delivery of genes coding for angiogenic factors and antigens. Delivery systems that are designed for localized delivery of plasmids to accessible tissues have shown excellent safety profiles in patients.

Although many more nonviral delivery systems have been designed and tested in limited ways for efficacy and safety *in vitro* and *in vivo* in animal models, they have not been administered to humans to date. Cationic polymers such as polyethylenimine, chitosan, poly(ortho esters) and other derivatives have been tested by several groups, but, so far only polyethylene glycol-poly-L-lysine has been introduced into patients [20].

While considerable effort in gene therapy has been directed towards the identification of therapeutic genes administered by multiple routes and delivery systems, the optimal expression of the delivered plasmid has also drawn substantial attention. Constitutive viral promoters have been used to obtain high levels of gene expression and tissue-specific promoters have been tested to restrict the expression to a desired tissue, such as skeletal muscle, prostate or liver. Additional enhancing elements have been incorporated to enable higher levels of expression, more accurate RNA splicing and longer duration of expression. More recently, regulation of expression has been investigated to allow gene expression to be turned on or off in response to systemic administration of low molecular weight regulatory molecules such as antibiotics or hormones. Such "gene switches" have not reached the clinic yet but have shown promise in preclinical studies with both viral vectors and nonviral systems.

Gene-based products, in particular plasmid-based vaccines, are technologies that have the potential to be cheaper, simpler and more convenient-to-use than existing vaccine products [21]. Initially, these so-called "disruptive technologies" tend to perform poorly compared to conventional technologies. Thus, performance enhancement is required to create innovative products for widespread application. Using the example of plasmid-based vaccines, Kaslow provides a unit operational analysis detailing the multi-step process from selection of candidate immunogen(s) to the generation of an effective immune response in humans [21]. Some of the steps are immunogen-dependent and relate, for instance, to the selection of the gene sequence, expression cassette, delivery system, device and route of administration, whereas other steps are immunogen-independent and generic, including, for example, selection of the plasmid backbone and bacterial host and manufacturing process. In this review, these unit operations, including the plasmid design, manufacturing, formulation and administration are analyzed in the context of gene-based products that have been tested in patients. The rational design of delivery and expression systems, coupled with the use of devices to control the location and functioning of synthetic gene-based medicines and vaccines, along with their manufacturing, are presented here. Future potential developments in the spatial and temporal modulation of gene function are briefly exemplified. A path from product concept to clinical trial is presented and exemplified with the case study of a formulated plasmid-based anthrax vaccine.

2 Clinical experience

Table 2 contains a list of some of the most recent company-sponsored clinical trials involving the delivery of plasmid into humans. The table lists trials

Table 2 Recent company-sponsored plasmid-based human clinical trials

Company [Ref.]	Gene(s)	Technology	Route	Indication	Phase	Back-bone	Promoter/ Enhancer	Intron	Term
CANCER VACCINES									
Corixa [212]	L523S	DNA + Ad boost; Biojector	IM	Lung	1	PVAX (pUC)	CMV/CMV	None	BGH
ImClone [212]	gp75	Naked DNA	IM	Melanoma	1	pUC	CMV/CMV	Intron A	BGH
INFECTIOUS DISEASE VACCINES									
Epimmune [212]	21 CTL epitopes	PVP	IM	HIV	1	Blue-script	CMV/CMV	Pre-pro insulin	SV40 early
IDM Pharma [212]	Various HIV	DNA PVP + Protein w/Alum	IM	HIV	1	Blue-script	CMV/CMV	Pre-pro insulin	SV40 early
IDM Pharma, Bav. Nord [212]	Various HIV	DNA + MVA boost	IM	HIV	1	Blue-script	CMV/CMV	Pre-pro insulin	SV40 early
Genencor (Innogenetics) [214]	HBSAg	PVP	IM	HBV	1	Blue-script	CMV/CMV	Pre-pro insulin	SV40 early
Chiron [212]	B Gag Env	PLGA, gp140 boost w/MF59	IM	HIV	1				
GeoVax [212, 213]	Various HIV	Naked DNA + MVA boost	IM	HIV	1		CMV/CMV	None	BGH
Merck [215]	Gag	Naked DNA	IM	HIV	1	pV1J	CMV/CMV	CMV	BGH
PowderMed [216, 217]	HBsAg	Gene gun	ID	HBV	1	pUC	CMV/CMV	CMV	BGH
PowderMed [218–221]	HA	Gene gun	ID	Influenza	1	pUC	CMV/CMV	CMV	BGH
CytRx, AdvBioLab [212]	Gag 5 env	Naked DNA + gp120 boost	IM	HIV	1	pUC	CMV/CMV	CMV	BGH

Table 2 (continued)

Company [Ref.]	Gene(s)	Technology	Route	Indication	Phase	Back-bone	Promoter/ Enhancer	Intron	Term
Vical [222]	PA/LF	Cationic Lipid	IM	Anthrax	1	pUC	CMV/CMV	CMV	BGH
Vical [222]	gB/pp65, IE1	Poloxamer + CRL1005	IM	CMV	1	pUC	CMC/CMV	CMV	mRBG
Vical/VRC [222]	Glycoprotein	Naked DNA	IM	Ebola	1	pUC	CMV/CMV	CMV	BGH
Wyeth [212]	B gag (p37)	IL-12/Bupivacaine	IM	HIV	1	pUC	CMV/CMV	none	BGH
Wyeth [212]	B gag	IL-12/Bupivacaine peptide boost	IM	HIV	1	pUC	CMV/CMV	none	BGH
Wyeth [212]	B gag	Bupivacaine,GM-CSF, peptide boost	IM	HIV	1	pUC	CMV/CMV	none	BGH
MGI Pharma [109, 223]	E6, E7	PLG	IM	HPV	2	NA	CMV/CMV	NA	BGH
THERAPEUTIC PROTEINS									
AnGes Inc [212]	Hep GF	Naked DNA	IM	PAD	2	NA	CMV/CMV	NA	BGH
Copernicus [20]	CFTR	PEG-PLL	IN	Cystic Fibrosis	2	PUC	CMV/CMV	CMV	HTLV1 + BGH
Corautus Genetics Inc [224]	VEGF-2	Naked DNA	IM	CAD	2	PUC	RSV/CMV	Rat insulin	Rat insulin
Centelion (Sanofi/Aventis) [224]	FGF-1	Naked DNA	IM	PAD	2	pUC	CMV/CMV	NA	NA
Protiva Biotherapeutics Inc [224]	HSV-TK	Cationic lipid	IV	Cancer	1	pUC	CMV/CMV	CMV	RBG
Valentis [224]	DEL-1	Poloxamer-188	IM	PAD	2	pUC	CMV/CMV	Chimera IVS8	BGH

using DNA encoding vaccine antigens (cancer and infectious diseases) and therapeutic proteins, and includes company sponsor, gene(s) encoded by the DNA, technology coupled with the DNA, route of administration, disease indication, clinical trial phase and plasmid components.

Three-quarters of current company-sponsored plasmid-based clinical trials involve intramuscular vaccination and over half of these involve vaccination against HIV-1. The remainder involve anti-cancer vaccination or immunotherapy and therapies for peripheral/coronary artery diseases. Most trials have recruited less than 200 patients and almost 90% are in clinical phase 1. Trials involve DNA vaccines encoding antigens for serious debilitating diseases (influenza, AIDS, malaria, CMV) and potential biodefense-related pathogens (anthrax and Ebola). The technologies used in clinical trials with plasmid involve the use of polymers, cationic lipid systems, anesthetics, viral or protein boosting and needle-free delivery devices.

The next section will detail the basic and applied technologies being employed to advance a plasmid-based drug product to market. Four basic unit operation technologies to be discussed are: 1) plasmid design; 2) manufacturing; 3) formulation; 4) administration. A brief analysis of potential areas of future optimization will follow each unit operation description.

3 Technology reviews

3.1 Plasmid design

3.1.1 Structural elements

Components for manufacturing

The structural components used for plasmid production in high-density fermentation growth generally include a bacterial origin of replication (*ori*) and a selectable marker.

Plasmid replication

The first component, the bacterial *ori*, functions to control plasmid copy number per bacterial cell. While early non-viral systems used in human clinical trials contained the pBR322 ColiE1 *ori* [22], virtually all plasmids entering the clinic over the past 2–3 years contain the modified *ori* initially developed in the pUC series of plasmids constructed by J. Messing [23]. The breakthrough in technology came when it was found that the plasmid gene, termed *rop*, limited the plasmid number to 20–40 copies per cell. The now well-characterized pUC family of plasmids lack *rop* and also have a single base change in the *ori* sequence encoding the RNA elements required for replication. These two modifications increased copy number by 3–4 fold without any apparent effect on plasmid stability [24]. Therefore, fermentation growth of pUC-derived plasmids encoding average size proteins (1500 bp) could replicate to more than 500 copies per cell. Plasmid yield per cell is important when calculating the eventual cost of goods and may change after fermentation scale-up to thousands of liters of bacterial culture. Thus, there is considerable effort to further increase the copy number for plasmids during scale-up manufacturing.

Plasmid selection

The second structural component usually necessary for bacterial cell production of plasmid is a selectable marker gene. This is required for positive selection of cells transformed with the plasmid encoding the marker. Typically, these markers have been antibiotic resistance genes since the growth of bacteria in media containing antibiotic drugs is convenient, inexpensive, and extremely efficient. The selection marker genes most commonly used in laboratory-scale plasmids have been β-lactamase (resistance to bacterial lytic penicillin-based drugs) and tet^{R} (resistance to the bacteriostatic drug tetracycline). However, with the development of plasmids for direct delivery to human tissue, the FDA resolved that the safest selectable marker gene would be one that is not commonly used in human health and would therefore not allow the possibility of conferring antibiotic resistance to human symbiotic bacteria [25]. In addition, there was also concern that the use of penicillin drug family members would induce anaphylactic shock in individuals allergic to these drugs if even a low contamination of drug was present in purified plasmid preparations. Therefore, nearly all plasmids used in clinical studies (and most animal models of human disease) have the transposable element, TN903 gene, an aminoglycoside enzyme conferring kanamycin resistance during bacterial cell growth. Kanamycin is not an antibiotic widely used in humans and perhaps for that reason there are essentially no known allergies to it. The FDA has thus declared that this is an acceptable selectable marker for use on plasmids destined for human trials (FDA Points to Consider for Plasmid DNA Vaccines for Preventive Infectious Disease Indications, 1996).

However, the introduction of any antibiotic resistance factor as a selectable marker on plasmids delivered at high dose for veterinary as well as human applications still has the theoretical risk of transferring a resistance activity into the human intestinal microenvironment. Therefore, non-antibiotic selection systems have been developed and one such system, using a pCOR host-vector system, has entered clinical trials. The pCOR plasmid carries a synthetic amber suppressor tRNA gene (*sup* Phe) which, when combined with a host cell carrying an amber mutation in the *argE* gene (arginine auxotroph), provides a means to select for plasmid propagation without antibiotics [26]. In addition, the pCOR plasmid carries the *cer* gene to decrease plasmid multimerization, and a modified origin of replication (R6Kγ) that is restricted to host cells carrying the *pir* (encoding the psi protein) gene. The host cell therefore must encode both the psi protein and a mutated *argE* gene. This host/plasmid system has been used to produce plasmids at a level comparable to pUC-based vectors (100 mg/liter of media) both at the laboratory scale and in 800 liter fermentors under fed-batch conditions [26].

3.1.2
Components for eukaryotic cell expression

Expression cassette

There are three basic domains controlling the eukaryotic expression cassette in plasmids being used for clinical studies: the eukaryotic enhancer/promoter domain, the transcribed region domain (including the sequence encoding the protein(s) of interest), and the transcription termination domain. The transcribed region also contains untranslated regions at both the 5′ and 3′ ends of the transcription product (mRNA) and therefore forward of, and downstream from, the coding region. Each of these domains contain regulatory elements favored due to their ability to maximize constitutive gene expression.

Enhancer/promoter

All plasmids that have been advanced into human clinical studies have enhancer/promoter elements derived from viral vectors. In an early plasmid construct, the Rous Sarcoma Virus promoter (from the long terminal repeat and termed "RSV promoter") was used to express the HLA-B7 class I heavy and light (β2-microglobulin) proteins. However, as shown in Table 2, the majority of plasmids entering clinical trials over the past 3–4 years use the human Cytomegalovirus Immediate Early (hCMV-IE) enhancer/promoter. The reason for this is based largely on a variety of comparative studies from transfected cells in culture and from animal studies suggesting that the CMV enhancer/promoter domain constitutively expresses the highest level of proteins [27–35]. This has resulted in several studies to screen regions of the CMV enhancer/promoter for the specific controlling elements responsible for such promiscuous expression. For example, well-spaced binding sites for the inflammation responsive transcription factor, NFκB, suggest that these elements may be responsible for high levels of expression during the early period when local inflammation occurs as a result of the injection of plasmid [36]. As described further below, NFκB sites have been manipulated in several experimental systems, both *in vitro* and *in vivo*, to map a correlation between these sites, levels of expression, and, in the case of antigenic protein, immunogenicity. It remains unclear what hCMV-IE enhancer/promoter elements contribute to expression in human tissue and how such elements could be exploited. Indeed, the high expression of

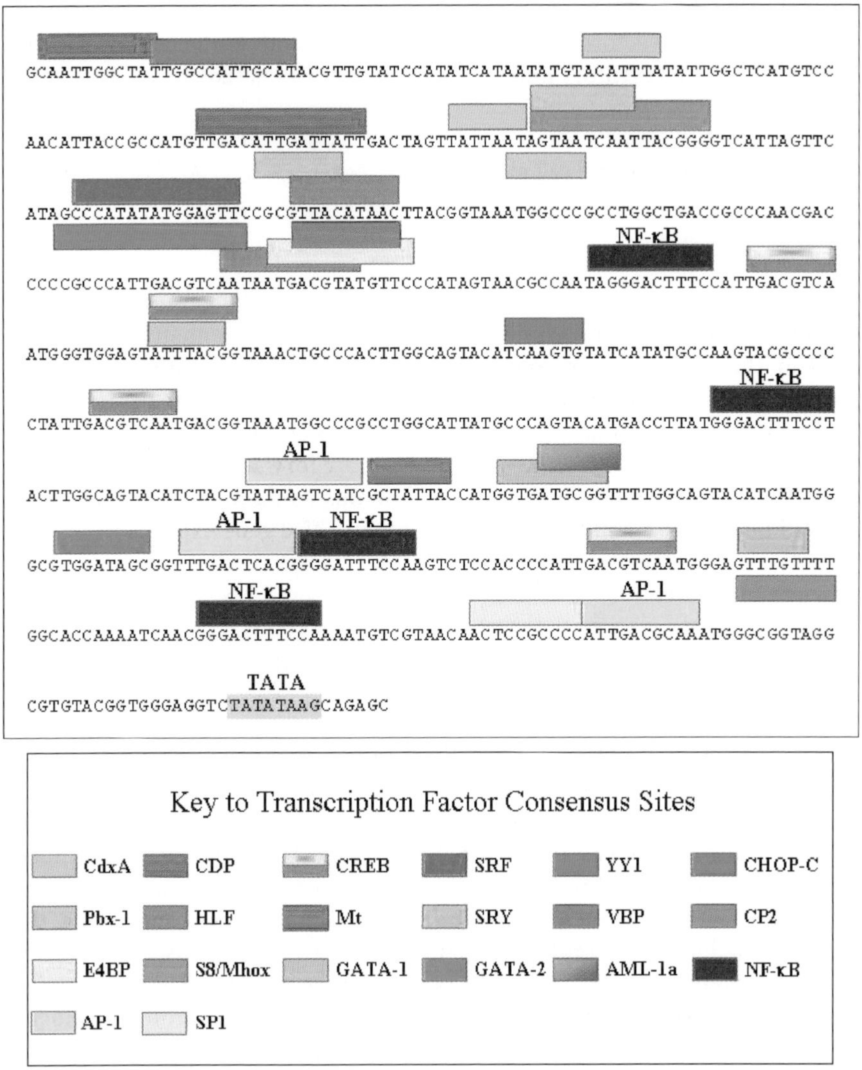

Fig. 1 The human CMV promoter contains numerous consensus sites for several known transcription factors. The diversity and number of putative factors both contribute to making the CMV promoter such a ubiquitous and potent driver of transgene expression. (Consensus site map was generated using the TFSEARCH program and database at http://molsun1.cbrc.aist.go.jp/research/db/TFSEARCH.html)

hCMV-IE enhancer/promoter may result from the large number of predicted eukaryotic transcription factor binding sites and the lack of a repressor site, as shown in Fig. 1.

Two human studies have employed a chimeric enhancer/promoter using hCMV-IE and RSV components [37, 38]. A chimeric enhancer/promoter was designed to deliver the therapeutic protein human Vascular Endothelial Growth Factor 2 (VEGF-2) in humans to reverse angina pectoris and promote angiogenesis in heart muscle [39–41]. Another trial used the same plasmid to deliver VEGF-2 to skeletal muscle to relieve arterial blockage in individuals with critical limb ischemia [41]. There is no published evidence indicating the preclinical evaluation of this chimeric enhancer/promoter and there is no indication from the design of the clinical trial that any comparative evaluation between the individual CMV and RSV components was done in humans.

Transcriptional termination domain

There has been little evidence to indicate the mechanism by which transcriptional termination is controlled in eukaryotic cells. However, since all vertebrate RNA polymerase II-dependent mRNA products are polyadenylated at sites 25–50 bases 3′ to an AAUAAA sequence, it is assumed that this signal sequence is coupled to termination and processing (addition of up to 250 adenosines to the 3′ end). Without mechanistic understanding, all plasmids taken into clinical trials to date have assumed that termination as well as polyadenylation signal sequences are normally contiguous to the AAUAAA signal sequence and that arbitrary addition of several hundred bases 3′ to the AAUAAA will suffice for this control element to function. It has also been assumed that the processing of mRNAs from an extrachromosomal plasmid element will be recognized and processed by the enzymatic activity normally used for transcripts derived from genomic DNA.

As seen in Table 2, the most commonly used AAUAAA-containing termination elements are derived from either SV40 T antigen, the rabbit beta-globin gene, or the bovine growth hormone gene comprising the contiguous 100–300 bases to each side of this sequence. The elements were found, along with their 3′ untranslated region (UTR) domains, to be efficient termination/polyadenylation sequences with a large number of heterologous 5′ coding sequences [42, 43]. Preclinical evidence has indicated that these termination sequences may affect the level and type of immune response [44]. However, this has yet to be demonstrated in humans.

Transcribed domain

As introduced above, the transcribed domain comprises: i) the region encoding the translated protein; ii) the 5′ flanking UTR; iii) the 3′ flanking UTR. When designing an expression plasmid for *in vivo* delivery of a protein to effect a significant biological response, these three components of the transcribed domain should control efficient transcription, nuclear transport, translation, mRNA stability and the intracellular transport of the expressed protein. The efficiency of eukaryotic transcription from bacterial plasmids remains largely unknown. This is a difficult problem to address in preclinical and clinical studies since the structure(s) of plasmid in the nucleus following cell entry is unclear and the best structure of the plasmid (covalently closed circle, nicked circle, or linear) for effecting the highest level of transcription is also unknown.

The ability of the transcribed mRNA to be transported from the nucleus to the cytoplasmic ribosome is known to be affected by the presence of an intron on the primary transcript and the strength of processing (intron removal). Consequently, many (but not all) of the plasmids used in human clinical trials have introns and all of these intronic sequences have been used based on preclinical animal studies showing their requirement for enhanced expression. The effective location of an intron can be either within the 5′ UTR, the coding region or the 3′ UTR. The hCMV-IE 5′ UTR, often used in conjunction with the hCMV-IE enhancer/promoter, can include the 856 bp intron A. The SV40 3′ UTR can also contain an intron sequence. Both have been shown to increase expression of a biological protein in transfected cells in culture as well as in animal models, therefore justifying their continued use in plasmids destined for clinical evaluation. Preclinical studies have shown that the hCMV-IE intron can be reduced in size and still be effective. It has also been shown that the intron can be made chimeric with alternative splice donor or splice acceptor sequences and still retain normal or even increased splicing activity. Only one plasmid with such chimeric introns has been taken into human trials [45, 46]. Finally, it should be added that some viral and prokaryotic sequences contain cryptic splice donor and/or acceptor sequences. These sequences, when present with either a 5′ or 3′ known intron, could cause differential splicing and, in turn, significantly diminish levels of protein expression. Therefore, such non-eukaryotic sequences should be carefully examined prior to construction of expression plasmids.

While preclinical data has shown that the introns present in coding regions can induce a higher level of biologically active protein, there is no evidence that introns within coding sequences have reached the clinic. This may be due to the concern that introns within some coding regions may not be active in all cell types. This has been seen in preclinical studies of two genes (human growth hormone; mouse alkaline phosphatase) where the cDNA is expressed at a level at least 10-fold higher in mammalian muscle relative to the complete gene sequence (P. Hobart, unpublished observations). In each case, the protein coding sequence within the gene contained two or more introns.

Translation of open reading frame (ORF) within mRNAs that reach the ribosome is known to be affected by the capping to modify the 5′ end, the presence of a consensus Kozak sequence to initiate translation and a polyadenylation-modified 3′ end. It has also been shown that there are elements in both the 5′ and 3′ UTRs that can confer either stability or instability based on the host cell type. These structural components of mRNAs and the post-transcription modifications to eukaryotic mRNAs are assumed to be assured when using the well characterized 5′ and 3′ UTRs. However, efficiency of translation and stability may be affected by the mRNA secondary structure and this may be significantly affected by the protein coding sequence. While there is good evidence that secondary structures within the 5′ UTRs of prokaryotic mRNAs affect expression by masking elements such as translation start sequences [47], such information is not yet available for eukaryotic mRNA, especially when dealing with *in vivo* expression in heterologous cell types. Therefore, UTRs used in clinical studies have been chosen because of their successful expression of heterologous coding sequences when tested in transfected cells in culture and in preclinical animal models.

Translation of mRNAs is also affected by codon usage [48, 49]. The frequency of codons used in eukaryotic ORFs can be significantly different from those used by prokaryotes. This difference is presumably also reflected in the expression of tRNAs and the tRNA synthetases, components required for translation of these codons. These differences can have the combined effect of limiting translational efficiency and reducing protein expression. Therefore, the most frequently used mammalian codons should be used in plasmid-based eukaryotic expression cassettes to encode all prokaryotic, viral and eukaryotic proteins in mammalian cells. This is now easy to do and cost-effective based on the greatly reduced base pair cost for chemical synthesis.

Plasmids that are now in clinical trials are constructed with synthetic sequences that have been optimized for human codon usage. While initially justified for increasing expression of prokaryotic or lower eukaryotic antigens, the design of synthetic sequences can also remove cryptic splice junctions, alternative open reading frames, unwanted restriction sites, and destabilizing sequences that may affect plasmid production in bacterial cells or the transcribed mRNAs when transfected into eukaryotic cells. It also provides the opportunity to introduce or reduce immune stimulatory CpG motifs from plasmid eukaryotic ORFs produced in bacterial cells. This depends on whether the plasmid is designed to express an antigen or a therapeutic protein [50, 51].

Intracellular targeting of expressed proteins to effect release from transfected cells has been accomplished by fusing ORFs to eukaryotic leader peptides [52]. This is necessary for both the efficient delivery of therapeutic proteins and induction of a robust humoral immune response to expressed antigens. The fusion of a leader peptide does not appear to be needed to enhance the cellular response to plasmid-expressed antigens where proteins are to target the TAP system for MHC presentation [53].

Finally, translation of the mRNA may involve a bicistronic transcript, enabling the expression of two ORFs from a single mRNA. Such bicistronic constructs have been used in clinical trials [9, 22]. In these constructs, the first ORF requires the 5′ cap of the mRNA for translation initiation whereas the second ORF is 3′ to a viral cap and requires a Cap-Independent Translational Entry (CITE) sequence. These sequences, also termed Internal Ribosome Entry Site (IRES), provide an alternative method for ribosomal recognition of a mRNA's translational initiation sites. CITE sequences have been used on plasmids to express both heavy and light chains of the Class I MHC protein HLA-B7 upon direct injection into tumor tissues [22]. They have also been used to deliver a patient-specific B-cell lymphoma light and heavy chain immunoglobulin as a tumor-specific vaccine injected into skeletal muscle [9].

3.1.3 Future plasmid design for human use

The inability of plasmids to achieve *in vivo* expression comparable to widely used viral vectors indicates that further improvements are needed to increase the efficiency of plasmid delivery and expression. This may be facilitated by plasmid formulations and/or devices, which will be presented below. However, structural components of plasmids may also contribute significantly to their ability to increase the biological activity of the expressed protein(s).

Plasmid entry into mammalian cells

The role of the primary, secondary, or tertiary structure of plasmids in effecting binding to and uptake by mammalian cells is not well understood. Plasmids are presumed to be foreign macromolecules within higher eukaryotes and are therefore recognized as pathogens and normally phagocytized and removed. The highly investigated role of unmethylated CpG motifs in the primary sequence of plasmids is presumed to contribute to the recognition as "foreign" DNA and has been seen as an advantage for enhancing protein-based vaccines. Recent evidence indicating that CpG motifs are recognized and signal through Toll-Like Receptor 9 (TLR-9), and that the motifs enable DNA to actually bind to this receptor, are significant developments in the understanding of how DNA plasmids affect mammalian cells [54]. However, the role of other primary structural sequences, secondary sequences (B vs. Z DNA), and tertiary sequences (covalently closed circles [CCC], duplexes, nicked circles, and linear DNA) on the recognition and uptake by mammalian cell surface proteins and subsequent transport of the plasmid to the mammalian cell nucleus are largely unknown. Most evidence demonstrating relative efficiencies of CCC vs. nicked circular vs. linear plasmid taken up by mammalian cells comes from cells in culture, which may not be predictive of plasmid behavior in cells *in vivo*.

The mechanism(s) of cell receptor uptake and intracellular transport used by viruses is one avenue that must be recognized as both highly efficient and highly evolved with regard to recognition by host cell surface protein receptors of viral particle coat proteins [55–58]. The role of the viral nucleic acid component in this process is not well described and is considered minor relative to coat proteins. However, the role of the viral nucleic acid sequence in subsequent intracellular trafficking may be instructive.

Plasmid nuclear targeting

Transport of plasmid across the cytoplasm and into the nucleus is an additional rate limiting step in DNA delivery and expression [59]. Recent experimental approaches have suggested that eukaryotic host cell DNA binding proteins with nuclear localization signal (NLS) sequences may provide one way to enhance nuclear transport. Such binding proteins, usually transcription factors, come equipped with a NLS sequence, and have demonstrated high affinity for targeted nucleic acid sequence motifs [60]. This concept is highly desirable because the targeted sequences can be easily inserted into a plasmid, are specific, and are usually short [61, 62]. However, to date, improvement of nuclear transfer of plasmids *in vivo* has been relatively modest and since plasmid delivery *in vivo* is the result of several sequential events, it is difficult to assess nuclear transport solely via measurement of gene expression. Nuclear transfer of plasmids *in vivo* remains a poorly understood and addressed step in the overall delivery of plasmids.

Eukaryotic promoter

Transcriptional promoters have been chosen for use in plasmids mainly for their ability to increase mRNA and subsequent protein expression per transfected cell. However, for vaccines, additional factors may affect the promoter's ability to enhance antigen immunogenicity. These include the type of cell that is transfected (such as structural vs. migratory dendritic and phagocytic cell) by the plasmid in the injected tissue such as muscle, skin or mucosa, and by the ability of the promoter to sustain expression over a long period of time. Promoters designed to elevate expression in dendritic or other antigen-presenting cells may be useful for plasmid based vaccines where targeted expression of the antigen may prove more important than promoter strength [63].

There is a long-standing concern that the expression of plasmids, even those reaching the nucleus, may be affected by bacterial sequences (plasmid backbone) commonly found in plasmids and required for growth in bacterial cells. The preclinical evidence supporting this concern includes transfection of cells (*in vitro* and *in vivo*) with supercoiled minicircles that lack all backbone sequences and have been purified away from CCC pDNAs using CsCl gradient centrifugation [64, 65]. However, a clinical evaluation of the backbone effects on expression in humans will require improvements in minicircle DNA manufacturing.

Untranslated regions

The analysis of UTR structural information and the mechanism by which this information is manifested in cells has been of interest since the structure of mRNA was first revealed. Over the past few years, with ever-increasing RNA sequence information and numbers of proteins found to bind to eukaryotic mRNAs, and with experimental analysis indicating that these protein levels change in normal, tumor, and pathogen-infected cells, the analysis has become more defined and may involve elements that can be exploited in plasmid-based gene delivery systems.

UTR functions have been grouped into three general categories [66, 67]: 1) subcellular transport and targeting; 2) regulation of translational efficiency of the mRNA; 3) sequences that affect promotion of RNA stability. Both 5′ and 3′ UTRs have a role in subcellular transport and targeting. The 5′ UTRs are generally looked upon as being primary or secondary sequences that regulate the translational efficiency of the mRNA. The 3′ UTRs (and some 5′ UTRs) are looked upon as domains that affect RNA stability. UTRs contain AU rich elements (ARE) that are generally viewed as conferring instability to mRNAs [66].

It now appears that the AU motifs are more likely to be indicators of a site binding a subclass of nucleic acid binding proteins (AUBPs) and that the type of protein can either stabilize or destabilize the bound mRNA [68, 69]. Indeed, the HuR protein, which stabilizes mRNAs, and AUBP 1 protein which destabilizes mRNAs, both bind to the same ARE motifs. These general characteristics have been accepted for more than 20 years with only a few new caveats. Therefore, future plasmids should include UTR elements that may be stabilized by AUBPs normally present in transfected cells in targeted tissues cells. Similarly, since the function of elements within 5′ UTRs that effect increased translation are so closely tied to species and cell-specific expression, it may be difficult to find domains that have cassette-like activity that can act promiscuously in muscle, tumor or antigen-presenting cells.

3.2 Plasmid manufacturing

Among all biologicals, plasmid manufacturing is an extremely convenient and generic process, independent of composition (plasmid backbone and gene sequence). Although products have been manufactured under GMP for Phase 3 clinical trials, process improvement is still an active area of development. The scope of this review does not allow a thorough analysis of plasmid manufacturing processes, but this topic has been well reviewed recently [24, 70]. Much of the literature in this field has outlined effective methods of manufacturing plasmid in sufficient quantities to meet clinical needs [71–73]. The major challenge remains to achieve scaleable and economical means of producing large quantities of plasmid. With human doses currently in the milligram range, the production of a plasmid-based product with broad indications could require tens to hundreds of kilograms of plasmid annually. In order to meet this demand, productivity improvements in all manufacturing unit operations are required. Almost all processes for the manufacture of bulk plasmid can be described by the following generic series of unit operations:

Fermentation → Harvest → Lysis → Isolation → Purification → Bulk preparation

Each of these unit operations is briefly reviewed below.

Fermentation

The production of plasmid is almost universally achieved using *Escherichia coli* (*E. coli*) bacterial fermentation. Competent *E. coli* cells are transformed with the plasmid of interest to create an *E. coli* master cell bank. *E. coli* are grown aerobically in either a rich or a defined medium. There is a large body of literature outlining methods to improve *E. coli* fermentation performance [74–76]. Much of this effort has focused on maintaining high plasmid copy numbers in the bacteria to maximize production of a bacterial recombinant protein of interest. Studies of the effects of bacterial fermentation media on the production of plasmid show that plasmid yield can be improved by supplementing media with amino acids and nucleosides [77]. The C : N ratio may be a critical media formulation parameter for maximizing specific plasmid productivity [78]. Plasmids made in *E. coli* typically contain an antibiotic resistance gene. Bacterial fermentation is completed in the presence of an antibiotic in order to promote the selective growth of bacteria containing plasmids encoding the antibiotic resistance gene. An alternative to the use of antibiotic genes [26, 79], which utilizes repressor titrations to favor growth of plasmid-containing cells, was discussed above (see "Plasmid selection").

Harvest

Bacterial cell harvest typically involves centrifugation or microfiltration. Large-scale centrifugation in a current Good Manufacturing Practices (cGMP) environment can be capital intensive. At large-scale, microfiltration may offer the advantage of lower capital cost, but this must be balanced against the cost of microfiltration membranes and the number of times a membrane can be used. Microfiltration also offers the ability to wash away spent media components and extracellular impurities prior to commencing purification.

Lysis

Since plasmids are intracellular, the *E. coli* host cells are lysed as a first step in their isolation. Lysis is a critical unit operation, as this step can affect the ratio of covalently closed circular (CCC) plasmid to other forms, and the amount and physicochemical characteristics of cellular impurities that must be removed downstream. Due to the complex nature of the process streams, this is a poorly understood and difficult to monitor unit operation. There are three methods that have been employed to lyse *E. coli* cells. *Chemical lysis* by alkaline detergent solution followed by precipitation of cell debris with acetate is the most widely used method for clinical-scale manufacturing [71]. This method is relatively simple and results in a significant fraction of cell impurities being removed when the lysed solution is clarified. The yield and purity profile resulting from this method is sensitive to conditions of mixing, reaction times and localized concentrations [80, 81]. Such sensitivity to conditions implies that alkaline lysis may reach a practical limit in scale. *Heat lysis* has been used in laboratory preparations for several years and has been applied to large-scale plasmid preparation [82]. This method offers the advantage of denaturation of proteins and host cell nucleic acid. *Mechanical lysis* by homogenization is a technique widely used for isolation of proteins. However, the usefulness of current mechanical lysis methods has been discounted because the larger size of the plasmids makes them more susceptible to damage from the high shear and potential cavitation events during processing. The addition of a condensing or protecting agent may serve to partially protect plasmid during mechanical lysis [83].

Isolation

Following lysis are step(s) designed to remove cell debris and isolate crude plasmid. This can be done by precipitation with detergent [84] or polymer and/or salt [85]. This step might be eliminated by taking a lysate directly to expanded bed chromatography [73, 86], but bed capacity may be limited and a precolumn precipitation may be required to make this approach economically feasible.

Purification

Purification is often accomplished by chromatography. Several common approaches utilize one or more of the following: 1) anion exchange [71, 87], size exclusion [71, 87], hydrophobic interaction [88], reverse phase [72, 89], affinity and hydroxyapatite chromatography [90]. Traditional chromatography resins have suffered from low capacity due to the lack of accessible surface area for binding the relatively large plasmid molecules. More recently, resins have been developed with larger pore structures, making them more appropriate for plasmid purification [90, 91].

A novel approach to plasmid purification has recently been described which utilizes the affinity binding of plasmid to calcium silicate as a purifying method [92]. This method is readily scaleable and replaces expensive chromatography resins with disposable bulk adsorbents. The economics will depend on the number of times chromatography resins can be reused versus the waste handling costs for the spent bulk absorbants.

Bulk Preparation

The next step is to place the bulk plasmid in the proper vehicle at the appropriate concentration. This can be done by precipitating the plasmid and dissolving the precipitate at the desired concentration. However, the most commonly recommended method is to exchange buffer and concentrate the product using a 50–100 kDa pore size ultrafiltration membrane.

Drug Product Preparation/Formulation

Although some clinical trials have been initiated with the drug product as naked DNA, there is considerable work in progress aimed at improving the therapeutic efficacy of plasmids by using adjuvants or delivery systems including cationic lipids or nonionic polymers. A formulation step with excipients must be completed prior to drug product filling.

Analytical testing

Plasmid used as a drug substance can be considered a well-characterized biological. The manufacturing methods described above are standard bioprocessing methods and are plasmid sequence-independent. Each plasmid product has a defined molecular weight and consistent structure, and the key impurities from the manufacturing process are well-known. The tests for identity, purity, and potency of the product have become well-established and routine. 1) The *identity* of the plasmid can be determined by sequencing. This, however, is not practical or necessary to complete on each production batch. Therefore, identity is typically established using gel electrophoresis to demonstrate that the overall size and restriction patterns conform to specifications. At some point in the development process the genetic sequence of both plasmid strands must be checked to verify that the processing does not significantly alter base sequence stability. Identity usually includes a characterization of the plasmid conformation(s) with the determination of the predominant species such as supercoiled, open circular, multimers and possibly linear. These forms can all be generated during fermentation. Supercoiled forms can be transformed into open circular and linear forms during the manufacturing process. Data on the relative *in vivo* potency of these forms is scarce, so the required approach for cGMP manufacture is to monitor the relative quantity of each of these forms and verify that the manufacturing process delivers batch-to-batch consistency in structure. 2) The *purity* of plasmid products is assayed by conventional, well-established methods. Typical impurities in *E. coli* fermentation products are host *E. coli* protein, DNA,

Table 3 Quality control assays for plasmid drug substance and drug product used in clinical trials

Test type	Test description
Identity	Total size; restriction map by gel electrophoresis Plasmid conformations by HPLC
Purity	Residual *E. coli* chromosomal DNA Residual RNA Residual *E. coli* protein or amino acids Endotoxin by LAL Sterility Appearance pH
Potency/Strength	*In vitro* potency by ELISA, FACS, or RT-PCR DNA concentration (A_{260}); Excipient concentration

RNA, and endotoxin. Plasmid purification is designed to reduce these impurities to below acceptable limits. 3) The *potency/strength* are typically verified *in vitro* using transfection of a cell line and measurement of the expressed gene products such as protein by ELISA or FACS, or mRNA by quantitative RT-PCR. Since plasmids are well-characterized biologicals with low and consistent impurity levels and a reproducible ratio of the different conformers, it should be possible in the future to relate the plasmid concentration to potency. Light absorbance (A_{260}) measurements are highly reproducible and not subject to the variability of biological assays. The analytical tests for plasmid drug substance and/or formulated drug product are summarized in Table 3.

3.3 Plasmid delivery systems and formulation

Relatively complex delivery systems have been investigated in preclinical models including condensing elements, targeting moieties, endosomal release agents and nuclear localization molecules [18]. However, to date, only the simpler first-generation formulations have entered clinical trials. These formulations include noncondensing polymeric delivery systems such as poloxamers, poly[lactide-co-glycolide], poly[vinyl pyrrolidone], condensing polymers such as poly-L-lysine and cationic lipid-based systems.

Poloxamer formulations

Poloxamers are surface active, water-soluble, nonionic tri-block copolymers composed of blocks of polyoxypropylene (POP) and polyoxyethylene (POE) and are often denoted as POE-POP-POE. Figure 2 shows the general struc-

$$HO\text{-}(CH_2CH_2O)_x\text{-}(CH_2\underset{}{\overset{CH_3}{\overset{|}{C}}}HO)_y\text{-}(CH_2CH_2O)_x\text{-}H$$

POE_x	POP_y	POE_x

Fig. 2 General structure of poloxamer

Table 4 Characterization of poloxamers being used in plasmid-based clinical trials

Poloxamer	MW	(POE)x	(POP)y
188	8400	$x = 52$	$y = 30$
CRL 1005	12 000	$x = 7$	$y = 207$

ture of these polymers and Table 4 describes the characteristics of two specific poloxamers being used in gene therapy clinical trials. Poloxamer 188 (CRL 5861 or FLOCOR from CytRx Corporation or Pluronic F-68 from BASF Corporation) are hydrophilic, water-soluble polymers, whereas CRL 1005 (CytRx Corporation) is a hydrophobic polymer that forms self-assembled particulate structures at room temperature [93].

Valentis' lead product candidate, Deltavasc (VLTS-589) [94], has been evaluated in a Phase 2 clinical trial for the treatment of peripheral arterial disease. Deltavasc uses a polymeric gene delivery system consisting of the synthetic, nonionic poloxamer 188 in Tris(hydroxymethyl)aminomethane hydrochloride (Tris-HCl) [94], to deliver the angiogenic DEL-1 (development-regulated endothelial locus-1) gene to oxygen-starved tissue in the lower extremities. All components are aseptically mixed using an in-line process, filter sterilized, lyophilized and stored at 2–8 °C. Poloxamer 188 is considered to be a protective, interactive, noncondensing polymer [95], which allows formulated plasmid to diffuse into the extracellular spaces of muscle tissue and increase the distribution of plasmid around muscle cells. In addition, the polymer is claimed to protect the plasmid from nuclease degradation. Valentis recently reported that the Deltavasc product did not meet its primary endpoint in a Phase 2 clinical trial in patients with intermittent claudication. Although a statistically significant increase in exercise tolerance and ankle brachial index was observed at 90 days compared to baseline, there was no difference from the control group that received the PINC polymer alone.

Merck and Company is investigating a prime-boost regimen utilizing poloxamer CRL1005 adjuvanted plasmid expressing the HIV-1 gag protein and an adenovirus type 5 vector boost [5]. Preclinical studies in nonhuman primates have been published describing the immunogenic potential of this approach [5, 96–98]. There are no published data on the exact plasmid formulation being used as the prime boost, but a recent publication [97] described the physicochemical characterization and immunological evaluation of an optimized poloxamer formulation consisting of CRL1005 and the cationic surfactant benzalkonium chloride (BAK). Suspensions of CRL1005 form 1–10 micron diameter particles above their phase transition temperature (cloud point, 4–8 °C) and the physical properties of the particles are altered by the addition of BAK. Binding of BAK to the hydrophobic surface of the poloxamer particles reduced particle size to 300 nm and rendered the surface charge positive. DNA has been shown to interact with these cationic particles, changing their surface charge from positive to negative. Between 1.5–3% of the plasmid was shown to be associated with the CRL1005/BAK particles (in other words 20–40 plasmids/CRL1005/BAK particle). The mechanism by which CRL1005/BAK enhanced plasmid vaccination has not been determined, but the investigators postulated that CRL1005/BAK particles act as an adjuvant and/or plasmid delivery enhancer.

Vical Incorporated also has a CRL1005 formulation for a plasmid-based vaccine in Phase 1 clinical trial for prevention of cytomegalovirus (CMV)-associated disease in the allogeneic hematopoietic cell transplant population. Details on this polymer-based formulation are forthcoming [99, 100].

Protective, interactive, noncondensing (PINC) polymers

Several gene therapy clinical trials have been conducted or are ongoing (see Table 2) using colloidal polymers such as poly(vinyl pyrrolidone) or PVP, which interacts with plasmid via hydrogen bonding and hydrophobic interactions. The interactions result in protection of the plasmid from nucleases, possibly by providing a hydrophobic coat around the plasmid [101–105]. The polymer does not condense DNA (plasmid structure remains flexible) and the resulting formulation can be hyperosmotic, resulting in improved dispersion of the plasmid through the extracellular matrix of the muscle and solid tumor tissues. The increased hydrophobicity of the complex compared to free plasmid may facilitate muscle cell uptake.

Poly(lactide-co-glycolide) (PLG)

A growing appreciation of the importance of delivering plasmid to antigen-presenting cells (APCs) has led to increased interest in the formulation of DNA vaccines with phagocytosable particles [106]. PLG is a synthetic polymer of lactic and glycolic acids readily available from GMP-grade suppliers (Birmingham Polymers Inc., a subsidiary of Durect Corporation; Boehringer Ingelheim, and Alkermes, Inc.). PLG has been approved by FDA for several therapeutic applications and exhibits biodegradability, biocompatibility and safety in humans. Zycos (a subsidiary of MGI Pharma, Inc.) has a novel plasmid-based immunogen (Zyc 300) in Phase 1 clinical trials [107], comprising PLG microparticles formulated with plasmid expressing a mutated CYP1B1 protein. The CYP1B1 protein is over-expressed in all primary human tumors but exhibits minimal expression in critical normal tissues, making this an attractive tumor specific target for immunotherapy. Zycos also has a therapeutic human papillomavirus vaccine (ZYC101a) in Phase 2 clinical trials [108], comprising plasmid encapsulated in PLG microparticles. The plasmid encodes multiple HLA-restricted epitopes derived from human papilloma virus (HPV-16 and 18) E6 and E7 proteins, which are HPV oncoproteins consistently expressed in neoplastic cells [109]. The standard method of encapsulating DNA in PLG microparticles is the formation of a water-oil-water double emulsion followed by solvent evaporation [110]. Plasmid in aqueous solution is homogenized with PLG in organic solvent to form a primary emulsion, which is then added to a second aqueous phase of polyvinyl alcohol and homogenized to form the double emulsion. The organic solvent is then evaporated and the PLG solidifies to form microparticles containing plasmids. The particles are filtered and washed and lyophilized to remove the water from the internal aqueous phase. In preclinical mouse studies, formulation of PLG with taurocholic acid (TA) or monomethoxyl polyethylene-glycol-distearoylphosphatidylethanolamine (PEG-DSPE) has been described [107]. The exact mechanism of action of this formulation is unknown, but particulates and their components may be pro-inflammatory and/or may target the DNA to APCs [107].

Chiron Corporation is in Phase 1 clinical trials with an HIV vaccine designed to produce HIV-specific humoral (B-cell) and cellular (T-cell) immune system responses by Clade B gag and envelope plasmid/PLG vaccination priming followed by MF59-adjuvanted recombinant envelope glycoprotein boosting. The manufacture and physical characterization of this type of plasmid delivery system has been well-described [111–114]. The PLG microparticles are coated with the cationic surfactant cetyltrimethylammonium bromide (CTAB) during preparation using a standard double emulsion followed by solvent evaporation technique [110–114]. A net positive surface charge allows plasmid to be adsorbed onto the particle surface via electrostatic interaction. Increased immunogenicity is described to result from plasmid persistence at the injection site, recruitment of mononuclear phagocytes and plasmid delivery to APCs [111, 114].

Polylysine and polyethylene glycol conjugates

A polymeric condensing system based on poly-L-lysine has been tested in a Phase 1–2 clinical trial to deliver a CFTR plasmid by the intranasal route [115]. The unimolecular/polymer complex consists of a single plasmid compacted with polyethylene glycol (PEG)-substituted 30-mer lysine polymer (CK_{30}). The polylysine 30-mer with a *N*-terminal cysteine was produced by solid-phase synthesis as the trifluoroacetate (TFA) salt; the cysteine residue was then conjugated with polyethylene glycol (average MW 10 000) to form PEGylated polysine CK_{30}PEG10k. The TFA counterion was exchanged with acetate by gel filtration. The plasmid (8.3 kbp) encodes the cystic fibrosis transmembrane regulator (CFTR) protein. When the acetate counterion of CK_{30}PEG10k is mixed with plasmid, the plasmid is condensed into rod-like nanoparticles with a diameter of 12–15 nm and a length of 100–150 nm [116]. These DNA nanoparticles are unique for a non-viral system in that they can be highly concentrated without aggregation in saline (12 mg/ml), are stable in serum and can be combined with liposomes and used to transfect post-mitotic cells. Once inside cells, they will directly cross into the nucleus via the 25 nm nuclear pore complex [117]. In the Phase 1–2 clinical trial, administration of compacted DNA nanoparticles to the nasal epithelium of CF patients was safe. There were no significant adverse events and cell transfection was detected in the nasal epithelium cells. There was evidence of partial CFTR chloride channel reconstitution and 8 out of 12 subjects had evidence of partial nasal potential difference corrections in response to isoproterenol.

Cationic lipid-based formulations

Cationic lipid-based formulations have been used as delivery system for pDNA-based therapeutics and vaccines for over a decade. Cationic lipids are typically first formulated with colipids to form liposomes. Incorporating positively charged amphiphilic cationic lipids into the liposomes imparts multiple positive charges to the hydrophilic surfaces of the liposomes that are exposed to the bulk aqueous environment. When cationic liposomes and DNA are mixed in aqueous solution, the positively charged amine groups in the cationic lipid headgroup bind the negatively charged phosphate groups of the plasmid, during which a structural rearrangement occurs in both the liposomes and DNA. The primary driving force for the association of cationic liposomes with DNA is electrostatic, but hydrophobic interactions and hydration forces are also involved in the structural rearrangement to spontaneously form plasmid/lipid complexes (so-called "lipoplexes") [118–120]. The detailed structure of the plasmid/lipid complex depends on several factors, including the initial liposome size, the DNA/cationic lipid ratio, the formulation vehicle, the formulation procedure and the structures of the cationic and neutral lipids comprising the initial liposomes.

Liposomes are typically prepared by hydration of dried films using agitation. Large, multilamellar vesicles (MLV) are formed upon hydration and can be used directly for formulation. MLV are typically in the micron size range. There are several procedures that reduce liposome size and enable formulation at relatively high lipid and DNA concentrations. The most common methods are extrusion and homogenization. Extrusion involves forcing liposomes through one or more filters with selected pore sizes. Homogenization and microfluidization involves forcing liposomes through orifices at high pressure, a process similar to milk homogenization. Homogenization and extrusion can also be combined, and both processes are used to produce liposomes on a large scale.

Lipoplexes can be prepared by in-line mixing, which involves aseptically combining sterile plasmid and sterile bulk liposomes at a defined rate and shear force, most commonly using a peristaltic pump and static mixer. The physical properties of the complexes prepared by in-line mixing depend principally on the cationic lipid/DNA molar ratio. When the cationic lipid is in molar excess, the lipid coats the DNA and positively charged complexes are formed. If the DNA is in molar excess, negatively charged complexes are formed and the DNA is either free in solution or extends from the liposome surface [121]. Encapsulation has also been proposed as another method for lipoplex preparation. DNA is entrapped within the aqueous compartments of the liposomes by freeze-drying followed by controlled rehydration and centrifugation to remove unencapsulated DNA [122].

A second method of encapsulation produces stabilized plasmid-lipid particles (SPLP) by a detergent dialysis method [123]. Plasmids are incubated with the cationic lipid DODAC in the presence of buffered octylglucoside containing 150 mm NaCl. The colipid DOPE and PEG analogs are then added. The PEG analogs are added to increase the circulating half-life of the complexes following IV injection. The plasmid-lipid mixture is then dialyzed extensively against buffered saline. Nonencapsulated plasmid is removed by anion exchange chromatography and empty vesicles can be removed by sucrose density centrifugation [124]. The principal aim of SPLP is to produce complexes that circulate long enough in the blood to allow accumulation at sites of inflammation and tumor growth. When SPLPs are administered intravenously (i.v.), about 10% of the injected dose remains in circulation twenty-five hours post injection [125, 126]. Conventionally prepared lipoplexes are cleared from the circulation within a few hours of injection. Due to the relatively long circulating half-life, plasmid delivered by SPLP can be taken up and expressed by the tumor cells, resulting in a local therapeutic effect [127, 128]. Following i.v. administration, about 80% of the injected dose is targeted to the tumor and tumor-specific protein expression has been demonstrated [124–126].

Lipoplexes have been shown to be therapeutically efficacious in a variety of animal tumor models (Table 5). As a result, cationic lipid/DNA complexes have been tested for a variety of cancer indications using different routes of administration *in vivo*. Plasmid encoding IL-2 or transplantation antigen has been used for intratumoral injection of lipoplexes [129–132]. Lipoplexes may stimulate the immune response by increasing cytokine levels. Human clinical trials have been completed for head and neck (Phase 1 and 2) and ovarian cancer (Phase 1) using the cationic lipid DC-Chol, with the co-lipid DOPE, complexed to pDNA encoding E1A, a tumor inhibitor gene. The SPLP technology for DNA encapsulation is being used in a Phase 1 trial to formulate a thymidine kinase (HSV-TK) gene product (Pro-1) for late stage metastatic melanoma [133]. Plasmid encoding the HSV-TK gene is encapsulated into liposomes and delivered IV in combination with the prodrug valcyclovir. The combination of HSV-TK with prodrug causes cell death by inhibiting the incorporation of dGTP into DNA and preventing DNA synthesis [134]. The toxic metabolite apparently spreads to surrounding tumor cells. Preclinical studies have shown that treatment of mice with Pro-1 inhibits tumor growth, causes tumor regression, and protects animals from tumor challenge [124, 135].

Lipoplexes have been used for a variety of other therapeutic indications. pDNA encoding genes to treat cystic fibrosis and alpha-1 antitrypsin deficiency has been delivered to the lung airways [136, 137], the nasal epithelium [138, 139], and lipoplexes have been used for delivering the angiogenic factor VEGF to the vasculature [140, 141].

Table 5 Cationic lipid formulations used for cancer indications

Lipids	Route	Model	Gene	Ref.
DMRIE:DOPE	IP	Mouse ovarian tumor	IL-2	[225]
DMRIE:DOPE	IT	Mouse renal cell carcinoma	IL-2	[226]
DMRIE:DOPE	IT	Mouse renal cell	IL-12	[227]
DMRIE:DOPE	IT	Mouse melanoma	hIL-2	[228]
TMAG/DLPC/DOPE	IVT	Human glioma in nude mice	hIFN-beta	[229]
LPD-PEG-Folate	IV	Mouse breast tumors	HSV-1 TK	[230]
DP3	IV	Mouse lung cancer	p53	[231]
DOTMA:Chol	IV	Mouse squamous cell	IL-12	[232]
DOTMA:Chol, DOTMA:DOPE	IV	Mouse metastatic lung	IL-2	[233]
DOTMA:Chol	IT	Head and neck mouse squamous cell	IL-2	[234–236]
DOTMA:Chol, DOTIM:DOPE	IT	Canine melanoma	Staph enterotoxin B, GM-CSF or IL-2	[237]
DOTAP:Chol/protamine	IV	Fibrosarcoma	IL-12, p53	[238]
DOTAP:Chol/protamine	IV	Mouse HPV	HPV16 E7	[239]
DOTAP:Chol	IV, IT	Mouse human lung xenografts	p53, FHIT	[240]
GL-67: DOPE: DMPE-PEG 5000, DC-Chol:DOPE	IP	Mouse mesothelioma (sarcoma)	hsp-65	[241]
DOTMA:Chol	Pulm	Mouse lung cancer	IL-12	[242]

Plasmid formulations and methods of generating humoral responses with DNA vaccines can be further improved. Formulation of DNA with cationic lipids has been shown in a variety of *in vivo* studies for several infectious disease targets to result in enhanced humoral responses against several antigen targets. Referenced preclinical studies using cationic lipids are summarized in Table 6.

Most of the preclinical studies listed in Table 6 employed intramuscular injection of lipoplexes to treat infectious diseases in rodent models. Plasmid doses vary considerably in the studies. For example, adult Welsh ponies were injected IM at Day 0 and 28 with 200 micrograms of plasmid encod-

Table 6 Cationic lipid formulations used for infectious diseases

Lipids	Route	Model	Gene	Ref.
Vaxfectin™	IM	Mice	NP (flu)	[158]
Vaxfectin™	IM	Rabbits	NP (flu)	[158]
Vaxfectin™	IM	Mice	5 antigens incl. NP (flu)	[243]
Vaxfectin™	IM	Mice	JE prM and E (Japanese encephalitis virus)	[244]
Vaxfectin™, GAP-DLRIE:DOPE	Salivary	Mice	NP (flu)	[245]
Vaxfectin™, GAP-DLRIE:DOPE	IM, IN	Mice	Ag85A, AG85B, PStS-3 (TB)	[246]
Vaxfectin™	IM, ID, IN	Mice	gp140 (HIV)	[247]
Vaxfectin™	IM, ID, IN	Baboons	tat, nef, gag/pro, env (HIV)	[248]
DMRIE:DOPE	IM	Ponies	glycoprotein G (rabies)	[249]
DMRIE:DOPE	IM	Dogs	glycoprotein HA and F (canine distemper)	[249]
Vaxfectin™, DMRIE:DOPE	IM	Rabbits	PA, LF (Anthrax)	[?]
DDAB:PC	IM	Mice	NS3 (hepatitis C)	[250]
DC-Chol:DOPE	IM, IP, IN, SC, ID	Mice, G-pigs	Env (HIV)	[251]
DC-Chol:DOPE	IN	Mice	env/rev (HIV)	[252]
DC-Chol:DOPE (± mannan)	IM, IN	Mice	rev (HIV)	[253]
DOTMA:DOPE	IM, IP, IV	Mice	NP (flu)	[254]
TM-TPS: DOPE (CellFECTIN)	IM, ID	Monkeys	HBsAg (HBV)	[255]
DOTAP:PC:DOPE	IM	Mice	HBsAg (HBV)	[256]
DODAC:DOPE: PEG_{2000}C8CER	IM, IN	Mice	HA (flu)	[257]
DOSPER	IN, Oral	Mice	HA (flu)	[?]

ing rabies glycoprotein-G [142]. The plasmid was formulated with either aluminum phosphate or as DMRIE-DOPE lipoplexes. All of the horses injected with cationic lipid-formulated plasmid (but none of the horses injected with plasmid adsorbed to aluminum phosphate) seroconverted by 14 days after the prime and developed anti-glycoprotein G titers indicative of protection against a virulent rabies challenge by 14 days after the boost. This study shows that even small doses of plasmid, properly formulated with DMRIE:DOPE cationic lipid system and injected into large animals, can elicit protective immunity against an infectious disease.

The efficacy of different particulate systems has been demonstrated in several animal models. However, the efficacy in humans for cancer and infectious disease indications is still under investigation. Improving particulate systems for vaccine and therapeutic applications remains an active area of research and development for several companies. It is likely that optimization will require particular combinations of formulations, regimens, and routes of delivery. A key advantage of particulate formulations is the ability to produce well-characterized, reproducible formulations. In addition to efficacy in humans, a critical component to the success of particulate formulations will be the ability to manufacture single vial formulations cost effectively. Enhanced stability and elimination of the cold chain will also be critical to the production of the next generation of vaccines.

3.4
Plasmid administration: Devices

In the global pharmaceutical market, the holy grail of drug delivery continues to be oral administration using pills, capsules or liquids. However, biotechnology-derived pharmaceuticals include large molecules such as proteins and DNA, and there are few technologies available to enable such macromolecules to be therapeutically delivered after oral administration to humans. Thus, most DNA-based drugs to date are administered into tissues, usually within the upper extremities.

Conventional needle-based drug injections were developed about 150 years ago, and became widely used in the 1920s for mass vaccinations, and even more widely used following the introduction of penicillin after World War II. Modern syringe-needle devices are individually wrapped, sterile, single-use, disposable, readily available and cheap (< 8 cents each). Their basic design and use are universal and administration compliance is high. Currently, the primary device used for DNA delivery is a standard, manually operated syringe-needle.

The World Health Organization (WHO) estimates that 25 billion needles are currently used per year, half of which are used for catheters, about 12 billion are used for injections, and 1 billion for vaccinations [143]. WHO predicts that by 2005, 2 billion vaccinations will be administered by needle and syringe. Within the past 25 years or so, it was realized that improper use and disposal of needles has resulted in the transfer of infections from one patient to another and from patient to the healthcare worker, annually causing 16 million cases of hepatitis B, 4.7 million cases of hepatitis C, 160 000 cases of HIV and 1.3 million deaths [144, 145]. These acquired diseases cost the world about $535 million per year and add a "hidden cost" of $0.125 per injection [146–148].

In the United States, 6 million healthcare workers suffer about 1 million needle injuries annually, causing 2400 cases of hepatitis B/C and many cases of HIV [146, 147]. Within the past decade, the number of injections required to fully immunize a child has risen from 8 to 14 and is expected to rise further. There is mounting concern that needle/syringe devices are increasingly becoming vectors for disease [149, 150].

In addition to the needle as disease vector, the use of standard needles and syringes for injections has other disadvantages. The training of a person to give injections is relatively complicated, but is required to ensure no inadvertent damage to arteries, lymphatic ducts and nerves. Some patients are needlephobic and may refuse care. The accuracy of administration can also be problematic since the tissue that is injected depends on how far the needle is pushed into the patient, meaning that so-called "subcutaneous" injections may be intramuscular or vice versa, resulting in different drug pharmacokinetics and distributions and thus possibly altering the therapeutic effects.

The needle and syringe industry is evolving to cope with the above problems. The first generation of syringes were made of glass and the needles were made of steel and both were typically autoclaved and reused. Becton Dickinson and Company (BD) has emerged as the world leader in the manufacture of second-generation, cheap, disposable syringe-needle injectors. Currently, third-generation needle-syringes are being developed as cheap, single-use, disposable, drug pre-filled or not, pen or syringe devices for insulin and other injectables [151]. These third generation devices are the so-called "auto-destruct" or "auto-disable" (AD) syringes that preclude reuse and minimize disease transmission. However, all of these third-generation devices still use a needle and it is the needle itself that is the vector for disease. Improper use or disposal may still result in disease dissemination.

An alternative to the use of needle-syringe devices is needle-free injection [151, 152]. Prototype technologies used metal gun- or pen-shaped devices hooked to gas propellant tanks. The drug vial is inserted onto the device and multiple doses of fluid containing the drug can be propelled at high speed through a narrow orifice in the injector tip. The resulting fluid stream would be injected with high velocity into the skin or muscle. This technology was used to vaccinate military personnel for decades. This basic first generation needle-free injection technology remained in use until it became apparent that using the same device for multiple patients could result in disease cross-contamination, just as with standard needles and syringes [153–155].

Second-generation needle-free injection devices emerged in the 1980s and comprised small pens or tubular devices with narrow drug delivery orifices.

The larger devices could be connected to a bulk propellant source (such as He, N_2, or CO_2 tanks) and some portable models contained a small propellant accessory. These devices were designed to propel the drug under high pressure through a narrow orifice. The portable devices, such as the Biojector 2000 (www.bioject.com), could only inject 10–15 doses before gas cartridge replacement, rendering them impractical for rapid and convenient mass immunization.

Third-generation needle-free devices have only emerged within the past few years and remain largely untested within the clinical arena [156]. Some of these devices are mass-produced, relatively cheap (< $1), small, self-contained, disposable, single-use, and can be stored either pre-filled with drug product or not. These devices do not have needles and can be discarded with no possibility of being reused and presumably less possibility of creating injuries or transmitting disease compared with needle and syringe-based systems.

Various needle-free injection devices have been used to deliver plasmid into tissues of animals and more recently into humans. A decade ago, it was shown that the Med-E-Jet pneumatic injection device does not significantly shear plasmid and is able to deliver a plasmid vaccine encoding influenza nucleoprotein (NP) into skeletal muscle to generate anti-NP antibody and T cell immune responses [157]. Later, it was demonstrated that rabbits injected once in a single muscle with naked DNA encoding NP using a needle-free Biojector 2000, AdvantaJet (www.advantajet.com) or Medi-Jector (www.antarespharma.com) device developed anti-NP antibody titers that were about 20-fold higher than when using syringe-needle and same dose, volume and injection regimen [158]. Rabbits given a boost with Biojector developed anti-NP antibody titers that were up to 100-fold higher than syringe-needle injections. Thus, in addition to the potential safety benefits of using needle-free injectors, plasmid vaccination effectiveness may actually be enhanced using a needle-free injection device.

Clinical trials have revealed that patients and health care workers prefer needle-free devices over needle and syringe injections [146, 159–167]. The ability of needle-free devices to enhance immune responses to plasmid vaccines in animals [158] has further encouraged the use of needle-free plasmid vaccinations in humans. Several clinical trials have used the Biojector 2000 device to administer plasmid vaccines, although none of these studies have compared, in detail, immune response differences from needle and syringe injections [6, 8, 9, 168, 169]. Various other needle-free devices are available beside the Biojector and their development is being monitored by the CDC [156]. No studies have yet been reported on the use of these devices for plasmid vaccination or plasmid therapeutics in humans. These devices operate under the same principle of rapid extrusion of liquid through a small orifice, but vary in propellant mechanism (mechanical, gas pressure or solid propellant). Since the propellant is separated from the injectable drug, one would expect that some of these devices would enhance plasmid vaccination just as well as the Biojector device.

Another set of needle-free devices called gene guns use biolistic technology that was initially developed in the 1980s to propel plasmid into plant and animal cells in culture dishes [170]. In the early 1990s, the devices were adapted to administer plasmid *in vivo*. Micron-sized inert gold or tungsten particles are coated with plasmid and propelled at high speeds into animal tissues such as skin [110, 171–176]. The particles carry the plasmid into cell nuclei where it is released by the particles and the encoded genes are expressed. The primary application has been plasmid vaccination and early clinical studies are encouraging [177]. Advantages of gene guns are the low plasmid dose required to elicit immune responses and the ability to skew the immune response toward a humoral (antibody) response. The major disadvantages are the need for a bulky machine, a helium tank and/or separate propellant cartridges and the limited particle penetration into tissues (restricting the nonsurgical injection route to skin). Gene guns have recently been evaluated for their utility in vaccinating humans using plasmid encoding hepatitis B virus surface antigen and malarial parasite circumsporozooite protein [7, 177–180]. These studies show gene guns to be safe and well-tolerated and able to generate protective antibody titers with as little as 4–12 μg of plasmid encoding hepatitis B antigen.

Hybrid needle/needle-free devices are becoming available for drug delivery. Disposable abrasion devices for skin vaccination are being developed whereby the device containing a vaccine is scratched across the skin multiple times in such a way that the vaccine is delivered at a modestly disrupted stratum corneum [181]. Another device consists of a microarray of needles, each about the diameter of a human hair, arranged in a patch that sticks to the skin in such a way that the microneedles penetrate the epidermis to deliver the encased vaccine or therapeutic drug [182].

The fundamental gaps in the development of needle-free technologies to deliver plasmid vaccines in humans are 1) lack of information on safety and efficacy relative to needle-syringe devices, 2) plasmid dosing and dose-sparing, 3) relevance of tissue route of injection, and the 4) relative cost vs. benefit compared to the established use of 8-cent needle-syringe devices. Although most of the applications of needle-free devices have focused on nonplasmid vaccines and therapeutics, their increasing availability and potential cost and efficacy advantages will encourage their evaluation for plasmid delivery.

4
Case study: From concept to clinic

The development of a plasmid drug product from concept to licensure is a generic process that is mostly independent of the gene encoded by the plasmid (see Fig. 3). Following conception of a drug product and indication is a **discovery phase** where the appropriate genes are chosen, obtained and cloned into a carefully chosen plasmid expression system, after which plasmid is formulated and an administration device is chosen, usually needle and syringe. Then the plasmid is tested for efficacy (proof of concept and dose-ranging studies), while at the same time a manufacturing process and analytical methods are initiated to ultimately confirm routine product identity, purity, potency and stability. During the latter three stages, a pre-Investigational New Drug (IND) application meeting with the FDA is generally held to ensure regulatory alignment. Following the pre-IND meeting, a decision is made to proceed to the **preclinical development phase** which involves confirming general safety and lack of toxicity of the product, the post-administration biodistribution and kinetics and possibly genomic integration studies. These studies provide data for the preparation and filing of an IND application. Even before the actual IND is filed, manufacturing validation and scale-up, product stability and analytical method qualification and validation must be initiated and will progress well into the the next phase. Once the IND is allowed, a **clinical development phase** is initiated and involves conducting Phase 1 through 3 human clinical trials. Successful clinical trials are then followed by the filing of a Biologics License Application

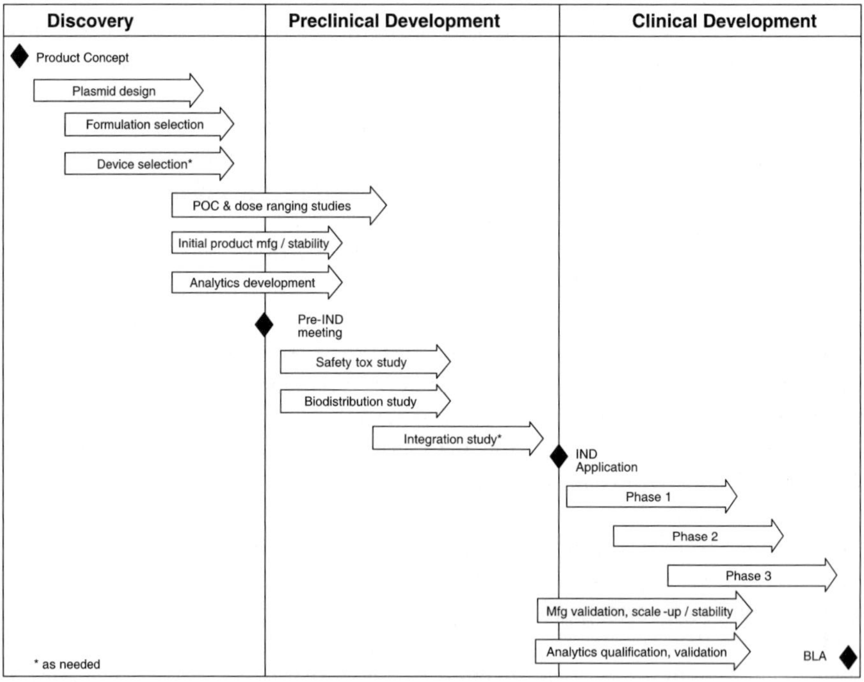

Fig. 3 Generic plasmid product development flowchart

(BLA). This three-phase product development method will be exemplified with a case study below.

4.1 Anthrax vaccine case study

Anthrax bacteria mainly infect livestock and are transmitted through food and soil. The aerosolized anthrax spore letter attacks following the terrorist attacks of 2001 generated public concern about the threat from anthrax used as a bioterror agent. Twenty-two cases of anthrax (11 inhalation and 11 cutaneous) were reported as a result of spores sent via the U.S. Postal Service, with five fatalities resulting from inhalation of spores [183]. These biological attacks resulted in a heightened awareness of bioterrorism threats and the U.S. Government responded with several measures to stimulate research and development, including the National Institute of Allergy and Infectious Diseases (NIAID) integrated programs for biodefense research (http://www2.niaid.nih.gov/biodefense/; NIH Biodefense Website).

4.2 Concept of an anthrax prophylactic vaccine with broad protection

Anthrax pathology in humans is primarily the result of a multi-component toxin secreted by *B. anthracis* bacterium [184]. The toxin derives from a combination of three separate gene products, designated Protective Antigen (PA), Lethal Factor (LF), and Edema Factor (EF). All three genes are encoded on a 184-kb plasmid designated pXO1 residing in *B. anthracis* bacteria [185]. PA83 is a 735 amino acid single chain protein that binds to a mammalian cell surface receptor [186]. Cleavage of PA83 by a furin-like enzyme yields PA63 [187, 188] which multimerizes and binds LF and EF. The PA63-LF complex (called Lethal toxin or Letx) or the PA63-EF complex is endocytosed by cells [189]. Once internalized, LF (a 776 amino acid zinc metalloprotease) cleaves several isoforms of MAP kinase kinases (Mek1, Mek2, MKK3) to disrupt host cell signal transduction events [190]. Letx causes extensive tissue hypoxia and pleural edema and is considered responsible for the rapid lethality of anthrax spore inhalation infection [184, 191].

Currently, the only licensed human anthrax vaccine is Anthrax Vaccine Adsorbed (AVA), prepared from a bacterial supernatant adsorbed onto aluminum hydroxide. Protection against anthrax infection is associated in large part with a humoral immune response directed against PA [192–195]. Passive infusion studies in animal challenge models have confirmed that PA antibody alone can be protective against Letx [196]. Some evidence suggests that EF and LF may also contribute to specific immunity [193, 196, 197]. There are several disadvantages of AVA vaccination. The Institute of Medicine recently concluded: "The current anthrax vaccine is difficult to standardize, is incompletely characterized, and is relatively reactogenic (probably even more so because it is administered subcutaneously), and the dose schedule is long and challenging. An anthrax vaccine free of these drawbacks is needed, and such improvements are feasible." [198].

DNA vaccines, many of which are in clinical trials, represent the next generation in the development of vaccines [41, 199]. The ability to conveniently deliver more than one antigen is one of the major advantages of plasmid over recombinant proteins, and has been demonstrated using a five-plasmid vaccine encoding genes for both blood-stage and liver-stage antigens of *P. falciparum* [200]. Multicomponent plasmid vaccines encoding four antigens from *B. anthracis*, Ebola, Marburg, and Venezuelan equine encephalitis viruses have also been tested [201].

DNA-based immunization has been shown to protect against a lethal challenge of anthrax in animal models [202, 203]. A plasmid vaccine encoding PA63 elicited protective immunity against a lethal toxin challenge in rabbits. A plasmid vaccine encoding a fragment of the LF gene product could provide partial protection against a lethal toxin challenge [204]. This DNA vaccine platform and development infrastructure needed to rapidly carry a DNA vaccine toward licensure (see Fig. 3) can be adapted to the development of a human prophylactic DNA vaccine for *B. anthracis*. The next section illustrates critical path activities for the development of a protective anthrax DNA vaccine.

4.3 Plasmid design

The choice of the right antigen(s) to target with a plasmid vaccine is critically important. For many disease targets, there is a long history of work that clearly defines which antigen(s) to choose, but the form of antigen to be delivered (such as secreted vs. cytoplasmic) is also essential. In the case of anthrax, the target protein is known to be PA, and passive anti-PA antibody transfer has been shown to provide complete protection. Based on this previous work with anthrax rPA protein and DNA vaccines indicating that LF immunogen may also enhance protection, a genetically detoxified bivalent PA and LF DNA vaccine was developed [205, 206].

Many vaccine target antigens have unwanted biological activities that would present a safety concern if injected into healthy or immunocompromised patients. To eliminate the zinc metalloprotease activity of LF described above, the entire LF domain IV was deleted. The ability of PA to bind LF and EF as well as to be internalized was eliminated by deleting the furin cleavage site of PA83. These deletions render both proteins biologically inactive but retain as much of the antigen sequence as possible.

Presentation of the expressed immunogen extracellularly, either by secretion or by transmembrane display, leads to increased humoral responses whereas immunogen designed to elicit a cellular response are best constructed to remain cytoplasmic [207]. Therefore, for the detoxified anthrax immunogen, the PA and LF coding sequences were modified by deleting the *B. anthracis* leader sequence and replacing it with the human TPA signal sequence for more efficient extracellular secretion.

The base sequences for the plasmid pX01 containing the PA and LF genes evolved to express optimally in the *B. anthracis* bacteria. But bacteria and mammalian cells differ in codon frequency usage during translation and bacterial codon usage can be a limiting factor in mammalian cell immunogen expression and immunogenicity from plasmid vaccines [208, 209]. There are several methods of optimizing plasmid expression for mammalian cell hosts. The method used for anthrax plasmids was to match the frequency of individual codon usage with that seen in the human genome. Other methods include changing only the most rare codons and using one or two of the most frequently used codons for a given amino acid [210, 211].

Since there is still much that is empirical about designing DNA constructs for expression and immunogenicity, it is prudent to construct several plasmids and test them first for expression *in vitro*, and then immunogenicity in a small mammal, usually mice. Multiple constructs can be rapidly screened to elicit the most effective immune response either *in vitro* if a neutralization assay exists, or *in vivo* with an animal challenge. This is extremely important for plasmid vaccines, since relying solely on *in vitro* expression as a selection criterion for plasmid constructs can be deceptive. Several studies have concluded that the amount of antigen expressed *in vitro* from a given plasmid does not necessarily determine relative immunogenicity *in vivo* [206].

In a series of preclinical studies designed with the goal of developing a safe and efficacious human anthrax vaccine, Hermanson et al. [206] constructed and screened six different PA and seven LF plasmid constructs for *in vitro* protein expression and for mouse humoral immunogenicity (as measured by antibody titers and *in vitro* Letx neutralization assay). All constructs included the human TPA leader sequence for efficient secretion from mammalian cells

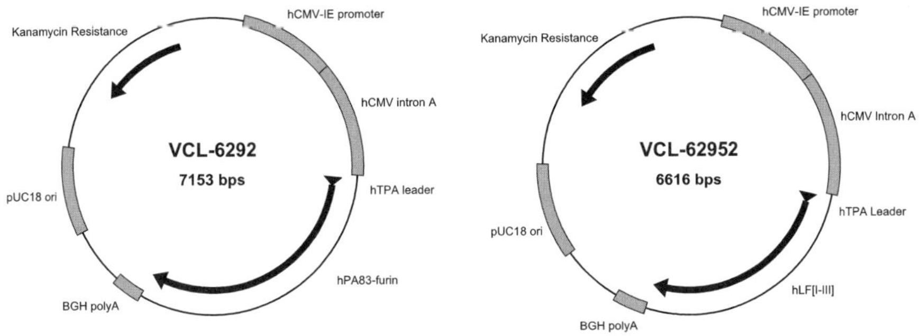

Fig. 4 Maps of the PA and LF plasmids

in order to induce the highest possible antibody response. Constructs synthesized with codons optimized for expression in humans were compared to constructs encoding *B. anthracis* wild-type sequences. Full-length PA83 (with the furin cleavage site deleted), the processed PA63, both humanized and *B. anthracis* wild type PA constructs, as well as various detoxification modifications were compared. LF constructs tested included humanized and wild-type full length versions and truncated versions in which the metalloprotease domain had been deleted. Based on the initial *in vivo* immunogenicity screen, the plasmids shown in Fig. 4 were chosen for formulation selection.

4.4 Formulation selection

As part of a large immunogenicity and efficacy study in rabbits, a PA construct (VCL-6292) was evaluated for the generation of neutralizing antibodies to *B. anthracis* Letx using four formulations and a needle-free delivery device [206]. The formulations and device were chosen based on their ability to move directly into the clinic, and the results of the rabbit immunization are shown in Fig. 5.

Groups of ten rabbits were immunized with 1 mg of formulated VCL-6292 at 0, 4, and 8 weeks. At 6 and 10 weeks, the Letx neutralization titer was determined for each animal as described [206].

Group	Formulation	Delivery
A	Vaxfectin™	i.m. needle
B	Poloxamer	i.m. needle
C	DMRIE/DOPE (DM-DP)	i.m. needle
D	Proprietary bacterial adjuvant	i.m. needle
E	Vaxfectin™	needle-free device

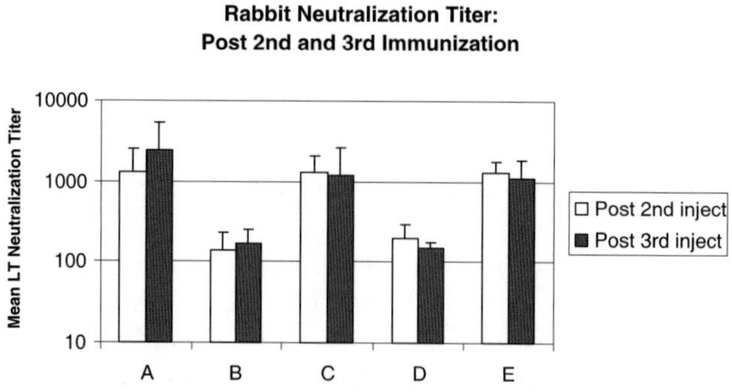

Fig. 5 Plasmid formulation and needleless device testing

The data showed that the groups injected with the cationic lipid formulations Vaxfectin™ and DMRIE/DOPE yielded statistically higher Letx neutralizing titers than groups injected with a poloxamer or proprietary bacterial adjuvant. There was no difference in immunogenicity seen with delivery by needle or needle-free device.

4.5 Device selection

To accelerate the anthrax DNA vaccine development, intramuscular injection using a standard syringe-needle was selected. Once the product configuration (plasmid, formulation, dose, injection regimen) was defined, needle-free jet injectors could be substituted for needle injections for preclinical and/or clinical evaluation.

4.6 Proof of concept (POC) and dose ranging studies

The plasmid design, formulation, injection route and device selected based on data from the mouse screening studies were advanced to a 140 rabbit immunogenicity and efficacy study in which it was demonstrated that DNA vaccines generated neutralizing titers to Letx that were as high as those induced by the licensed AVA [206]. Based on the rabbit immunogenicity data, a subset of immunized rabbits was exposed to a lethal aerosolized anthrax bacterial spore challenge. The 1-month and 7.5-month (Group 4) challenge data from this rabbit experiment are shown in Table 7.

The data indicate that 50 out of 50 animals that had received the cationic lipid-formulated PA-containing vaccine were protected from exposure to a lethal dose of anthrax spores, whereas all 22 spore-challenged control rabbits died. Injection of LF DNA (no PA DNA) protected 5 of 9 rabbits and the licensed AVA protected 4 of 4 rabbits. The LF vaccine elicited prolonged survival even in the four animals that died (4–7 vs. 2–4 days). Thus, a proof of concept was achieved in the rabbit challenge model and a product configuration was selected to advance an anthrax plasmid-based vaccine candidate to preclinical safety studies.

4.7 Initial product manufacturing, stability and analytics

Once a product candidate is established, manufacturability, scalability and stability studies need to be undertaken. For these studies, the majority of the analytical assays for the plasmid drug substance and in-process testing are generic. For the final drug product, specific assays typically need to be developed for the selected formulation (such as particle size), expres-

Table 7 *B. anthracis* inhalation spore challenge results

Group	Vaccination	Survival	Pre-challenge Letx Neut Titer Median ± S.D.	Inhaled Spore Dose Median LD_{50} *± S.D. (LD_{50} Range)	Days to Death Median ± S.D.
1	PA83Δfurin	8/8	5120 ± 6110	98 ± 28 (56–129)	> 21
2	PA83Δfurin/DM-DP	8/8	2560 ± 3055	116 ± 69 (56–238)	> 21
3	PA83Δfurin-2 inj.	8/8	1076 ± 634	94 ± 74 (52–252)	> 21
4 †	PA83Δfurin	10/10	844 ± 843	93 ± 40 (27–155)	> 21
5	PA83Δfurin+LF[I]–2inj.	8/8	415 ± 199	95 ± 43(72–205)	> 21
6	PA83Δfurin+LF[I-III]	8/8	3948 ± 1325	113 ± 47 (65–192)	> 21
7	LF[I-III]	5/9	453 ± 339	112 ± 67 (46–241)	5.3 ± 1.5
8	Vector	0/5	< 20	90 ± 35 (58–144)	2.6 ± 0.5
9	AVA-2 inj.	4/4	1280 ± 805	135 ± 34 (107–176)	> 21
	Naïve rabbits	0/12	ND	110 ± 49 (57–208)	2.6 ± 0.7
†	Naïve rabbits	0/5	ND	112 ± 72 (34–198)	2.8 ± 1.4

Rabbits were selected for challenge on week 13 by calculating the mean Letx neutralization titer for each group and excluding 1 rabbit with a higher and lower titer than the mean but always including the highest and lowest titer animals. * $1LD_{50}$ = 105 000 Ames spores. † All ten Group 4 rabbits and five naïve animals were challenged on week 40.

sion (such as potency) and immunogenicity testing (such as neutralizing antibodies).

4.8 Pre-IND meeting and preclinical studies and IND application

After a pre-IND meeting with the FDA to obtain concurrence on preclinical safety studies, clinical trial design and the regulatory path, standard preclinical safety studies were carried out. These studies included 1) plasmid repeat-dose GLP toxicology in rabbits, 2) tissue biodistribution in mice, and 3) plasmid integration into genomic DNA in rabbits. Upon completion of the preclinical studies and allowance of the filed IND, a Phase 1 clinical trial was initiated.

4.9 Conclusion of the case study

The anthrax plasmid vaccine described above progressed in just 13 months from concept to the start of preclinical safety studies in support of IND filing. This rapid development was facilitated by the large body of work on *B. anthracis* regarding i) the mechanism of toxicity, ii) antigens that provide a protective immune response, iii) *in vitro* assay for toxin neutralizing antibody response, and iv) animal models for critical *in vivo* challenge studies. However, even for less well-characterized targets than anthrax, plasmid vaccine technology can significantly shorten the time needed to select a vaccine candidate and begin preclinical studies. This case study demonstrates the ease with which one can do translational research using plasmid vaccine technology. The substantial cumulative safety database for DNA vaccines, the well-characterized nature of manufactured product and the speed with which new vaccine targets can be introduced into the clinic are strong factors for continuing the development of plasmid vaccine technology.

5 Conclusions

Plasmid-based pharmaceutical products remain in the early stages of clinical testing. Plasmids have been shown to be safe and well-tolerated in humans, are easy and relatively inexpensive to manufacture and have been explored in clinical trials to treat a wide variety of acquired diseases. The path from concept to clinical trial has been mapped in detail and well-vetted in regulatory agencies, and provides a generic infrastructure for the development of future plasmid-based products. The extent of future success in using plasmids as pharmaceutical drugs for treating the human condition will depend on the results of the many current clinical trials and on the continued development of improved plasmid expression and delivery systems, stable and cost-effective formulations, improved manufacturing processes and optimized administration methods.

Acknowledgements The authors wish to thank Drs. Thomas G. Evans, David C. Kaslow, Ruth Vahle, Jukka Hartikka and Mr. Alan Engbring for their helpful editorial comments, and we thank Dr. Rohit Mahajan for contributing Fig. 1.

References

1. Rolland A (ed) (1999) Advanced gene delivery: From concepts to pharmaceutical products. Drug Targeting and Delivery Series. Harwood, Academic Publishers, Amsteldjik, The Netherlands
2. Rolland A, Sullivan SM (eds) (2003) Pharmaceutical gene delivery systems. Marcel Dekker, Cold Spring Harbor, NY
3. Friedmann TE (1999) The development of gene therapy. Cold Spring Harbor Laboratory Press, Cold Spring Harbor, NY
4. Gurunathan S, Klinman DM, Seder RA (2000) Annu Rev Immunol 18:927–974
5. Shiver JW, Fu TM, Chen L, Casimiro DR, Davies ME et al. (2002) Nature 415:331–335
6. Wang R, Epstein J, Baraceros FM, Gorak EJ, Charoenvit Y et al. (2001) Proc Natl Acad Sci USA 98:10817–10822
7. Rottinghaus ST, Poland GA, Jacobson RM, Barr LJ, Roy MJ (2003) Vaccine 21:4604–4608
8. Epstein JE, Charoenvit Y, Kester KE, Wang R, Newcomer R et al. (2004) Vaccine 22:1592–1603
9. Timmerman JM, Singh G, Hermanson G, Hobart P, Czerwinski DK et al. (2002) Cancer Res 62:5845–5852
10. Edelstein ML, Abedi MR, Wixon J, Edelstein RM (2004) J Gene Med 6:597–602
11. Peng Z (2005) Current status of Gendicine in China: Recombinant human Ad-p53 agent for treatment of cancer. Hum Gene Ther 16:1016–1027
12. Wilson JM (2005) Gendicine: The first commercial gene therapy product. Hum Gene Ther 16:1014
13. Blaese RM, Culver KW, Miller AD, Carter CS, Fleisher T et al. (1995) Science 270:475–480
14. Berns A (2004) N Engl J Med 350:1679–1680
15. Cavazzana-Calvo M, Hacein-Bey S, de Saint Basile G, Gross F, Yvon E et al. (2000) Science 288:669–672
16. Cohen J (1993) Science 259:1691–1692
17. Rolland A (2001) Curr Opin Investig Drugs 2:1767–1769
18. Rolland AP (1998) Crit Rev Ther Drug 15:143–198
19. Wolff JA, Malone RW, Williams P, Chong W, Acsadi G et al. (1990) Science 247:1465–1458
20. Copernicus Software (2004) Website. See www.copernicus.com, last accessed 13th September 2005
21. Kaslow DC (2004) T Roy Soc Trop Med H 98:593–601

22. Nabel GJ, Nabel EG, Yang ZY, Fox BA, Plautz GE et al. (1993) Proc Natl Acad Sci USA 90:11307–11311
23. Yanisch-Perron C, Vieira J, Messing J (1985) Gene 33:103–119
24. Prather KJ, Sagar SL, Murphy JC, Chartrain M (2003) Enzyme Microb Tech 33:865–883
25. Smith HA (1994) Vaccine 12:1515–1519
26. Soubrier F, Cameron B, Manse B, Somarriba S, Dubertret C et al. (1999) Gene Ther 6:1482–1488
27. Tong X, Engehausen DG, Freund CT, Agoulnik I, Guo Z et al. (1998) Anticancer Res 18:719–725
28. Sutherland LC, Williams GT (1997) J Immunol Meth 207:179–83
29. Christenson SD, Lake KD, Ooboshi H, Faraci FM, Davidson BL et al. (1998) Stroke 29:1411–1415; discussion 1416
30. Teramoto S, Ito H, Ouchi Y (1999) Thromb Res 93:35–42
31. Thompson TA, Gould MN, Burkholder JK, Yang NS (1993) In Vitro Cell Dev Biol 29A:165–170
32. Zarrin AA, Malkin L, Fong I, Luk KD, Ghose A et al. (1999) Biochim Biophys Acta 1446:135–139
33. Smith RL, Traul DL, Schaack J, Clayton GH, Staley KJ et al. (2000) J Virol 74:11254–11261
34. Maass A, Langer SJ, Oberdorf-Maass S, Bauer S, Neyses L et al. (2003) J Mol Cell Cardiol 35:823–831
35. Hartikka J, Sawdey M, Cornefert-Jensen F, Margalith M, Barnhart K et al. (1996) Hum Gene Ther 7:1205–1217
36. Leung TH, Hoffmann A, Baltimore D (2004) Cell 118:453–464
37. Cullen BR, Raymond K, Ju G (1985) Mol Cell Biol 5:438–447
38. Boshart M, Weber F, Jahn G, Dorsch-Hasler K, Fleckenstein B et al. (1985) Cell 41:521–530
39. Losordo DW, Vale PR, Hendel RC, Milliken CE, Fortuin FD et al. (2002) Circulation 105:2012–2018
40. Khan TA, Sellke FW, Laham RJ (2003) Gene Ther 10:285–291
41. Makinen K, Manninen H, Hedman M, Matsi P, Mussalo H et al. (2002) Mol Ther 6:127–133
42. Yeung G, Choi LM, Chao LC, Park NJ, Liu D et al. (1998) Mol Cell Biol 18:276–289
43. Pfarr DS, Rieser LA, Woychik RP, Rottman FM, Rosenberg M et al. (1986) DNA 5:115–122
44. Zinckgraf JW, Silbart LK (2003) Vaccine 21:1640–1649
45. Mendiratta SK, Quezada A, Matar M, Wang J, Hebel HL et al. (1999) Gene Ther 6:833–839
46. Abruzzese R (2003) US Pat No. 20030220286
47. Miller JH, Reznikoff WS (1980) The operon. Cold Spring Harbor Laboratory, New York
48. Gustafsson C, Govindarajan S, Minshull J (2004) Trends Biotechnol 22:346–353
49. Plotkin JB, Robins H, Levine AJ (2004) Proc Natl Acad Sci USA 101:12588–12591
50. Hemmi H, Takeuchi O, Kawai T, Kaisho T, Sato S et al. (2000) Nature 408:740–745
51. Aderem A, Hume DA (2000) Cell 103:993–996
52. Luke CJ, Carner K, Liang X, Barbour AG (1997) J Infect Dis 175:91–97
53. Pertmer TM, Eisenbraun MD, McCabe D, Prayaga SK, Fuller DH et al. (1995) Vaccine 13:1427–1430

54. Cornelie S, Hoebeke J, Schacht AM, Bertin B, Vicogne J et al. (2004) J Biol Chem 279:15124–15129
55. Greber UF, Willetts M, Webster P, Helenius A (1993) Cell 75:477–486
56. Wickham TJ, Roelvink PW, Brough DE, Kovesdi I (1996) Nat Biotechnol 14:1570–1573
57. Douglas JT, Rogers BE, Rosenfeld ME, Michael SI, Feng M et al. (1996) Nat Biotechnol 14:1574–1578
58. Fasbender A, Zabner J, Chillon M, Moninger TO, Puga AP et al. (1997) J Biol Chem 272:6479–6489
59. Zelphati O, Liang X, Hobart P, Felgner PL (1999) Hum Gene Ther 10:15–24
60. Mesika A, Grigoreva I, Zohar M, Reich Z (2001) Mol Ther 3:653–657
61. Young JL, Benoit JN, Dean DA (2003) Gene Ther 10:1465–1470
62. Dean DA, Dean BS, Muller S, Smith LC (1999) Exp Cell Res 253:713–722
63. Ross R, Sudowe S, Beisner J, Ross XL, Ludwig-Portugall I et al. (2003) Gene Ther 10:1035–1040
64. Bigger BW, Tolmachov O, Collombet JM, Fragkos M, Palaszewski I et al. (2001) J Biol Chem 276:23018–23027
65. Chen ZY, He CY, Ehrhardt A, Kay MA (2003) Mol Ther 8:495–500
66. Pesole G, Liuni S, Grillo G, Licciulli F, Mignone F et al. (2002) Nucleic Acids Res 30:335–340
67. Chen CY, Shyu AB (1995) Trends Biochem Sci 20:465–470
68. Lu JY, Schneider RJ (2004) J Biol Chem 279:12974–12979
69. Fan XC, Steitz JA (1998) Embo J 17:3448–3460
70. Shamlou PA (2003) Biotechnol Appl Bioc 37:207–218
71. Horn NA, Meek JA, Budahazi G, Marquet M (1995) Hum Gene Ther 6:565–573
72. Ballantyne J, Klocke D, Smiley L (2002) http://www.aldevron.com/pdf/WestNilevaccine.pdf
73. Varley DL, Hitchcock AG, Weiss AM, Horler WA, Cowell R et al. (1999) Bioseparation 8:209–217
74. Chen W, Graham C, Ciccarelli RB (1997) J Ind Microbiol Biot 18:43–48
75. Riesenberg D, Schulz V, Knorre WA, Pohl HD, Korz D et al. (1991) J Biotechnol 20:17–27
76. Shiloach J, Kaufman J, Guillard AS, Fass R (1996) Biotechnol Bioeng 49:421–428
77. Wang Z, Lee G, Shi Y, Wegrzyn G (2001) Process Biochem 36:1085–1093
78. O'Kennedy RD, Baldwin C, Keshavarz-Moore E (2000) J Biotechnol 76:175–183
79. Williams SG, Cranenburgh RM, Weiss AM, Wrighton CJ, Sherratt DJ et al. (1998) Nucleic Acids Res 26:2120–2124
80. Ciccolini LA, Ayazi PS, Titchener-Hooker NJ, Ward JM, Dunnill P (1999) Bioprocess Eng 21:231–237
81. Clemson M, Kelly WJ (2003) Biotechnol Appl Bioc 37:235–244
82. Lee AL, Sagar SL (2002) Merck & Co, Patent No US 2002/0001829-A1
83. Wilson RC, Murphy JC (2002) USA, Patent No 2002 0197637
84. Lander RJ, Winters MA, Meacle FJ, Buckland BC, Lee AL (2002) Biotechnol Bioeng 79:776–784
85. Horn N, Budahazi G, Marquet M (1998) Vical Incorporated, United States, Patent No 5707812
86. Ferreira GN, Cabral JM, Prazeres DM (2000) Bioseparation 9:1–6
87. Ferreira GN, Cabral JM, Prazeres DM (1999) Biotechnol Prog 15:725–731
88. Diogo MM, Ribeiro SC, Queiroz JA, Monteiro GA, Tordo N et al. (2001) J Gene Med 3:577–584
89. Lee SY (1996) TIBTECH 14:98–105

90. Wils P, Escriou V, Warnery A, Lacroix F, Lagneaux D et al. (1997) Gene Ther 4:323–330
91. Stadler J, Lemmens R, Nyhammar T (2004) J Gene Med 6(Suppl 1):S54–66
92. Winters MA, Richter JD, Sagar SL, Lee AL, Lander RJ (2003) Biotechnol Prog 19:440–447
93. Alexandridis P, Hatton TA (1995) Colloid Surface A 96:1–46
94. Quezada A, Larson J, French M, Ponce R, Perrard J et al. (2004) J Pharm Pharmacol 56:177–185
95. Lowrie DB, Whalen R (eds) (1999) DNA vaccines. Humana, Totowa, NJ
96. Casimiro DR, Chen L, Fu TM, Evans RK, Caulfield MJ et al. (2003) J Virol 77:6305–6313
97. Evans RK, Zhu DM, Casimiro DR, Nawrocki DK, Mach H et al. (2004) J Pharm Sci 93:1924–1939
98. Caulfield MJ, Wang S, Smith JG, Tobery TW, Liu X et al. (2002) J Virol 76:10038–10043
99. Vilalton A, Mahajan RK, Hartikka J, Rusalov D, Martin T et al. (2005) Hum Gene Ther 16:1143–1150
100. Selinsky C, Luke C, Wloch M, Geall A, Hermanson G et al. (2005) Hum Vaccines 1:16–23
101. Mumper RJ, Rolland AP (1998) Adv Drug Deliver Rev 30:151–172
102. Mumper RJ, Duguid JG, Anwer K, Barron MK, Nitta H et al. (1996) Pharm Res 13:701–709
103. Mumper RJ, Ledebur HCJr (2001) Mol Biotechnol 19:79–95
104. Mumper RJ, Wang J, Klakamp SL, Nitta H, Anwer K et al. (1998) J Control Release 52:191–203
105. MacLaughlin F, Rolland A (1999) In: Mathiowitz E (ed) Encyclopedia of controlled drug delivery. Wiley, New York, pp 874–889
106. Donnelly J, Berry K, Ulmer JB (2003) Int J Parasitol 33:457–67
107. McKeever U, Barman S, Hao T, Chambers P, Song S et al. (2002) Vaccine 20:1524–1531
108. Klencke B, Matijevic M, Urban RG, Lathey JL, Hedley ML et al. (2002) Clin Cancer Res 8:1028–1037
109. Garcia F, Petry KU, Muderspach L, Gold MA, Braly P et al. (2004) Obstet Gynecol 103:317–326
110. Cui Z, Mumper RJ (2003) Crit Rev Ther Drug 20:103–137
111. Denis-Mize KS, Dupuis M, Singh M, Woo C, Ugozzoli M et al. (2003) Cell Immunol 225:12–20
112. Briones M, Singh M, Ugozzoli M, Kazzaz J, Klakamp S et al. (2001) Pharmaceut Res 18:709 712
113. Singh M, Briones M, Ott G, O'Hagan D (2000) Proc Natl Acad Sci USA 97:811–816
114. Denis-Mize KS, Dupuis M, MacKichan ML, Singh M, Doe B et al. (2000) Gene Ther 7:2105–2112
115. Konstan MW, Wagener JS, Hilliard KA, Kowalczyk TH, Hyatt SL et al. (2003) In: American Society of Gene Therapy, Sixth Annual Meeting, 4–8 June 2003, Washington D.C.
116. Ziady AG, Gedeon CR, Miller T, Quan W, Payne JM et al. (2003) Mol Ther 8:936–947
117. Liu G, Li D, Pasumarthy MK, Kowalczyk TH, Gedeon CR et al. (2003) J Biol Chem 278:32578–32586
118. Duguid J, Durland R (1999) In: Rolland A (ed) Advanced gene delivery: From concepts to pharmaceutical products. Harwood, Amsterdam, pp 45–63

119. Choosakoonkriang S, Wiethoff CM, Koe GS, Koe JG, Anchordoquy TJ et al. (2003) J Pharm Sci 92:115–130
120. Hirsch-Lerner D, Barenholz Y (1998) Biochim Biophys Acta 1370:17–30
121. Eastman SJ, Siegel C, Tousignant J, Smith AE, Cheng SH et al. (1997) Biochim Biophys Acta 1325:41–62
122. Gregoriadis G, Bacon A, Caparros-Wanderley W, McCormack B (2002) Vaccine 20(Suppl 5):B1–9
123. Wheeler JJ, Palmer L, Ossanlou M, MacLachlan I, Graham RW et al. (1999) Gene Ther 6:271–281
124. MacLachlan I, Cullis PR, Graham RW (2000) In: Smythe Templeton N, Lasic D (eds) Gene therapy: Therapeutic mechanisms and strategies. Marcel Dekker, New York, pp 267–290
125. Fenske DB, MacLachlan I, Cullis PR (2001) Curr Opin Mol Ther 3:153–158
126. Fenske DB, MacLachlan I, Cullis PR (2002) Methods Enzymol 346:36–71
127. Monck MA, Mori A, Lee D, Tam P, Wheeler JJ et al. (2000) J Drug Target 7:439–452
128. Tam P, Monck M, Lee D, Ludkovski O, Leng EC et al. (2000) Gene Ther 7:1867–1874
129. Galanis E, Hersh EM, Stopeck AT, Gonzalez R, Burch P et al. (1999) J Clin Oncol 17:3313–3323
130. Hoffman DM, Figlin RA (2000) World J Urol 18:152–156
131. Morse MA (2000) Curr Opin Mol Ther 2:448–452
132. Rini BI, Selk LM, Vogelzang NJ (1999) Clin Cancer Res 5:2766–2772
133. Thompson J (2003) Hum Gene Ther 14:310–312
134. Mesnil M, Yamasaki H (2000) Cancer Res 60:3989–3999
135. Protiva Biotherapeutics (2004) Website. Protiva Biotherapeutics, Seattle, WA (see www.protivabio.com, last accessed 13th September 2005)
136. Scheule RK, Cheng SH (1998) Adv Drug Deliver Rev 30:173–184
137. Brigham KL, Lane KB, Meyrick B, Stecenko AA, Strack S et al. (2000) Hum Gene Ther 11:1023–1032
138. Hyde SC, Southern KW, Gileadi U, Fitzjohn EM, Mofford KA et al. (2000) Gene Ther 7:1156–1165
139. Noone PG, Hohneker KW, Zhou Z, Johnson LG, Foy C et al. (2000) Mol Ther 1:105–114
140. Morse MA (2001) Curr Opin Mol Ther 3:97–101
141. Yla-Herttuala S, Martin JF (2000) Lancet 355:213–222
142. Fischer L, Minke J, Dufay N, Baudu P, Audonnet JC (2003) Vaccine 21:4593–4596
143. Voelker R (1999) JAMA 281:1879–1881
144. Kane A, Lloyd J, Zaffran M, Simonsen L, Kane M (1999) Bull World Health Organ 77:801–807
145. Hutin YJ, Chen RT (1999) Bull World Health Organ 77:787–788
146. Kermode M (2004) Health Promot Int 19:95–103
147. Simonsen L, Kane A, Lloyd J, Zaffran M, Kane M (1999) Bull World Health Organ 77:789–800
148. Miller MA, Pisani E (1999) B World Health Organ 77:808–811
149. Aylward B, Lloyd J, Zaffran M, McNair-Scott R, Evans P (1995) Bull World Health Organ 73:531–540
150. Levine MM (2003) Nat Med 9:99–103
151. Clements CJ, Larsen G, Jodar L (2004) Vaccine 22:2054–2058
152. Levine MM, Sztein MB (2004) Nat Immunol 5:460–464
153. Sweat JM, Abdy M, Weniger BG, Harrington R, Coyle B et al. (2000) Ann NY Acad Sci 916:681–682

154. CDC (1986) MMR Weekly 35:373–376
155. Mollohan AB (1998) Letter to DoD regarding the DoD's use of jet injectors. Online at: http://www.cdc.gov/nip/dev/bailey-mollohan-letter-DoD-1998Jul09.htm (last accessed 13th September 2005)
156. CDC (2004) Needle-free injection technology website. Centers for Disease Control and Prevention, Atlanta, GA (online at http://www.cdc.gov/nip/dev/jetinject.htm, last accessed 13th September 2005)
157. Vahlsing HL, Yankauckas MA, Sawdey M, Gromkowski SH, Manthorpe M (1994) J Immunol Methods 175:11–22
158. Hartikka J, Bozoukova V, Ferrari M, Sukhu L, Enas J et al. (2001) Vaccine 19:1911–1923
159. Dialynas M, Hollingsworth SJ, Cooper D, Barker SG (2003) J Am Podiatr Med Assoc 93:23–26
160. Silverstein JH, Murray FT, Malasanos T, Myers S, Johnson SB et al. (2001) Endocrine 15:15–17
161. Munshi AK, Hegde A, Bashir N (2001) J Clin Pediatr Dent 25:131–136
162. Verrips GH, Hirasing RA, Fekkes M, Vogels T, Verloove-Vanhorick SP et al. (1998) Acta Paediatr 87:154–158
163. Haiduven D, Ferrol S (2004) AAOHN J 52:102–108
164. Wig N (2003) Indian J Med Sci 57:192–198
165. Murakami H, Kobayashi M, Zhu X, Li Y, Wakai S et al. (2003) Soc Sci Med 57:1821–1832
166. Ernest SK (2002) West Afr J Med 21:70–73
167. Porta C, Handelman E, McGovern P (1999) AAOHN J 47:237–244
168. Epstein JE, Gorak EJ, Charoenvit Y, Wang R, Freydberg N et al. (2002) Hum Gene Ther 13:1551–1560
169. Boyer JD, Cohen AD, Vogt S, Schumann K, Nath B et al. (2000) J Infect Dis 181:476–483
170. Chen D, Maa YF, Haynes JR (2002) Expert Rev Vaccines 1:265–276
171. Bennett AM, Phillpotts RJ, Perkins SD, Jacobs SC, Williamson ED (1999) Vaccine 18:588–596
172. Benvenisti L, Rogel A, Kuznetzova L, Bujanover S, Becker Y et al. (2001) Vaccine 19:3885–3895
173. Mahvi DM, Sheehy MJ, Yang NS (1997) Immunol Cell Biol 75:456–460
174. Weiss R, Scheiblhofer S, Freund J, Ferreira F, Livey I et al. (2002) Vaccine 20:3148–3154
175. Oran AE, Robinson HL (2003) J Immunol 171:1999–2005
176. Tachedjian M, Boyle JS, Lew AM, Horvatic B, Scheerlinck JP et al. (2003) Vaccine 21:2900–2905
177. Swain WE, Heydenburg Fuller D, Wu MS, Barr LJ, Fuller JT et al. (2000) Dev Biol (Basel) 104:115–119
178. McConkey SJ, Reece WH, Moorthy VS, Webster D, Dunachie S et al. (2003) Nat Med 9:729–735
179. Roy MJ, Wu MS, Barr LJ, Fuller JT, Tussey LG et al. (2000) Vaccine 19:764–778
180. Tacket CO, Roy MJ, Widera G, Swain WF, Broome S et al. (1999) Vaccine 17:2826–2829
181. Mikszta JA, Alarcon JB, Brittingham JM, Sutter DE, Pettis RJ et al. (2002) Nat Med 8:415–419
182. Prausnitz MR (2004) Adv Drug Deliver Rev 56:581–587
183. Jernigan DB, Raghunathan PL, Bell BP, Brechner R, Bresnitz EA et al. (2002) Emerg Infect Dis 8:1019–1028

184. Dixon TC, Meselson M, Guillemin J, Hanna PC (1999) N Engl J Med 341:815–826
185. Mock M, Mignot T (2003) Cell Microbiol 5:15–23
186. Bradley KA, Mogridge J, Mourez M, Collier RJ, Young JA (2001) Nature 414:225–229
187. Klimpel KR, Molloy SS, Thomas G, Leppla SH (1992) Proc Natl Acad Sci USA 89:10277–10281
188. Singh Y, Chaudhary VK, Leppla SH (1989) J Biol Chem 264:19103–19107
189. Mogridge J, Cunningham K, Lacy DB, Mourez M, Collier RJ (2002) Proc Natl Acad Sci USA 99:7045–7048
190. Duesbery NS, Webb CP, Leppla SH, Gordon VM, Klimpel KR et al. (1998) Science 280:734–737
191. Moayeri M, Haines D, Young HA, Leppla SH (2003) J Clin Invest 112:670–682
192. Ivins B, Fellows P, Pitt L, Estep J, Farchaus J et al. (1995) Vaccine 13:1779–1784
193. Ivins BE, Welkos SL (1988) Eur J Epidemiol 4:12–19
194. Pitt ML, Little S, Ivins BE, Fellows P, Boles J et al. (1999) J Appl Microbiol 87:304
195. Pitt ML, Little SF, Ivins BE, Fellows P, Barth J et al. (2001) Vaccine 19:4768–4773
196. Kobiler D, Gozes Y, Rosenberg H, Marcus D, Reuveny S et al. (2002) Infect Immun 70:544–560
197. Pezard C, Weber M, Sirard JC, Berche P, Mock M (1995) Infect Immun 63:1369–1372
198. Joellenbeck L, Zwanziger LL, Durch JS, Strom BL (eds) (2002) The anthrax vaccine, is it safe? Does it work? National Academies Press, Washington, DC
199. Donnelly JJ, Ulmer JB, Liu MA (1998) Dev Biol Stand 95:43–53
200. Hedstrom RC, Doolan DL, Wang R, Gardner MJ, Kumar A et al. (1997) Springer Semin Immun 19:147–159
201. Riemenschneider J, Garrison A, Geisbert J, Jahrling P, Hevey M et al. (2003) Vaccine 21:4071–4080
202. Friedlander AM, Welkos SL, Ivins BE (2002) Curr Top Microbiol 271:33–60
203. Gu ML, Leppla SH, Klinman DM (1999) Vaccine 17:340–344
204. Price BM, Liner AL, Park S, Leppla SH, Mateczun A et al. (2001) Infect Immun 69:4509–4515
205. Ferrari M, Hermanson G, Rolland A (2004) Curr Opin Mol Ther 6:506–512
206. Hermanson G, Whitlow V, Parker S, Tonsky K, Rusalov D, Ferrari M, Lalor P, Komai M, Mere R, Bell M, Brenneman K, Mateczun A, Evans T, Kaslow D, Galloway D, Hobart P (2004) Proc Natl Acad Sci USA 101:13601–13606
207. Boyle J, Koniaras C, Lew A (1997) Int Immunol 9:1897–1906
208. Narum DL, Kumar S, Rogers WO, Fuhrmann SR, Liang H et al. (2001) Infect Immun 69:7250–7253
209. Deml L, Bojak A, Steck S, Graf M, Wild J et al. (2001) J Virol 75:10991–11001
210. Bulmer M (1987) Nature 325:728–730
211. Haas J, Park EC, Seed B (1996) Curr Biol 6:315–324
212. US National Institutes of Health (2004) ClinicalTrials.gov website. US NIH, Bethesda, MD (see http://www.clinicaltrials.gov, last accessed 13th September 2005)
213. Powell K (2004) Nat Biotechnol 22:799–801
214. Genencor Int. (2004) Genencor to start Phase I study for therapeutic vaccine for Hepatitis B (online Press Release). Genencor International, Palo Alto, CA (see http://genencor.com/wt/gcor/pr_1075498454, last accessed 13th September 2005)
215. IAVI (2004) Ongoing trials of preventive HIV vaccines (online table). International AIDS Vaccine Initiative, New York (see http://www.iavireportonline.org/specials/OngoingTrialsofPreventiveHIVVaccines.asp, last accessed 13th September 2005)

216. PowderJect Pharmaceuticals Plc (2001) Interim report (online). PowderJect Pharmaceuticals Plc, Oxford, UK (see http://www.corporate-ir.net/media_files/lse/pjp.uk/reports/pj_interim_2001.pdf, last accessed 13th September 2005)
217. PowderJect Pharmaceuticals Plc (2004) PowderMed launched in spin out of powder injection DNA vaccine programmes from Chiron Vaccines (online Press Release). PowderJect Pharmaceuticals Plc, Oxford, UK (see http://www.powdermed.com/pdf/PowderMed_launch_press_release.pdf, last accessed 13th September 2005)
218. PowderJect Pharmaceuticals Plc (2002) Memorandum (online). House of Commons, London (see http://www.parliament.the-stationery-office.co.uk/pa/ld200203/ldselect/ldsctech/23/23w48.htm, last accessed 13th September 2005)
219. Biotechnology Investment Today (2004) PowderMed launched in spin out of powder injection DNA vaccine programmes from Chiron Vaccines (online article). Biotechnology Investment Today, London (see http://www.zerostart.net/pressroom_release.php?id=1526, last accessed 13th September 2005)
220. Guardian Newspaper (2001) DNA shot fails to hit market spot (online article). Guardian Newspapers Limited, Manchester, UK (see http://www.guardian.co.uk/print/0%2C3858%2C4175367-103676%2C00.html, last accessed 13th September 2005)
221. Conry RM, Curiel DT, Strong TV et al. (2002) Clin Cancer Res 8:2782–2787
222. Vical Inc. (2004) Website. Vical Incorporated, San Diego, CA (see www.vical.com, last accessed 13th September 2005)
223. Crum CP, Beach KJ, Hedley ML, Yuan L, Lee KR et al. (2004) J Infect Dis 189:1348–1354
224. US National Institutes of Health (2004) Clinical trials in human gene transfer (online). US NIH, Bethesda, MD (see http://www4.od.nih.gov/oba/rac/clinicaltrial.htm, last accessed 13th September 2005)
225. Horton HM, Dorigo O, Hernandez P, Anderson D, Berek JS et al. (1999) J Immunol 163:6378–6385
226. Saffran DC, Horton HM, Yankauckas MA, Anderson D, Barnhart KM et al. (1998) Cancer Gene Ther 5:321–330
227. Shi F, Rakhmilevich AL, Heise CP, Oshikawa K, Sondel PM et al. (2002) Mol Cancer Ther 1:949–957
228. Clark PR, Stopeck AT, Ferrari M, Parker SE, Hersh EM (2000) Cancer Gene Ther 7:853–860
229. Yagi K, Ohishi N, Hamada A, Shamoto M, Ohbayashi M et al. (1999) Hum Gene Ther 10:1975–1982
230. Bruckheimer E, Harvie P, Orthel J, Dutzar B, Furstoss K et al. (2004) Cancer Gene Ther 11:128–134
231. Zou Y, Zong G, Ling YH, Perez-Soler R (2000) Cancer Gene Ther 7:683–696
232. Anwer K, Meaney C, Kao G, Hussain N, Shelvin R et al. (2000) Cancer Gene Ther 7:1156–1164
233. Anwer K, Kao G, Proctor B, Rolland A, Sullivan S (2000) J Drug Target 8:125–135
234. Li D, Zeiders JW, Liu S, Guo M, Xu Y et al. (2001) Laryngoscope 111:815–820
235. O'Malley BW Jr, Couch ME (2000) Adv Oto-rhino-laryngol 56:279–288
236. Bishop JS, Thull NM, Matar M, Quezada A, Munger WE et al. (2000) Cancer Gene Ther 7:1165–1171
237. Dow SW, Elmslie ER, Willson AP, Roche L, Gorman C, Potter TA (1998) J Clin Invest 101:2406–2414
238. Whitmore M, Li S, Huang L (1999) Gene Ther 6:1867–1875
239. Dileo J, Banerjee R, Whitmore M, Nayak JV, Falo LDJr et al. (2003) Mol Ther 7:640–648

240. Ramesh R, Saeki T, Templeton NS, Ji L, Stephens LC et al. (2001) Mol Ther 3:337–350
241. Lanuti M, Rudginsky S, Force SD, Lambright ES, Siders WM et al. (2000) Cancer Res 60:2955–2963
242. Blezinger P, Freimark BD, Matar M, Wilson E, Singhal A et al. (1999) Hum Gene Ther 10:723–731
243. Reyes L, Hartikka J, Bozoukova V, Sukhu L, Nishioka W et al. (2001) Vaccine 19:3778–3786
244. Nukuzuma C, Ajiro N, Wheeler CJ, Konishi E (2003) Viral Immunol 16:183–189
245. Sankar V, Baccaglini L, Sawdey M, Wheeler CJ, Pillemer SR et al. (2002) Oral Dis 8:275–281
246. D'Souza S, Rosseels V, Denis O, Tanghe A, De Smet N et al. (2002) Infect Immun 70:3681–3688
247. Locher CP, Witt SA, Ashlock BM, Levy JA (2004) DNA Cell Biol 23:107–110
248. Locher CP, Witt SA, Ashlock BM, Polacino P, Hu SL et al. (2004) Vaccine 22:2261–2272
249. Fischer L, Tronel JP, Minke J, Barzu S, Baudu P et al. (2003) Vaccine 21:1099–1102
250. Jiao X, Wang RY, Feng Z, Alter HJ, Shih JW (2003) Hepatology 37:452–460
251. Ishii N, Fukushima J, Kaneko T, Okada E, Tani K et al. (1997) AIDS Res Hum Retrov 13:1421–1428
252. Okada E, Sasaki S, Ishii N, Aoki I, Yasuda T et al. (1997) J Immunol 159:3638–3647
253. Toda S, Ishii N, Okada E, Kusakabe KI, Arai H et al. (1997) Immunol 92:111–117
254. Yokoyama M, Zhang J, Whitton JL (1996) FEMS Immunol Med Mic 14:221–230
255. Gramzinski RA, Millan CL, Obaldia N, Hoffman SL, Davis HL (1998) Mol Med 4:109–118
256. Perrie Y, Frederik PM, Gregoriadis G (2001) Vaccine 19:3301–3310
257. Wong JP, Zabielski MA, Schmaltz FL, Brownlee GG, Bussey LA et al. (2001) Vaccine 19:2461–2467

Adv Biochem Engin/Biotechnol (2005) 99: 93–118
DOI 10.1007/10_004

Published online: 1 November 2005

Nonviral Delivery of Cancer Genetic Vaccines

Steven R. Little · Robert Langer (✉)

Department of Chemical Engineering and Center for Cancer Research,
Massachusetts Institute of Technology, Cambridge, MA 02142, USA
rlanger@mit.edu

Abstract The potential use of genetic vaccines to address numerous diseases including cancer is promising, but currently unrealized. Here, we review advances in the nonviral delivery of antigen-encoded plasmid DNA for the purpose of treating cancer through the human immune system, as this disease has drawn the most attention in this field to date. Brief overviews of dendritic cell immunobiology and the mechanism of immune activation through genetic vaccines set the stage for the desirability of delivery technology. Several promising nonviral delivery techniques are discussed along with a mention of targeting strategies aimed at improving the potency of vaccine formulations. Implications for the future of genetic vaccines are also presented.

Keywords Cancer · Dendritic cells · DNA vaccines · Genetic vaccines · Nonviral gene delivery

1 Introduction

Shortly after the demonstration by Wolff and colleagues that plasmid DNA injected into the muscle of mice led to the expression of a gene [1], it was shown that plasmid encoding an antigen could produce antibodies [2], cytotoxic T-cell responses, and protection from a corresponding disease [3, 4]. This elicitation of an immune response using plasmid DNA encoding an antigen, rather than administration of antigen itself, is called genetic vaccination. This concept not only allows for the possibility of therapeutically treating many incurable diseases, but has also found application in the screening of pathogenic genome libraries for the determination of protective antigens [5], and the high-throughput generation of high specificity monoclonal antibodies [6]. However, it is generally accepted that safely eliciting the proper immune responses with genetic vaccines is currently unrealizable due to the potency and/or toxicity of these formulations. This review will focus on the advancements in targeting and delivery aimed at improving the potency of DNA vaccines.

2 Immunobiology of genetic vaccines

2.1 The role of dendritic cells in immunity

An understanding of the complex multicellular milieu of the immune system and the genetic vaccine mechanism is critical to the development of successful DNA vaccine delivery technologies. The primary activators of adaptive cell mediated immunity are professional antigen presenting cells (APC) called dendritic cells (DC). DCs, depending on their state, are the most powerful APC, capable of inducing naïve T-cell activation, or "priming". Only one DC is required to activate 100–3000 T-cells [7]. In the absence of antigen, DCs have been shown to contact approximately 500 individual T-cells per hour and can interact with over ten antigen-specific T-cells simultaneously [8]. However, DCs are initially thought to exist in an "immature" state in which they are relatively incapable of activating T-cells, yet fully capable of surveying the peripheral environment by capturing antigen at the tremendously high rate of approximately one cell volume per hour [9]. Upon capture of antigen, these cells are thought to go through phenotypic changes in which their rate of antigen uptake briefly increases and then almost completely ceases. The cell then goes through morphological changes and down-regulation of chemokine receptors responsible for the migration of the cell to the site of inflammation, and up-regulation of receptors that would mediate translocation to the lymph nodes where T-cells are waiting for activation. Surface expression of "co-stimulatory" molecules is up-regulated as well. These receptors (Signal 2) are required for the activation of T-cells in tandem with the presentation of antigen on MHC Class I and II (Signal 1) which is also up-regulated upon "maturation" of a DC.

Once a DC is in the lymph nodes, where the majority of T-cells reside, surface co-stimulatory and secreted cytokine signals between the APC and T-cell convey the location and type of antigen that was captured. T-cells, specific to the peptide-MHC complex through a highly variable T-cell receptor (TCR), will then proliferate into clones capable of either destroying a target cell which expresses that same complex on its surface ($CD8^+$ T_c), or in the case of helper T-cells ($CD4^+$ T_H), capable of secreting soluble signals to B and T_c cells which direct the type of immune response elicited. This complex dialogue between lymphocytes can polarize the immune response to react through a cytotoxic cell-mediated response, called a Th1 response (characterized by secretion of cytokines such as IFN-γ, IL-2, and IL-12 along with IgG2a antibody), or by B-cell secretion of antibody, called a Th2 response (characterized by secretion of IL-4 and IL-10 along with IgG1 antibody). The rapid generation of $CD8^+$ T_c and $CD4^+$ T_H clones is thought to involve genetic alterations in certain persisting T-cells that enhance the magnitude of cellular proliferation in response to a second encounter with a pathogen. This effect is called a memory response, the basis of prophylactic immunity.

2.2
Genetic vaccine mechanism

The mechanism of immune induction for a traditional vaccine depends upon its classification. Attenuated bacterial or viral vaccines are still competent at infecting target cells, but are extremely unlikely to induce a full-blown infection in persons with normal immune systems. Because of this, antigenic peptide is produced intracellularly and processed by the MHC Class I pathway to elicit a Th1 response. At the same time, the pathogen can be taken up and degraded by the MHC Class II pathway for elicitation of a Th2 response. These vaccines are extremely effective at inducing memory responses and usually result in lifetime immunity. However, for particularly dangerous pathogens, the risk of reversion to a virulent state is too great, especially in immunocompromised individuals. Therefore, a vaccine with fewer safety concerns is required. Completely inactivated or subunit vaccines cannot infect a cell, and are mainly processed by the MHC Class II pathway. This results in activation of only the humoral arm of the immune system (characterized by an antibody response), and the lack of necessary T_c immunity required by certain infections. Currently, there are no vaccines capable of safely stimulating the immune system to battle certain diseases such as HIV, hepatitis B and C, herpes, malaria, and cancer.

Genetic vaccines, however, can enter a cell and cause expression of a target gene intracellularly, mimicking viral infection, and allowing both arms of the immune system to be activated (Fig. 1). Most importantly, plasmid DNA vaccines achieve this feat without the use of any infectious agent. However, just like an inactivated or subunit antigen, one would surmise (as the scientists that pioneered genetic vaccination in the late 1980s did) that plasmid DNA alone would not be able to gain entry into the cytoplasm of a cell. However, it seems that this does actually occur, by some unknown mechanism, albeit to an extremely small extent. The exact mechanism for immune induction via genetic vaccines is still unknown (Fig. 2).

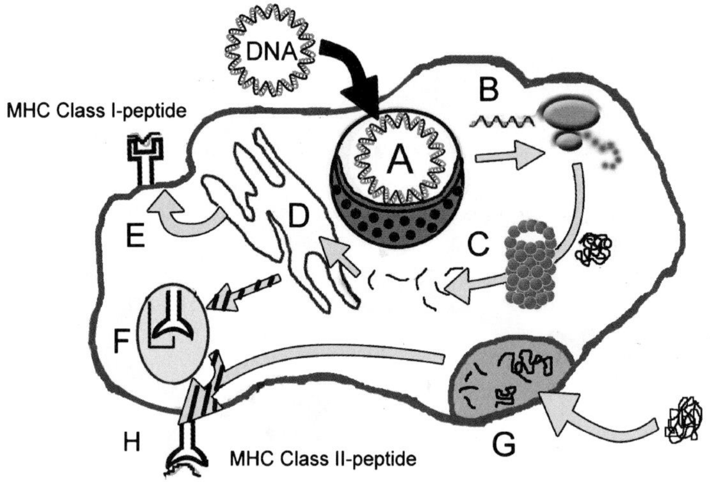

Fig. 1 Antigen processing and presentation in a cell. Upon cellular uptake of DNA from a genetic vaccine or viral infection (A), the gene is transcribed and translated (B) intracellularly to create a full length protein. This "antigen" is subsequently marked for degradation into smaller peptide fragments by the proteosome (C), and then transported into the endoplasmic reticulum for loading onto MHC Class I molecules (D). This MHC-peptide complex is then transported to the surface for presentation to antigen-specific CTLs (E). This complex, along with the appropriate co-stimulatory signals on the surface of an APC, would result in the activation and proliferation of the CTL (T_c). Alternatively, in the case of somatic cell presentation of the antigen fragment (thereby marking it as a target), this CTL would mediate lysis of the target cell. In APCs only, a MHC Class II complex is transported from the endoplasmic reticulum (along with an invariant chain to forbid Class I restricted peptide complexation) (F) in vesicles destined to fuse with endosomes containing exogenously captured antigen (G). The antigen is degraded in the late endosomes along with the invariant chain to allow MHC Class II – peptide loading, which is then transported for expression on the surface in tandem with co-stimulatory molecules for activation of helper T-cells (T_H) (H). The exclusivity of these pathways for processing and presentation of an endogenous or exogenous antigen to a T_c for a cell-mediated response (Th1) or a T_H for an antibody response (Th2), respectively, is under debate due to the apparent ability of some exogenous antigens to bypass endosomal trafficking into the cell cytoplasm

Speculation has occurred as to the potential route of plasmid internalization of myocytes (the cell primarily transfected using i.m. vaccination) such as the linking of membrane invaginations called caveolae to plasmid uptake [10]. However, myocytes themselves do not express the co-stimulatory molecules that are necessary (Signal 2) to induce activation of naïve T-cells upon expression of processed peptide associated with surface MHC Class I (Signal 1). It seems as though these cells may have a greater capacity to to-

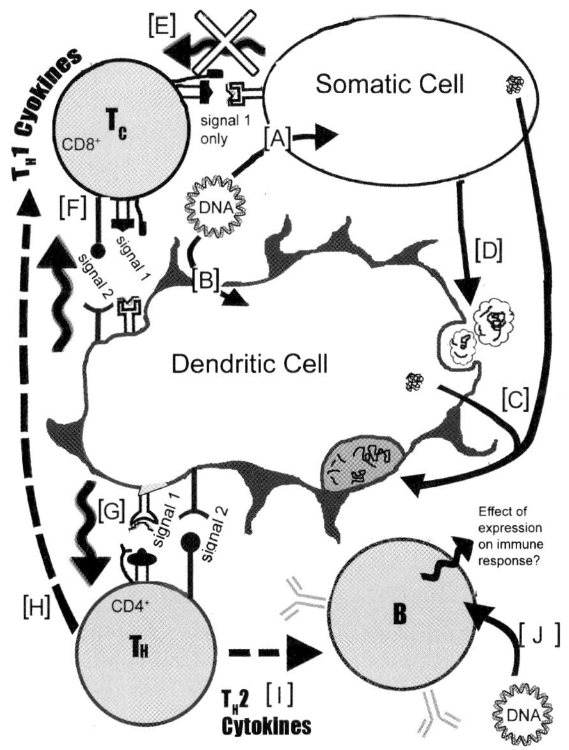

Fig. 2 Activation of cellular immunity by genetic vaccines. Plasmid DNA is taken up by somatic cells [A] and/or DCs [B] for intracellular expression of antigen. This antigen can be secreted [C] and subsequently taken up by DCs. Somatic cells may also, in certain circumstances, secrete or present antigen in such a way that it is taken up into the cytoplasm of a DC [D]. Importantly, there is no empirical evidence that somatic cells can activate $CD8^+$ T-cells due to a lack of co-stimulatory expression [E]. Antigen expressed intracellularly by a DC [B] or taken up through cross-priming [D] can be presented by MHC Class I along with co-stimulatory signal 2 to a $CD8^+$ T-cell to initiate priming [F]. Furthermore, antigen taken up exogenously [C] may be processed by the MHC Class II pathway and presented alongside signal 2 to $CD4^+$ T_H cells [G] which can subsequently secrete soluble cytokine signals such as IL-12 back to the DC, proliferative signals such as IL-2 and IFN-γ to T_c cells [H], or signal directed toward B-cells such as IL-4 to induce B-cell proliferation and antibody secretion [I]. The role of directly transfected B-cells is still yet to be determined [J]

lerize an antigen-specific T-cell due to the lack of Signal 2 in the presence of Signal 1 [11–13] (with the exception of their presence in the lymphoid organ where another cell may deliver Signal 2 [14]). Furthermore, studies have shown that immune responses are restricted to the haplotype of parent bone marrow-derived cells and not to the haplotype of transplanted somatic cells which express antigen from a genetic vaccine [15–18]. These somatic cells may serve as "antigen factories" to produce and secrete large amounts of antigen which then may be taken up and processed by DCs. However, this secreted antigen, under normal circumstances, is only thought to be processed by the MHC Class II pathway, similar to the way an inactivated or subunit vaccine is only capable of eliciting a Th2-type response.

Another possibility is that the APC itself takes up the antigen encoding DNA and presents the peptides on its surface in the context of both Signal 1 and Signal 2. This seems to be the simplest explanation and has been backed up empirically by several groups who have demonstrated APC expression of a reporter gene administered in a DNA vaccine by several delivery routes in the draining lymph nodes [19–24] in very small numbers of cells. However, some have questioned the extent to which this limited number of cells contributes to the observed immune response. It should be noted that as few as 500–1000 DCs tranfected in vitro and administered to a mouse have induced immune responses equivalent to standard genetic immunization (comparable to transfecting only 0.5–1% of the cells in the target area) [25]. Another consideration is that dendritic cells have proven to be particularly difficult to transfect with plasmid DNA, as will be discussed later in this review. This cell seems to be a logical target for delivery technology.

A more complex mechanism of immune activation using genetic vaccines involving both somatic and APCs has also been proposed [16, 17]. The theory of "cross-presentation" suggests that antigen secreted under special circumstances or associated with proteins that mediate cytoplasmic transport (such as heat shock proteins, apoptotic/necrotic bodies) could be processed by the MHC Class I pathway of an APC. Evidence of this cross-priming mechanism has been demonstrated [26–28] and seems to involve phagocytosis by an immature DC through $\alpha_V\beta_5$ integrins and CD36. Furthermore, studies have shown that this process requires $CD4^+$ T-cell help [29], either through secretion of soluble T_c proliferation signals such as IL-2 and/or interacting with the DC in such a way as to make it more capable of mediating T_c priming. Importantly, this is most likely in the context of some sort of "danger signal", otherwise the body would presumably react to self antigens that it would theoretically capture and survey at all times [30].

Finally, a recent report by Coelho-Castelo et al. demonstrated that B-cells express a reporter gene after i.m. injection of plasmid DNA, contrary to previous reports [31]. The contribution of transfected B-cells to the immune responses elicited by DNA vaccines remains uncertain; however, B-cells may be an interesting future target for DNA vaccine delivery.

3 Using Genetic Vaccines to Treat Cancer

Inflammation is a direct result of tissue damage due to the invasive growth typical of cancer. Therefore, it is logical to think that the immune system would be alerted to malignant tumor cells. Apparently, however, cancer is capable of evading this immune recognition. Cancer cells may accomplish this feat through several mechanisms [32]: 1) down-regulated presentation of certain tumor-associated antigens (TAA) that may alert the immune system, 2) complete loss of some TAA, and 3) secretion of soluble signals that can diminish the ability of APCs to initiate immune rejection. Examples of such secreted factors include IL-10 and TGF-β, which have both shown to dampen immune activation [32, 33]. Though it is still unclear as to what extent these mechanisms contribute to tumor persistence, it is certainly clear that tumor cells have developed ways to avoid recognition by the immune system.

Since tumor cells seem to evade immune recognition, it is believed that activation of these so-called "nonresponsive" TAA-specific T-cells is the key to effective cancer immunotherapeutics. The choice of antigen in which the immune system should be alerted is a major consideration in this process and is different for different types of cancer. This process of choosing the appropriate TAAs is beyond the scope of this review and is described in detail elsewhere [34, 35]. Despite the optimization of responses to different TAAs and the large number of clinical trials of genetic vaccine therapeutics, DNA vaccination has not produced the caliber of immune responses in humans that has been shown in small animals. For instance, as much as 5 mg of plasmid is required for effective DNA vaccination without a delivery vehicle (naked DNA vaccination) in nonhuman primates [36] while only 50–300 μg are required in mice [4]. Although large amounts of DNA have been shown to be tolerated by humans [37], technologies to increase the potency of DNA vaccines are certainly welcome.

4 DNA vaccine delivery

The inherent ability of naked plasmid DNA to gain access to the nucleus of a cell seems counterintuitive. Early experiments that led to the discovery of naked plasmid expression in mouse muscle were initially intended to demonstrate the superior ability of liposomal delivery systems over plasmid alone [38]. Surprisingly, this naked plasmid was actually capable of inducing more target gene expression than the liposomal formulations. However, the amount of uptake and expression, especially in muscle, is dwarfed by the total amount of plasmid administered (uptake is less than 0.01% of total) [10, 39, 40]. The transition from mice to primates that results in lower transfection may be due to an apparent difference in muscle structure with thicker perimysium [41]. A primary focus of delivery systems has been to increase this level of expression, and perhaps even more importantly, to target its delivery to the appropriate cell type.

The first attempt to increase potency in DNA vaccines by plasmid DNA delivery was the gene gun. This method involves delivery of DNA coated onto tiny gold beads which are accelerated to high velocities in order to penetrate into the skin [42]. Compared to direct i.m. injection of plasmid, gene gun vaccination requires 2–3 orders of magnitude less DNA [4, 43]. This may be due to the direct insertion of a plasmid-coated bead into skin cells, bypassing the need for uptake, but may also be partially due to the large amount of dendritic cells present in the dermis (Langerhans cells) [44]. However, a downside to gene gun vaccination is that the immune responses elicited are usually Th2-polarized.

Viral vectors are by nature far more effective as delivery vehicles and can be powerful activators of the immune system. However, potential toxicity and immune rejection issues related to viral vectors can be restrictive, especially in immunocompromised patients such as persons undergoing chemotherapy and the elderly population. An example of this shortcoming was apparent in research which demonstrated that restimulation with DCs transduced with an adenoviral genetic vaccine decreased the antigen-specific immune response in favor of a overbearing anti-adenovirus response in human melanoma patients [45]. Pre-existing immunity to a viral delivery vector can also be a limiting factor. Attenuated bacterial vectors have also shown promise for DNA delivery, but fall under similar safety constraints [46].

Synthetic, nonviral DNA vaccine delivery avoids many of the potential downsides of viral vaccines related to safety and immune rejection. However, these vaccines are far less effective at eliciting prophylactic immunity. Importantly, several key advances have been made which have significantly increased the therapeutic potential of nonviral genetic vaccines.

4.1 Electroporation

Electroporation is most commonly known as an effective in vitro transfection procedure. Electric pulses temporarily disrupt the cellular membrane and also physically translocate plasmid DNA into the cell due to its anionic charge (a process known as ionophoresis). The application of electroporation, in vivo, requires the use of probes or clamp electrodes to the site of DNA administration, resulting in a nonspecific, post-injection delivery method to the surrounding cells. Modification of process parameters from low voltage, long pulses to high voltage can significantly increase the effectiveness of this procedure for gene delivery [47].

Research into the delivery of TAA genes by application of electroporation has shown promise, but has its downsides. Electroporation-enhanced vaccination with both plasmid-encoded human GP100 and mouse TPR2 antigen was shown to elicit protection from melanoma challenge in mice [48]. It was also demonstrated that introduction of plasmid encoding inflammatory cytokine signals by electroporation at tumor sites induced transduction and tumor growth inhibition without the systemic cytokine levels seen using adenoviral delivery methods [49]. Futhermore, Kalat et al. used electroporation to facilitate the discovery of novel TAA plasmid construct variations by optimization of tyrosinase-related protein-2 antigens, leading to inhibition of tumor growth in two separate models [50]. This same group later demonstrated that electroporation could induce immune responses that were comparable to those induced by viral infection [51].

Although it has been suggested that the increase in transfection efficiency afforded by electroporation is responsible for the observed immune activation, it is also possible that tissue damage caused by electroporation at the immunization site may be responsible for the recruitment of APCs and inflammation, leading to a more effective response [52]. Electroporation can be destructive to tissues, and it has been reported that pain has occurred in patients during administration of electroporation during clinical trials [53].

4.2 Lipoplexes and liposomes

Cationic lipids are arguably the most common nonviral transfection reagent. These lipid formulations condense anionic plasmid DNA through the formation of lipid bilayers and through their cationic charges, leading to the formation of a lipoplex. Examples of common lipids used in these formulations are DOPE and cholesterol, both of which can also have endosomal disruption properties [54]. The mechanism of cellular uptake of lipoplexes was originally believed to be that of cell membrane fusion, but it is now widely accepted that this process occurs through endocytosis or phagocytosis of the lipoplex. Access to the cytoplasm is thought to be mediated by destabilization of the endosomal membrane [55].

Additionally, plasmid DNA can be encapsulated within a lipid vesicle called a dehydrated-rehydrated vesicle (DRV) [56]. DRVs are produced through the process of freeze drying lipoplexes (which is thought to increase the association of plasmid DNA with the flattened liposomal vesicles). Subsequent rehydration leads to the formation of a DRV with entrapped plasmid (Fig. 3). DRVs, compared to naked DNA and cationic lipoplexes, have shown the ability to generate improved cellular immunity and secretion of greater IgG1 levels in mice [56]. Various lipids have been shown to increase vaccine potency when added to these formulations [57] and have also been shown to enhance oral delivery of DRV formulations [58].

Another strategy using cationic lipids involves the incorporation of viral fusogenic peptides. These added components may impart advantageous viral properties to the complexes and have enhanced responses against tumor-

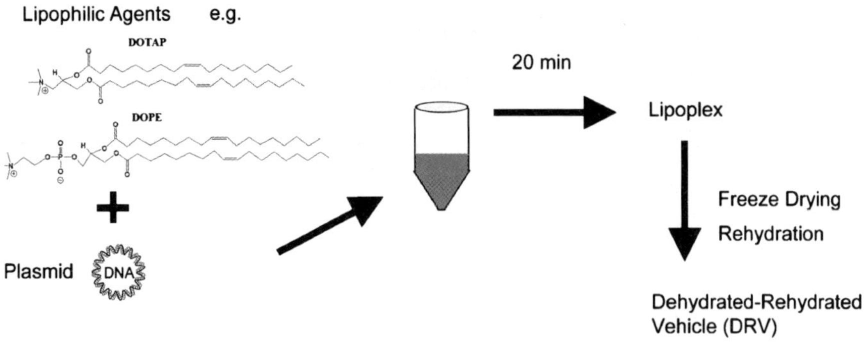

Fig. 3 Formation of lipoplexes and DRVs

associated antigens. Okamoto et al. demonstrated increased antibody response when using Hemaglutinating Virus Japan (HVJ) fusogenic peptides to deliver plasmid DNA i.m. in a mouse model while naked DNA was ineffective [59]. Zhou et al. later demonstrated that these same HVJ peptides used with cationic lipids i.m. could induce cellular immune responses and protection against melanoma along with antibody responses in a gp100 TAA model [60]. This delivery vector has also been used with i.n. administration but has shown to be less effective than i.m. injection in a gp100/TRP2 TAA model [61]. Influenza fusogenic peptides have been used for the treatment of prostrate carcinoma with the successful induction of CTL responses using a parathyroid hormone-related peptide [62].

Cationic lipid formulations are easy to prepare, can protect the plasmid DNA from nucleases in the extracellular environment, and are promising candidates for genetic vaccination in vivo. However, the polycationic nature of the lipoplex/liposomal formulations impart a degree of cellular promiscuity and a pronounced tendency to bind to proteins in serum. This can severely limit the ability to target particular cells of the immune system and maintain stable plasmid/lipid formulations. Creating targeted liposome vectors with improved serum stability could be the key to a successful cationic lipid delivery vehicle for genetic vaccines.

4.3 Particulate encapsulation and adsorption

Polymeric encapsulation or binding of DNA vaccines may be one of the most promising nonviral delivery methods for several reasons: 1) these particles protect plasmid payload from extracellular degradation [63], 2) these formulations are capable of carrying large payloads, making co-delivery of many plasmids and other immunostimulatory agents possible, 3) these particles can offer a phagocytosis-based passive targeting to APCs in the range of 1–10 μm (Fig. 4), 4) surface modifications are easily made which can further

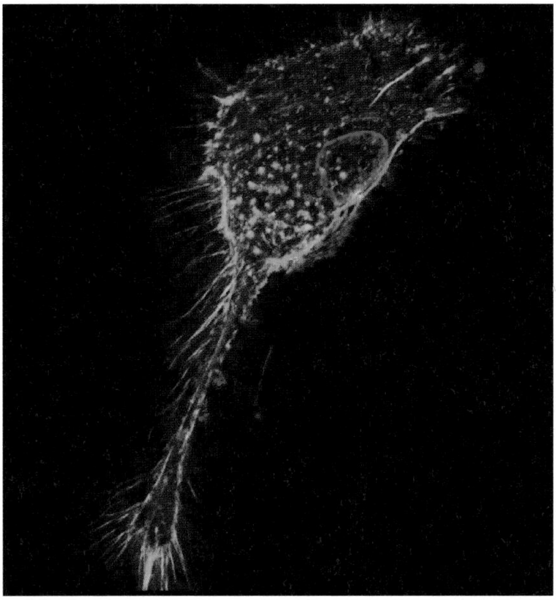

Fig. 4 Phagocytosed 5 μm fluorescently labeled microparticles (*red*) by a primary, human peripheral blood mononuclear derived DC that has been fixed and stained for actin (*green*) and nuclear material (*blue*)

enhance targeting and uptake [64], 5) these particles are widely believed to be adjuvants themselves.

The reason for the adjuvancy of microparticles is unknown, but several mechanisms have been proposed. The physical sizes of microparticles are characteristic of some pathogens [65] and the immune system may somehow mistake a particle for a potentially dangerous microorganism. It has been demonstrated that uptake of 1 μm latex microparticles by monocytes leads to the differentiation of DCs and their migration to lymph nodes [66]. Futhermore, phagocytosis of latex beads by DCs in vitro induces phenotypic maturation of the DCs, as shown by CD83 up-regulation [67]. Another potential explanation is the microparticle-mediated controlled release of antigen or plasmid DNA containing immunostimulatory bacterial CpG motifs that are recognized by TLR9 on the surface of a DC [68].

4.4 Microencapsulation of plasmid using PLGA

Traditionally, the polymer most often utilized in antigen and plasmid encapsulation is polylactic-co-glycolic acid. This biodegradable and biocompatible polymer is already approved by the US Food and Drug Administration, making it more easily advanced to the clinic than new polymers. It was originally used for degradable sutures [69] but has since found application in everything from delivery of narcotic agonists [70], contraceptives [71], pesticides [72], to the healing of bone fractures [73] and ligaments [74]. It decomposes by acid and base hydrolysis or enzyme-catalyzed degradation [75] and ultimately leaves the body as carbon dioxide. This polymer's application to the delivery of protein or peptide antigen vaccines comprehensive is covered in detail in several reviews [65, 76].

The most common methods used for encapsulating drugs with PLGA are the double-emulsion procedure and spray drying. Double emulsion (reviewed in [77]) is the most common method used for the encapsulation of plasmid (Fig. 5) while spray drying is used less frequently [78]. In both procedures an aqueous solution containing plasmid is emulsified with an organic phase containing the polymer to serve as the controlled release agent (for example PLGA). The release of plasmid from PLGA microparticles typically occurs through burst phases due to a property of PLGA and other similar polymers that allow diffusion of water into the interior of the particle (aptly named "bulk erosion") [79, 80]. This plasmid release is tunable using various molecular weights and polymer monomer ratios made up of hydrophobic lactide and hydrophilic glycolide.

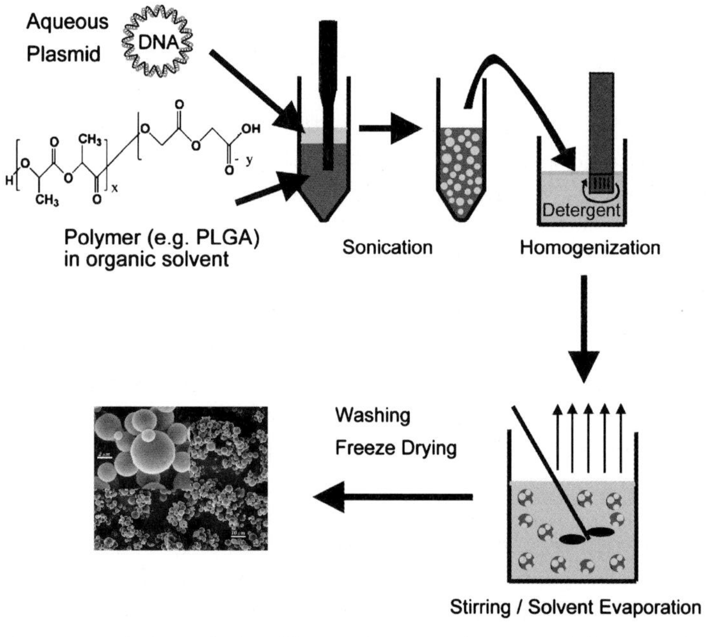

Fig. 5 Microencapsulation of plasmid DNA into a polymer particle using the double emulsion procedure. *Bottom left* photograph is a scanning electron microscopy image of 1000× (*body*) and 5000× (*inset*) magnifications of freeze-dried PLGA microparticles

The first attempt to deliver plasmid as a vaccine using PLGA microencapsulation was made by Hedley et al., who showed that stronger CTL responses were elicited using microencapsulated plasmid delivery s.c. and i.p. compared to naked plasmid injections using a VSV antigen system [81]. Clinical trials using the PLGA microparticle delivery system showed that 83% of patients demonstrated an immune response that persisted for six months using a plasmid encoding (HPV-16 E7) [82]. Furthermore, Phase 1 clinical trials for cervical intraepithelial neoplasia using PLGA delivery of plasmid indicated that no adverse side effects were observable, and 73% of patients exhibited a specific immune response [83]. Furthermore, 33% of patients exhibited a complete histologic response [83]. Later, it was demonstrated that oral immunization with plasmid DNA led to protective immunity against rotavirus challenge when PLGA microparticles were used for delivery [84].

The attractiveness of PLGA has generated a great deal of interest into the advancement of this technology to the clinic. Although these studies demonstrate the ability of PLGA to function as a genetic vaccine delivery vehicle, this polymer was never designed for this particular application and has several limiting disadvantages. Acidic degradation products that build up in the microparticle interior can severely stunt or permanently damage the activity of plasmid DNA. This can be attributed to PLGA ester bond degradation leading to acids that cannot easily diffuse out and away from the particle interior. It has been demonstrated using pH sensitive fluorescent probes and microscopy that the pH can drop to as low as 2 after three days of incubation [85]. Although this internal pH microclimate can stabilize some drugs [86], low pH has been shown to completely abolish plasmid transfection activity below a pH of 4 [80]. In addition, the amount of time needed for quantitative release of plasmid DNA from even low molecular weight PLGA microparticles is on the order of two weeks [87], while the lifespan of the majority of DCs after activation is approximately ten days [88].

Addition or replacement of PLGA with agents aimed at enhancing the immunogenicity of the formulations, such as lipophilic molecules (taurocholic acid (TA) and polyethylene-glycol-distearoylphosphatidylethanolamine (PEG-DSPE)), can increase both CTL and antibody response and can protect mice against tumor challenge i.v. [89]. Potential mechanisms for the observed heightened activity as a vaccine could involve membrane disruption upon uptake by a cell, or plasmid binding that may protect plasmid inside the particle or after release.

Besides microclimate pH deactivation of supercoiled plasmid, the process of fabrication itself can substantially damage DNA because of high sheer stresses encountered during sonication and homogenization, as these are required in the double emulsion procedure. It is also possible that organic/aqueous interfaces, which tend to denature proteins, have a deleterious effect. Furthermore, freeze-drying is commonly used prior to isolation of powdered microparticles and can also damage plasmid. To address these issues, Ando et al. put forth a technique for fabrication of plasmid microparticles which virtually eliminated the loss of supercoiled plasmid during fabrication [90]. This process involves the freezing of the internal aqueous phase containing plasmid DNA to shield it from sheer stress [90]. Stabilization agents, such as lactose, have also been shown to eliminate damage to plasmid during freeze-drying.

4.5
Cationic particle adsorption

To completely avoid the deleterious effects of processing plasmid DNA during the double emulsion procedure, cationic microparticles can be fabricated which retain a cationic surface to which polyanionic plasmid DNA could be bound. Addition of a cationic surfactant called cetyltrimethylammonium bromide (CTAB) produces this positively charged surface in contrast to the use of conventional detergent such as polyvinyl alcohol which imparts a negatively charged surface (PVA). These cationic microparticles are capable of eliciting humoral responses 250× greater than naked DNA and heightened CTL responses using a HIV p55 gag model with a relatively small dose of DNA (1 μg) [91]. Furthermore, these microparticles can transfect primary DCs, albeit to a low extent [92], and they have been found in draining lymph nodes three hours post injection [93]. Further studies showed that although naked DNA works better at higher dosages, this response is diminished upon injection of lower amounts of DNA [94]. Particles with surface-adsorbed plasmid, however, maintain high levels of Ab and CTL response with 1000× less plasmid [94]. The exact mechanism of these cationic microparticles and the effects of CTAB are unknown, but this system may allow for greater uptake by APCs, faster release of plasmid DNA, and endosomal release properties imparted by the cationic detergent.

Application of this genetic vaccine delivery system to cancer was first directed toward delivery of carcinoembryonic antigen (CEA)-encoded plasmid. This formulation inhibited CEA expressing adenocarcinoma cell growth in a population of vaccinated mice when used as a vaccine [95]. Furthermore, a boosting regimen with naked DNA i.m. encoding GM-CSF (a potent immunomodulatory cytokine), results in an increased frequency of responders and inhibition of tumor growth [95].

Another method of creating cationic nanoparticles using a hot cetyl alcohol-polysorbate 80 wax/aqueous emulsion formed by adding cationic surfactant was recently described by Cui et al. [96]. Cooling of this mixture leads to the formation of cationic microparticles which are 100 nm in diameter. The advantages of this method include uniformity of size, cationic surfaces capable of binding plasmid DNA, and elimination of organic solvents. Studies by this group have shown that cationic nanoparticles produced using this simple method elicit high antibody and Th1 cell mediated responses using a variety of administration routes [97–100].

4.6 Cationic chitosan nanoparticles

Chitosan is a biodegradable polymer derived from chitin in the shells of crustaceans, the second most abundant polymer on Earth behind cellulose (Fig. 6). Chitosan is the deacetylated version of chitin, and has a variety of potential uses in textiles, water treatment, and biodegradable films, to name just a few [101]. More importantly for genetic vaccine purposes, the mannose receptor commonly found on APCs interacts with acetylglucosamine, which is a repeating unit in chitosan [102].

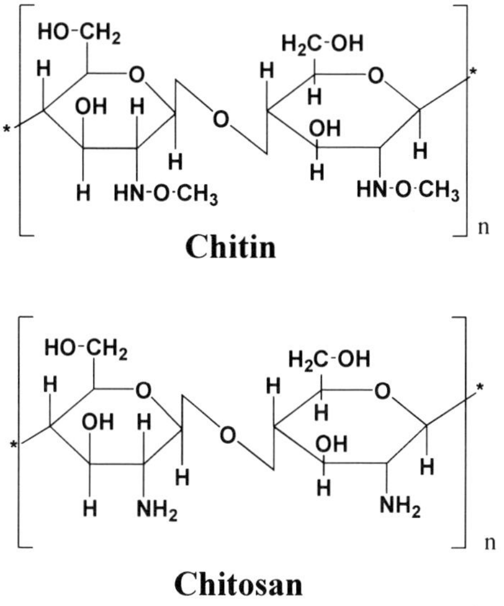

Fig. 6 Structures of chitin and chitosan

Similarly to cationic lipoplexes, the simplest way to create plasmid/chitosan complexes is through simple incubation and condensation through charge neutralization.

Alternatively, cationic nanoparticles can be prepared either by a coacervate method [103], or by addition of carboxymethylcellulose [104]. Generally, chitosan particles are directed toward the delivery of genetic vaccines to the mucosal tissue and, to our knowledge, have not been used for delivery of cancer genetic vaccines. However, Kabbaj et al. used chitosan nanoparticles to deliver tumor-inhibitory mycobacterial DNA to inhibit nuclease degradation [105]. The ability of this tumor inhibitor/chitosan formulation was 20× more effective than tumor inhibitor alone for inhibition of melanoma tumor cell growth.

4.7
pH-Sensitive microparticles

Wang et al. recently illustrated the use of polyorthoesters (POE) for fabrication of microparticulate genetic vaccines (Fig. 7). Unlike polymers which bulk erode such as PLGA, POEs degrade by surface erosion, eliminating the accumulation of acidic byproducts in the particle's interior. These polymers also degrade more rapidly at endosomal pH than at physiologic pH. Synthesis and testing of a series of these polymers revealed a POE that generated higher levels of secreted antibody and greater $CD8^{+}$ T-cell responses when compared to PLGA microparticle delivery. This POE, when administered as a genetic vaccine in mice, was shown to inhibit the growth of tumor cells expressing the Class I restricted epitope. It is speculated that differences in the immunogenicities of POE polymers are due to the timing of the release plasmid, which may correspond to the course of immune induction [106].

Furthermore, we have recently demonstrated the use of a pH-sensitive, degradable poly-β-aminoester (Fig. 8) in tandem with low molecular weight PLGA as a DNA vaccine delivery system [107]. This particular PBAE was discovered in a small library, and was capable of binding and condensing plasmid DNA [108]. It can also be used to form microparticles using fluores-

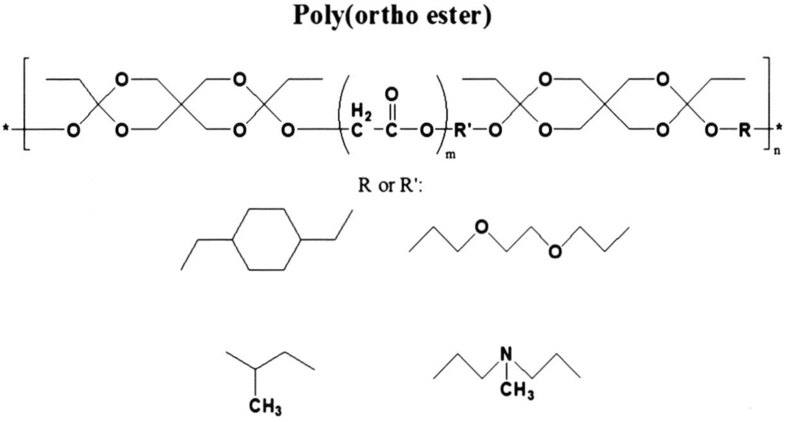

Fig. 7 Polyorthoesters [162]. R or R′ is shown *below* the polymer chain

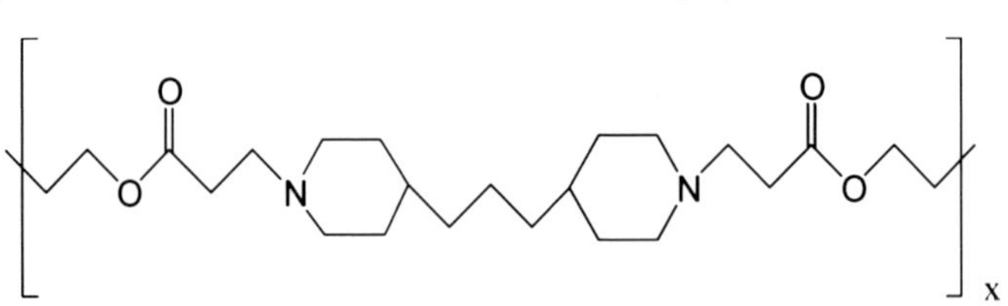

Fig. 8 A pH-sensitive poly-β-aminoester

cently labeled sugar and can quantitatively release this encapsulated material in the range of endosomal pH due to a unique balance of hydrophobicity and charge-inducible tertiary amines [109]. These tertiary amines in the PBAE backbone may also act as a weak base (or proton sponge), which may mediate phagosomal/lysosomal release [110]. These same groups act to buffer the internal aqueous environment due to ester bond degradation, and plasmid extracted from these particles after incubation in aqueous media has been shown to have higher integrity than pure PLGA microparticles [163]. These hybrid PBAE/PLGA microparticles have exhibited enhanced delivery of plasmid DNA to APCs when compared with PLGA alone and have strong adjuvant effects on DCs in vitro. Using a model antigen system, we demonstrated that mice vaccinated with these PBAE microparticle formulations were capable of exhibiting antigen-specific rejection of tumor cells expressing the target antigen. Other evidence suggests that the observed response was due to a polyclonal $CD8^+$ response, but the possibility of $CD4^+$ T-cell helper response cannot be ruled out.

4.8
Genetic engineering of DCs in vitro

An alternative strategy to targeting transfection of DCs in vivo is the isolation of immature DCs from a patient for antigen loading in vitro. The cells could then be injected back into the patient to allow the DCs to prime naïve T-cells specific to the antigen. One method to pulse TAA to DCs is through tumor antigens or tumor lysates [111–114]. These have the disadvantages of limited duration of antigen expression [115] and, in the case of tumor antigen pulsing, restriction of therapy to the haplotype of the antigen. Some groups have fused DCs with tumor cells by PEG co-culture to gain both the expression of correct tumor antigens with the costimulatory competency of DCs [116]. It has been demonstrated that these cells express surface molecules from both cells, and introduction back into mice results in antigen-specific CTLs and rejection of established metastasis. Others have recently employed electrofusion techniques to allow fusion of directly isolated tumor cells with DCs without the need for extended in vitro culture periods used with PEG co-culture [117].

Attempts to transfect DCs in vitro have resulted in only low levels of DC transfection using nonviral gene delivery. Primary DCs in culture have proven to be particularly difficult cells to transfect [92, 118–122]. In one particular instance it was necessary to use RT-PCR to detect the low levels of transfection inducible using nonviral means [92]. However, some progress has been made in sufficiently transducing a DC in vitro to render it capable of activating naïve, antigen-specific T-cells in vivo. One study employed the use of a cationic peptide called CL22, which demonstrated the ability to transfect DCs in vitro and protected mice against melanoma challenge while peptide pulsing was ineffective [119]. Another group reported a peptide containing ornithine and histidine DNA binding amino acids which were capable of transfecting a dendritic cell line. Injection of cationic peptide-transduced cells induced secretion of IFN-γ, while naked DNA-transduced cells were ineffective [123]. PEI complexes with mannose and adenoviral particle moieties have also been shown to transfect DCs in vitro and to stimulate proliferation in allogenic and autologous mixed lymphocyte reactions upon reinjection [118].

Transfection of DCs with RNA encoding reporter genes may be a more efficient alternative to plasmid, because unlike DNA transfection, RNA only needs to be delivered to the cytoplasm to be effective rather than the nucleus. In support of this concept, Strobel et al. reported the use of RNA to transduce primary human DCs, which resulted in twofold higher expression than DNA when using liposome delivery [124]. Furthermore, these cells, when reinjected, elicited stronger antigen-specific influenza matrix protein antigen memory T cell responses than DNA-transfected cells [124]. Importantly, DCs retained their immunological phenotype after transfection, which may be crucial factor in DC migration to the lymph nodes upon reintroduction.

Use of "whole tumor RNA" to transfect DCs via electroporation may be the best way to allow for natural immunodominance in processing and presenting the antigen optimally. This obviates the need for discovery of haplotype-restricted antigens in each patient. Resected tumors can be sampled and RNA can be amplified without loss of function to obtain complete RNA from tumor cells for transfection of DCs [125]. Using this technology, studies have been performed using whole tumor RNA for myeloma [126], breast carcinoma [127], colorectal cancer (renal cell carcinoma) [128–130], and chronic lymphocytic leukemia [131]. In addition, antisense oligonucleotides specific to inhibition of invariant chain expression have been delivered to DCs in vitro. Theoretically, loss of invariant chain expression would lead to the complexation of otherwise MHC Class I restricted epitopes on MHC Class II molecules. Using this technique, increased magnitude of immune response was observed along with persistence of $CD8^+$ T-cell responses in an ovalbumen model system [132].

5 Targeting genetic vaccines

The targeting of an antigen to a particular cell or organ can be accomplished by either targeting the delivery system for uptake by a specific cell, physically attaching the antigen to a targeting protein, or by using DNA that is transcriptionally regulated and only active in the target cell or cells. These methods will be briefly mentioned.

5.1
Targeting uptake using ligands

Our knowledge of the surface receptors and the various ligands involved in the uptake and activation (or suppression) of APCs is ever expanding. Fc receptors are believed to bind immune complexes and are involved in opsonization. The binding of Fc receptors to a DC surface results in this cell's maturation, as shown by up-regulation of co-stimulatory molecules [133]. CTLA-4 binds to B7.1/B7.2 at a high affinity and is also used to target DCs. Certain chemokines such as RANTES [134], IP-10 [135] and MCP-3 [135] can bind to the surface of DC through receptors as well. CD36 and $\alpha\beta5$ integrins have been implicated in receptor-mediated phagocytosis [136], and DEC205 (or the human homolog LY75) and DC-SIGN mediate endocytosis through surface receptors [137]. Another good example of DC targeting is through the use of mannose or mannan with a delivery system which binds to the APC mannose receptor (MMR). This strategy has produced a twofold increase in phagocytosis-mediated uptake of particle formulations by APCs in vitro [138]. The use of mannose has also been shown to increase transfection of cultured DCs [118] along with antibody and cell-mediated immune responses in vivo [98].

5.2
Targeting intracellular antigen processing pathways

Modifications can also be made to the gene delivered to physically link the post-transcription antigen to a molecule meant to target a specific processing pathway within a cell. Targeting of an antigen to a specific cellular compartment may change the way it is processed and presented to the immune system. For instance, if an antigen is normally processed by the exogenous pathway (MHC Class II), but is targeted to the endoplasmic reticulum, or ER (where antigen is loaded onto MHC Class I), it could conceivably be processed by the endogenous pathway. This could lead to immune responses that primarily elicit CTL activity instead of antibody secretion (called isotype switching), and possibly a more relevant therapy.

The most commonly used antigen fusion partner for targeting the MHC Class I pathway is ubiquitin. This protein marks antigen for degradation by the proteosome into peptides destined for the ER for Class I loading. This addition to plasmid constructs has shown to increase CTL responses, but at the cost of humoral responses [139–143]. Interestingly, in one study, the addition of ubiquitin in a fusion construct led to a decrease in humoral response while the CTL response observed remained unchanged [144]. Further studies are required before ubiquitin can be generalized for use in genetic vaccines. One particularly promising cancer vaccine fusion gene is calreticulin (CRT). This protein can not only target the MHC Class I pathway, but it also demonstrates anti-angiogenesis properties (the ability to inhibit blood vessel growth to the site of a tumor) [145–148]. CRT fusion modifications have been shown to exhibit marked anti-tumor activity as a DNA vaccine alongside a HPV-16 E7 antigen. It is likely that the anti-angiogenesis properties of CRT are at least partially responsible for the response observed [149].

Conversely, targeting the MHC Class II pathway can also be a valid way to increase vaccine potency if a humoral response is primarily desired. Fusion genes that can target this pathway include LAMP-1 [150–152] and LIMP II [153] which can direct antigen toward the lysosome. Furthermore, antigen targeted to the cell surface can also lead to exogenous processing [154]. Targeting an antigen for secretion by the cell would supposedly lead to this same surface uptake by another cell or even by the secreting cell. This antigen would then be endocytosed or pinocytosed by a DC and bound to MHC Class II in the lysosome. Intriguingly, both humoral and/or cell-mediated immunity are increased using these strategies [155, 156]. Could this observation provide support for a predominant cross-priming mechanism using these vaccines? Further investigation is clearly warranted.

5.3 Targeting through transcription

The cytomegalovirus promoter, or pCMV, is a strong viral promoter with the ability to elicit high levels of gene expression in a variety of cell types. Thus, it has found wide application in genetic vaccines and gene therapy. However, for some gene products such as immunomodulating cytokines (which can become extremely toxic if present in large quantities), uncontrollably high levels of gene expression is disadvantageous. Furthermore, antigen that is expressed by pCMV may persist at low levels after the vaccination and could even induce tolerance to that antigen through regulatory T-cell pathways [157]. One strategy to avoid this is to use transient promoters to target only a subset of cells, such as DCs. An example of this type of transcriptional regulation is the use of the lectin promoter. When used in a GFP plasmid to demonstrate transfection of DCs in vivo, anti-GFP CTL responses were observed [158]. Another candidate for transcriptional regulation is the mature DC specific fascin promoter. This promoter was shown to induce distinct Th1 responses as compared to the Th2 responses seen when using pCMV [159, 160]. This isotype switching seen when using transcriptional targeting could prove to enable some technologies where specific diseases require strong Th1 responses.

5.4 The future of nonviral delivery

Thus far, genetic vaccines for cancer have focused primarily on inciting "tolerized" T-cells to respond to an invading cancer cell. Major steps have been taken through new delivery technologies aimed at delivering plasmid-encoded antigen to be processed and presented by DCs in the presence of appropriate activation signals. Understanding the primary mechanisms and potential targets for genetic vaccine-induced immunity should lead to both prophylactic and therapeutic cancer therapies. Moreover, understanding how to activate a DC using a delivery vehicle, conceivably by imitating a pathogen, should also lead to the ability to bypass these mechanisms so as to avoid alerting the immune system. This should lead to the suppression of the same cells that seem to so readily attack our own bodies in autoimmunity and graft rejection.

The main focus of nonviral gene delivery research in the near future will be centered around increasing the potency further toward that of viral vectors. Research in the field has already come a long way by making improvements in both gene delivery and immune system activation. For example, the use of viral fusogenic peptides and alphaviral replicons [161] has allowed fundamental viral mechanisms to be implemented in completely synthetic delivery systems. With further understanding of viral gene delivery mechanisms, future synthetic delivery systems will mimic viral functionality even more, and we may someday witness a fully synthetic genetic vaccine delivery system with minimal safety concerns and the full potency of a virus.

References

1. Wolff JA, Malone RW, Williams P, Chong W, Acsadi G et al. (1990) Science 247:1465–1468
2. Tang DC, De Vit M, Johnston SA (1992) Nature 356:152–154
3. Ulmer JB, Donnelly JJ, Parker SE, Rhodes GH, Felgner PL et al. (1993) Science 259:1745–1749
4. Fynan EF, Webster RG, Fuller DH, Haynes JR, Santoro JC et al. (1993) P Natl Acad Sci USA 90:11478–11482
5. Johnston SA, Barry MA (1997) Vaccine 15:808–809
6. Chambers RS, Johnston SA (2003) Nat Biotechnol 21:1088–1092
7. Banchereau J, Steinman RM (1998) Nature 392:245–252
8. Bousso P, Robey E (2003) Nat Immunol 4:579–585
9. Dallal RM, Lotze MT (2000) Curr Opin Immunol 12:583–588
10. Wolff JA, Dowty ME, Jiao S, Repetto G, Berg RK et al. (1992) J Cell Sci 103:1249–1259
11. Knechtle SJ, Wang J, Jiao S, Geissler EK, Sumimoto R et al. (1994) Transplantation 57:990–996
12. Nickoloff BJ, Turka LA (1994) Immunol Today 15:464–469
13. Warrens AN, Zhang JY, Sidhu S, Watt DJ, Lombardi G et al. (1994) Int Immunol 6:847–853
14. Kundig TM, Bachmann MF, DiPaolo C, Simard JJ, Battegay M et al. (1995) Science 268:1343–1347
15. Corr M, Lee DJ, Carson DA, Tighe H (1996) J Exp Med 184:1555–1560
16. Ulmer JB, Deck RR, Dewitt CM, Donnhly JI, Liu MA (1996) Immunology 89:59–67
17. Doe B, Selby M, Barnett S, Baenziger J, Walker CM (1996) P Natl Acad Sci USA 93:8578–8583

18. Iwasaki A, Torres CA, Ohashi PS, Robinson HL, Barber BH (1997) J Immunol 159:11–14
19. Condon C, Watkins SC, Celluzzi CM, Thompson K, Falo LD Jr (1996) Nat Med 2:1122–1128
20. Casares S, Inaba K, Brumeanu TD, Steinman RM, Bona CA (1997) J Exp Med 186:1481–1486
21. Chattergoon MA, Robinson TM, Boyer JD, Weiner DB (1998) J Immunol 160:5707–5718
22. Porgador A, Irvine KR, Iwasaki A, Barber BH, Restifo NP et al. (1998) J Exp Med 188:1075–1082
23. La Cava A, Billetta R, Gaietta G, Bonnin DB, Baird SM et al. (2000) J Immunol 164:1340–1345
24. Chun S, Daheshia M, Lee S, Eo SK, Rouse BT (1999) J Immunol 163:2393–2402
25. Timares L, Takashima A, Johnston SA (1998) P Natl Acad Sci USA 95:13147–13152
26. Falo LD Jr, Kovacsovics-Bankowski M, Thompson K, Rock KL (1995) Nat Med 1:649–653
27. Kovacsovics-Bankowski M, Rock KL (1995) Science 267:243–246
28. Suto R, Srivastava PK (1995) Science 269:1585–1588
29. Bennett SR, Carbone FR, Karamalis F, Miller JF, Heath WR (1997) J Exp Med 186:65–70
30. Kurts C, Kosaka H, Carbone FR, Miller JF, Heath WR (1997) J Exp Med 186:239–245
31. Coelho-Castelo AA, Santos Junior RR, Bonato VL, Jamur MC, Oliver C et al. (2003) Hum Gene Ther 14:1279–1285
32. Pardoll D (2003) Annu Rev Immunol 21:807–839
33. Marincola FM, Jaffee EM, Hicklin DJ, Ferrone S (2000) Adv Immunol 74:181–273
34. Houghton AN (1994) J Exp Med 180:1–4
35. Pardoll DM (2002) Nat Rev Immunol 2:227–238
36. Barouch DH, Santra S, Schmitz JE, Kuroda MJ, Fu TM et al. (2000) Science 290:486–492
37. Le TP, Coonan KM, Hedstrom RC, Charoenvit Y, Sedegah M et al. (2000) Vaccine 18:1893–1901
38. Felgner PL (2003) Methods 31:181–182
39. Manthorpe M, Cornefert-Jensen F, Hartikka J, Felgner J, Rundell A et al. (1993) Hum Gene Ther 4:419–431
40. Winegar RA, Monforte JA, Suing KD, O'Loughlin KG, Rudd CJ et al. (1996) Hum Gene Ther 7:2185–2194
41. Jiao S, Williams P, Berg RK, Hodgeman BA, Liu L et al. (1992) Hum Gene Ther 3:21–33
42. Williams RS, Johnston SA, Riedy M, DeVit MJ, McElligott SG et al. (1991) P Natl Acad Sci USA 88:2726–2730
43. Pertmer TM, Eisenbraun MD, McCabe D, Prayaga SK, Fuller DH et al. (1995) Vaccine 13:1427–1430
44. Bergstresser PR, Fletcher CR, Streilein JW (1980) J Invest Dermatol 74:77–80
45. Tuettenberg A, Jonuleit H, Tuting T, Bruck J, Knop J et al. (2003) Gene Ther 10:243–250
46. Dietrich G, Spreng S, Favre D, Viret JF, Guzman CA (2003) Curr Opin Mol Ther 5:10–19
47. Vicat JM, Boisseau S, Jourdes P, Laine M, Wion D et al. (2000) Hum Gene Ther 11:909–916
48. Mendiratta SK, Thai G, Eslahi NK, Thull NM, Matar M et al. (2001) Cancer Res 61:859–863

49. Lohr F, Lo DY, Zaharoff DA, Hu K, Zhang X et al. (2001) Cancer Res 61:3281–4
50. Kalat M, Kupcu Z, Schuller S, Zalusky D, Zehetner M et al. (2002) Cancer Res 62:5489–5494
51. Paster W, Zehetner M, Kalat M, Schuller S, Schweighoffer T (2003) Gene Ther 10:717–724
52. Babiuk S, Baca-Estrada ME, Foldvari M, Storms M, Rabussay D et al. (2002) Vaccine 20:3399–3408
53. Rodriguez-Cuevas S, Barroso-Bravo S, Almanza-Estrada J, Cristobal-Martinez L, Gonzalez-Rodriguez E (2001) Arch Med Res 32:273–276
54. Zabner J (1997) Adv Drug Deliver Rev 27:17–28
55. Xu Y, Szoka FC Jr (1996) Biochemistry 35:5616–5623
56. Gregoriadis G, Saffie R, de Souza JB (1997) FEBS Lett 402:107–110
57. Perrie Y, Frederik PM, Gregoriadis G (2001) Vaccine 19:3301–3310
58. Perrie Y, Obrenovic M, McCarthy D, Gregoriadis G (2002) J Lipos Res 12:185–197
59. Okamoto T, Kaneda Y, Yuzuki D, Huang SK, Chi DD et al. (1997) Gene Ther 4:969–976
60. Zhou WZ, Kaneda Y, Huang S, Morishita R, Hoon D (1999) Gene Ther 6:1768–1773
61. Tanaka M, Kaneda Y, Fujii S, Yamano T, Hashimoto K et al. (2002) Mol Ther 5:291–299
62. Correale P, Cusi MG, Sabatino M, Micheli L, Pozzessere D et al. (2001) Eur J Cancer 37:2097–2103
63. Jones DH, Corris S, McDonald S, Clegg JC, Farrar GH (1997) Vaccine 15:814–817
64. Kempf M, Mandal B, Jilek S, Thiele L, Voros J et al. (2003) J Drug Target 11:11–18
65. O'Hagan DT, Singh M, Gupta RK (1998) Adv Drug Deliver Rev 32:225–246
66. Randolph GJ, Inaba K, Robbiani DF, Steinman RM, Muller WA (1999) Immunity 11:753–761
67. Thiele L, Rothen-Rutishauser B, Jilek S, Wunderli-Allenspach H, Merkle HP et al. (2001) J Control Release 76:59–71
68. Hemmi H, Takeuchi O, Kawai T, Kaisho T, Sato S et al. (2000) Nature 408:740–745
69. Frazza EJ, Schmitt EE (1971) J Biomed Mater Res 5:43–58
70. Woodland JH, Yolles S (1973) J Med Chem 16:897–901
71. Beck LR, Cowsar DR, Lewis DH, Gibson JW, Flowers CE (1979) Am J Obstet Gynecol 135:419–426
72. Sinclair RG (1973) Environ Sci Technol 7:955–956
73. Rokkanen P, Bostman O, Vainionpaa S, Vihtonen K, Tormala P et al. (1985) Lancet 1:1422–1424
74. Bercovy M, Goutallier D, Voisin MC, Geiger D, Blanquaert D et al. (1985) Clin Orthop Relat R 196:159–168
75. Williams DF, Mort E (1977) J Bioeng 1:231–238
76. Langer R, Cleland JL, Hanes J (1997) Adv Drug Deliver Rev 28:97–119
77. Odonnell PB, McGinity JW (1997) Adv Drug Deliver Rev 28:25–42
78. Gander B, Wehrli E, Alder R, Merkle HP (1995) J Microencapsul 12:83–97
79. Wang D, Robinson DR, Kwon GS, Samuel J (1999) J Control Release 57:9–18
80. Walter E, Moelling K, Pavlovic J, Merkle HP (1999) J Control Release 61:361–374
81. Hedley ML, Curley J, Urban R (1998) Nat Med 4:365–368
82. Klencke B, Matijevic M, Urban RG, Lathey JL, Hedley ML et al. (2002) Clin Cancer Res 8:1028–1037
83. Sheets EE, Urban RG, Crum CP, Hedley ML, Politch JA et al. (2003) Am J Obstet Gynecol 188:916–926
84. Chen SC, Jones DH, Fynan EF, Farrar GH, Clegg JC et al. (1998) J Virol 72:5757–5761
85. Fu K, Pack DW, Klibanov AM, Langer R (2000) Pharmaceut Res 17:100–106

86. Shenderova A, Burke TG, Schwendeman SP (1999) Pharmaceut Res 16:241–248
87. Walter E, Dreher D, Kok M, Thiele L, Kiama SG et al. (2001) J Control Release 76:149–168
88. Garg S, Oran A, Wajchman J, Sasaki S, Maris CH et al. (2003) Nat Immunol 4:907–912
89. McKeever U, Barman S, Hao T, Chambers P, Song S et al. (2002) Vaccine 20:1524–1531
90. Ando S, Putnam D, Pack DW, Langer R (1999) J Pharm Sci 88:126–130
91. Singh M, Briones M, Ott G, O'Hagan D (2000) P Natl Acad Sci USA 97:811–816
92. Denis-Mize KS, Dupuis M, MacKichan ML, Singh M, Doe B et al. (2000) Gene Ther 7:2105–2112
93. Denis-Mize KS, Dupuis M, Singh M, Woo C, Ugozzoli M et al. (2003) Cell Immunol 225:12–20
94. O'Hagan D, Singh M, Ugozzoli M, Wild C, Barnett S et al. (2001) J Virol 75:9037–9043
95. Luo YP, O'Hagan D, Zhou H, Singh M, Ulmer J et al. (2003) Vaccine 21:1938–1947
96. Cui Z, Mumper RJ (2002) Bioconjug Chem 13:1319–1327
97. Cui Z, Mumper RJ (2002) J Control Release 81:173–184
98. Cui Z, Mumper RJ (2002) Pharmaceut Res 19:939–946
99. Cui Z, Mumper RJ (2002) J Pharm Pharmacol 54:1195–1203
100. Cui Z, Baizer L, Mumper RJ (2003) J Biotechnol 102:105–115
101. Kumar M (2000) React Funct Polym 46:1–27
102. Stahl PD (1992) Curr Opin Immunol 4:49–52
103. Roy K, Mao HQ, Huang SK, Leong KW (1999) Nat Med 5:387–391
104. Cui Z, Mumper RJ (2001) J Control Release 75:409–419
105. Kabbaj M, Phillips NC (2001) J Drug Target 9:317–328
106. Wang C, Ge Q, Ting D, Nguyen D, Shen HR et al. (2004) Nat Mater 3:190–196
107. Little SR, Lynn DM, Ge Q, Anderson DG, Puram SV et al. (2004) P Natl Acad Sci USA 101:9534–9539
108. Lynn DM, Anderson DG, Putnam D, Langer R (2001) J Am Chem Soc 123:8155–8156
109. Lynn DM, Amiji MM, Langer R (2001) Angew Chem Int Edit 40:1707–1710
110. Boussif O, Lezoualch F, Zanta MA, Mergny MD, Scherman D et al. (1995) P Natl Acad Sci USA 92:7297–7301
111. Fields RC, Shimizu K, Mule JJ (1998) P Natl Acad Sci USA 95:9482–9487
112. Alters SE, Gadea JR, Sorich M, O'Donoghue G, Talib S et al. (1998) J Immunother 21:17–26
113. Tjandrawan T, Martin DM, Maeurer MJ, Castelli C, Lotze MT et al. (1998) J Immunother 21:149–157
114. Celluzzi CM, Mayordomo JI, Storkus WJ, Lotze MT, Falo LD Jr (1996) J Exp Med 183:283–287
115. Amoscato AA, Prenovitz DA, Lotze MT (1998) J Immunol 161:4023–4032
116. Gong J, Chen D, Kashiwaba M, Kufe D (1997) Nat Med 3:558–561
117. Siders WM, Vergilis KL, Johnson C, Shields J, Kaplan JM (2003) Mol Ther 7:498–505
118. Diebold SS, Lehrmann H, Kursa M, Wagner E, Cotten M et al. (1999) Hum Gene Ther 10:775–786
119. Irvine AS, Trinder PK, Laughton DL, Ketteringham H, McDermott RH et al. (2000) Nat Biotechnol 18:1273–1278
120. Walter E, Merkle HP (2002) J Drug Target 10:11–21
121. Walter E, Thiele L, Merkle HP (2001) STP Pharma Sci 11:45–56
122. Arthur JF, Butterfield LH, Roth MD, Bui LA, Kiertscher SM et al. (1997) Cancer Gene Ther 4:17–25

123. Chamarthy SP, Kovacs JR, McClelland E, Gattens D, Meng WS (2003) Mol Immunol 40:483–490
124. Strobel I, Berchtold S, Gotze A, Schulze U, Schuler G et al. (2000) Gene Ther 7:2028–2035
125. Boczkowski D, Nair SK, Nam JH, Lyerly HK, Gilboa E (2000) Cancer Res 60:1028–1034
126. Milazzo C, Reichardt VL, Muller MR, Grunebach F, Brossart P (2003) Blood 101:977–982
127. Muller MR, Grunebach F, Nencioni A, Brossart P (2003) J Immunol 170:5892–5896
128. Nencioni A, Muller MR, Grunebach F, Garuti A, Mingari MC et al. (2003) Cancer Gene Ther 10:209–214
129. Heiser A, Maurice MA, Yancey DR, Coleman DM, Dahm P et al. (2001) Cancer Res 61:3388–3393
130. Su Z, Dannull J, Heiser A, Yancey D, Pruitt S et al. (2003) Cancer Res 63:2127–2133
131. Muller MR, Tsakou G, Grunebach F, Schmidt SM, Brossart P (2004) Blood 103:1763–1769
132. Zhao Y, Boczkowski D, Nair SK, Gilboa E (2003) Blood 102:4137–4142
133. Regnault A, Lankar D, Lacabanne V, Rodriguez A, Thery C et al. (1999) J Exp Med 189:371–380
134. Kim JJ, Yang JS, Dang K, Manson KH, Weiner DB (2001) Clin Cancer Res 7:882s–889s
135. Biragyn A, Tani K, Grimm MC, Weeks S, Kwak LW (1999) Nat Biotechnol 17:253–258
136. Albert ML, Pearce SF, Francisco LM, Sauter B, Roy P et al. (1998) J Exp Med 188:1359–1368
137. Jiang W, Swiggard WJ, Heufler C, Peng M, Mirza A et al. (1995) Nature 375:151–155
138. Cui Z, Hsu CH, Mumper RJ (2003) Drug Dev Ind Pharm 29:689–700
139. Rodriguez F, Harkins S, Redwine JM, de Pereda JM, Whitton JL (2001) J Virol 75:10421–10430
140. Rodriguez F, An LL, Harkins S, Zhang J, Yokoyama M et al. (1998) J Virol 72:5174–5181
141. Rodriguez F, Zhang J, Whitton JL (1997) J Virol 71:8497–8503
142. Delogu G, Howard A, Collins FM, Morris SL (2000) Infect Immun 68:3097–3102
143. Tobery T, Siliciano RF (1999) J Immunol 162:639–642
144. Fu TM, Guan L, Friedman A, Ulmer JB, Liu MA et al. (1998) Vaccine 16:p1711–1717
145. Sadasivan B, Lehner PJ, Ortmann B, Spies T, Cresswell P (1996) Immunity 5:103–114
146. Spee P, Neefjes J (1997) Eur J Immunol 27:2441–2449
147. Pike SE, Yao L, Jones KD, Cherney B, Appella E et al. (1998) J Exp Med 188:2349–2356
148. Pike SE, Yao L, Setsuda J, Jones KD, Cherney B et al. (1999) Blood 94:2461–2468
149. Cheng WF, Hung CF, Chai CY, Hsu KF, He L et al. (2001) J Clin Invest 108:669–678
150. Rowell JF, Ruff AL, Guarnieri FG, Staveley-O'Carroll K, Lin X et al. (1995) J Immunol 155:1818–1828
151. Ji H, Wang TL, Chen CH, Pai SI, Hung CF et al. (1999) Hum Gene Ther 10:2727–2740
152. Wu TC, Guarnieri FG, Staveley-O'Carroll KF, Viscidi RP, Levitsky HI et al. (1995) P Natl Acad Sci USA 92:11671–11675
153. Rodriguez F, Harkins S, Redwine JM, de Pereda JM, Whitton JL (2001) J Virol 75:10421–10430
154. Forns X, Emerson SU, Tobin GJ, Mushahwar IK, Purcell RH et al. (1999) Vaccine 17:1992–2002
155. Geissler M, Bruss V, Michalak S, Hockenjos B, Ortmann D et al. (1999) J Virol 73:4284–4292
156. Rice J, King CA, Spellerberg MB, Fairweather N, Stevenson FK (1999) Vaccine 17:3030–3038

157. Steinman RM, Hawiger D, Nussenzweig MC (2003) Annu Rev Immunol 21:685–711
158. Morita A, Ariizumi K, Ritter R 3rd, Jester JV, Kumamoto T et al. (2001) Gene Ther 8:1729–1737
159. Ross R, Sudowe S, Beisner J, Ross XL, Ludwig-Portugall I et al. (2003) Gene Ther 10:1035–1040
160. Sudowe S, Ludwig-Portugall I, Montermann E, Ross R, Reske-Kunz AB (2003) Mol Ther 8:567–575
161. Schlesinger S, Dubensky TW (1999) Curr Opin Biotechnol 10:434–439
162. Wang B, Merva M, Dang K, Ugen KE, Boyer J et al. (1994) AIDS Res Hum Retrov 10:S35–41
163. Little SR, Lynn DM, Puram SV, Langer R (2005) J Con Rel 107(3):449–462

Adv Biochem Engin/Biotechnol (2005) 99: 119–145
DOI 10.1007/10_005

Published online: 25 October 2005

Adeno-associated Virus as a Gene Therapy Vector: Vector Development, Production and Clinical Applications

Joshua C. Grieger · Richard J. Samulski (✉)

University of North Carolina Gene Therapy Center,
Thurston Bowles Bldg. Rm. 7119 CB 7352, Chapel Hill, NC 27599, USA
jgrieger@med.unc.edu, rjs@med.unc.edu

Abstract Adeno-associated virus (AAV) has emerged as an attractive vector for gene therapy. AAV vectors have successfully been utilized to promote sustained gene expression in a variety of tissues such as muscle, eye, brain, liver, and lung [1–7]. As the significance of AAV as a gene therapy vector has been realized over the past years, recent developments in recombinant AAV (rAAV) production and purification have revolutionized the AAV field. It is now possible to produce high yields of vector (10^{12}–10^{13} genome-containing particles per mL) that are free of contaminating cellular and helper virus proteins. Such vectors have been successfully used in preclinical applications in animal models such as those of hemophilia, lysosomal storage diseases and vision deficiency, all of which have shown therapeutic benefits from rAAV treatment. Clinical trials using rAAV2 for the treatment of hemophilia B, cystic fibrosis, alpha-1-antitrypsin deficiency, and Canavan disease have begun, and reports from these phase I trials support the safety seen in preclinical trials. Eventually, tissue-specific vectors that can potentially evade the immune system will be required to optimize success in gene therapy. In recent years, this has led to the development of retargeted rAAV2 vectors and the identification and characterization of new serotypes from human and nonhuman primates that could potentially achieve these goals. AAV virologists and gene therapists alike have just begun to scratch the surface in terms of the utility of this small virus in a clinical setting. In this chapter, we will provide a comprehensive overview of the recent advances in rAAV vector production and purification, vector development, and clinical applications.

Keywords Adeno-associated virus · Adenovirus · Heparan sulfate proteoglycan · Transfection · Herpes simplex virus

1
AAV Biology

AAV is a member of the parvovirus family. Parvoviruses are among the smallest of the DNA animal viruses with a virion of approximately 25 nm in diameter composed entirely of protein and DNA. AAV has been classified as a dependovirus because it requires coinfection with helper viruses such as adenovirus (Ad) or herpes simplex virus (HSV) for productive infection in cell culture [8, 9].

The AAV2 genome is a linear, single-stranded DNA molecule containing 4679 bases [10]. The wild-type (wt) AAV genome is made up of two genes that encode four replication proteins and three capsid proteins respectively and is flanked on either side by 145 bp inverted terminal repeats [10, 11], as shown in Fig. 1. The larger replication proteins, Rep 78 and 68, are splice variants originating from the p5 promoter. Rep 68 and 78 are multifunctional and play a role in almost every aspect of the life cycle of AAV such as transcription, viral DNA replication and site-specific integration into human chromosome 19. The small replication proteins, Rep 40 and 52, have been characterized as important for DNA packaging into the preformed viral capsid within the nucleus [12, 13]. The virion is made up of the three capsid proteins Vp1, Vp2 and Vp3 in a ratio of 1 : 1 : 8 respectively. The capsid proteins are produced from the same open reading frame (ORF) but utilize different translational start sites. Vp3 is the most abundant subunit in the virion and is used for receptor recognition at the surface of cells defining the tropism of AAV. A phospholipase domain essential for viral infectivity has been identified in the unique *N*-terminus of Vp1 [14, 15]. The functional significance of Vp2 remains ambiguous, but recent studies have shown that it is possible to make viable virus lacking Vp2 in a recombinant background [16]. The biology of AAV2 has been extensively reviewed elsewhere [17, 18]. The current chapter is limited to issues concerning the use of AAV as a gene therapy vector.

One of the distinctive features of AAV replication has been the requirement for coinfection of the cell with a helper virus. In vitro, in the absence of helper virus, AAV can establish a latent infection by integrating into a unique site on human chromosome 19 [19–21]. Infection of cells latently infected with AAV by Ad or HSV will rescue the integrated genome and begin a productive infection [21–23]. While Ad early gene products serve as helper functions for AAV replication, late Ad gene products have been found to be unnecessary.

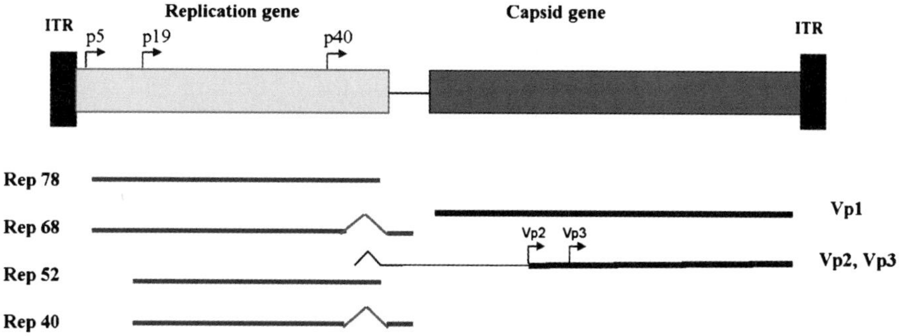

Fig. 1 Genomic structure of wild-type AAV. The p5 promoter initiates transcripts for Rep 78 (*when unspliced*) and Rep 68 (*spliced*). The p19 promoter initiates transcripts for Rep 52 (*unspliced*) and Rep 40 (*spliced*). Both Rep 52 and 40 products are in-frame with the Rep 78 and 68 products. The p40 promoter initiates all three capsid mRNAs. Vp1 is produced by an alternative splicing event. Using the same mRNA as Vp3, Vp2 is produced from an upstream alternative ACG start codon

The four Ad proteins required for helper function are the early region E1A, E1B, E4, and E2A. The synthesis of the Ad virus-associated (VA) RNAs is also required. The E1A protein is responsible for transactivation of AAV gene expression [24] and the E4 and E1B regulate the expression of AAV genes by aiding in mRNA transport to the cytoplasm [25] and aid in DNA replication by stopping the cell cycle in the G2 or S phase [25–28]. The E2A DNA binding protein aids in activating transcription from AAV promoters [29, 30] and may be involved in AAV DNA replication. By preventing the interferon-induced host cell shutoff mechanism of translation, Ad VA RNA assists the initiation of AAV protein synthesis.

Herpesviruses have also been identified as helper viruses for productive AAV replication. In order to determine the HSV genes essential for helper activity, Weindler et al. (1999) produced different HSV mutants defective in individual HSV replication genes [31]. Similar to Ad, AAV replication is not dependent upon the expression of late HSV genes. Genes encoding the helicase-primase complex (UL5, UL8, and UL52) and the DNA-binding protein (UL29) have been found to be sufficient to mediate the helper effect.

2
Vector production

AAV has developed into one of the most important and safest viral gene delivery vectors in the field of gene therapy. The simplicity of the AAV genome allows for the straightforward design of rAAV vectors to deliver transgenes of interest. The inverted terminal repeats flanking either side of the genome are the only cis-acting elements necessary for genome replication, integration and packaging into the capsid. rAAV can therefore be produced by replacing the replication and capsid genes with a promoter and gene of interest (vector DNA). The rep and cap genes that encode the replication and structural proteins are expressed in trans from a different plasmid lacking ITRs. The separation of these genes from the vector DNA (containing the ITR packaging signal) is critical for circumventing the formation of wtAAV. As mentioned previously, helper functions from adenovirus or herpes simplex virus are necessary for rAAV production.

In the past, the inability to produce high-titer stocks of rAAV without contamination from helper virus and wtAAV impeded the use of AAV for clinical applications. The initial protocol for production of rAAV involved co-transfection of cells with vector DNA and a packaging plasmid expressing the AAV rep and cap genes followed by infection with Ad at a low multiplicity of infection (moi) [32]. Yields of between 10^2 and 10^3 viral particles/cell are characteristic of this method and the stocks are often contaminated with helper virus. Several improvements on this method have been made over the years that have resulted in increased vector yields. The first was the recognition that low levels (below a certain threshold) of capsid proteins can decrease rAAV production [33]. The second was that reduced levels of Rep 78 expression result in higher capsid protein synthesis, increasing rAAV titers [33–36]. The third was replacement of the standard adenovirus infection by introducing essential Ad helper genes on a plasmid for DNA transfection [37, 38]. The fourth and fifth were the introduction of stable packaging cell lines [39] and the baculovirus system [40] to produce rAAV that are amenable to scaling up production. The majority of these improvements resulted in rAAV preparations that were free of contaminating Ad and its structural proteins. The next few sections will describe the aforementioned improvements used to increase rAAV titers and scale-up production in detail.

2.1
Transient transfection methods

The standard transient transfection methods described here utilize the calcium phosphate precipitation protocol reported by Chen and Okayama in 1987 [41]. The first demonstration that purified Ad DNA could supply helper functions to rAAV production was provided by Ferrari et al. in 1996 [42] and this finding led to a number of plasmid-based methods to produce high titer rAAV.

Xiao et al. introduced the triple transfection method in 1998 [38]. They generated plasmids pXX2 and pXX6. The pXX2 helper plasmid contains the wtAAV rep and cap genes essential for AAV production. The pXX6 Ad helper plasmid contains the Ad genes necessary for rAAV production and lacks the genes necessary for Ad virion production and the ability to produce Ad structural proteins. The ATG start site of Rep 78 and 68 was mutated to ACG to down-regulate their expression [35] and an additional p5 promoter element was cloned downstream of the Cap gene. It has been shown by a number of studies [43–46] that the lack of the p5 promoter cis-acting element causes down-regulation of the p19 and p40 promoters. They reasoned that an additional copy of the p5 promoter element may act as an enhancer for transcription

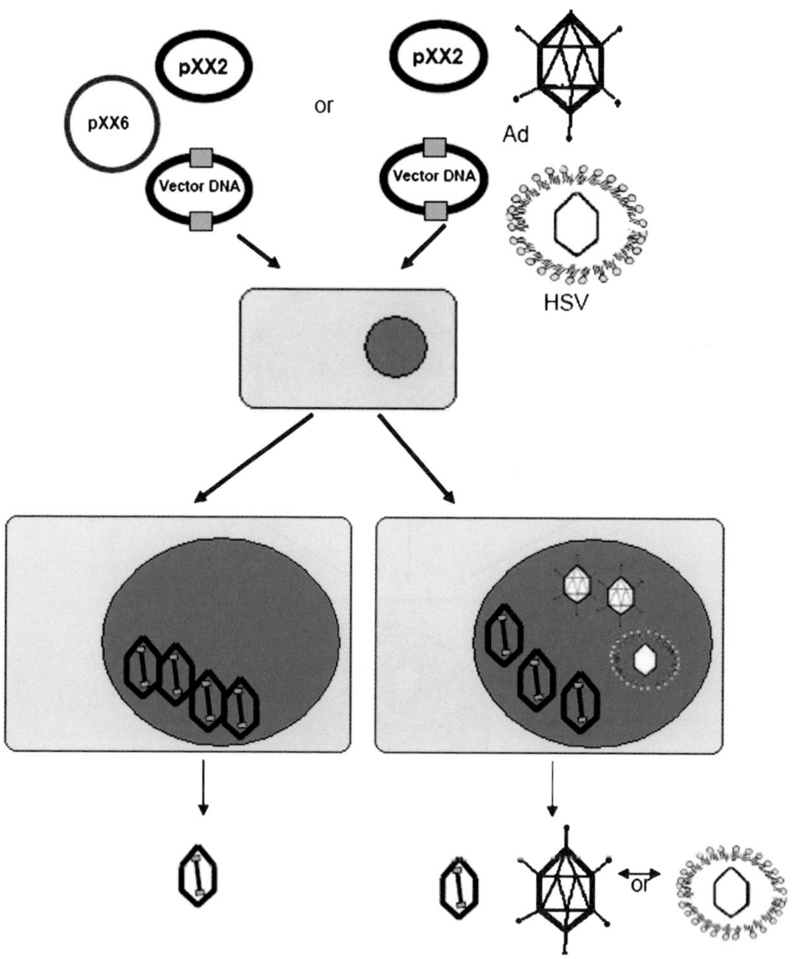

Fig. 2 **A** A schematic diagram of transfection methods used to produce rAAV with and without the use of helper virus. The triple transfection method developed by Xiao et al is depicted in the *left* of the transfection scheme. Three plasmids pXX6 containing the Ad helper genes, pXX2 containing the replication and capsid genes, and vector DNA (containing the ITRs flanking the transgene of interest) are transfected into cells. Cells are incubated 48–72 hours prior to isolating rAAV. The triple transfection method produces rAAV only while helper virus is found with rAAV when infected into cells for helper function

from the p19 and p40 promoters, which was illustrated in their studies. Utilizing the pXX2 plasmid and vector DNA followed by Ad infection, rAAV titers were increased 15-fold over its parental plasmid, pAAV/Ad [32]. Transfection of the pXX6, pXX2 and the vector plasmid (Fig. 2A) resulted in high-titer rAAV vectors with yields of approximately 1000 transducing units or 10^5 viral particles/cell without detectable contaminating Ad proteins and wtAAV.

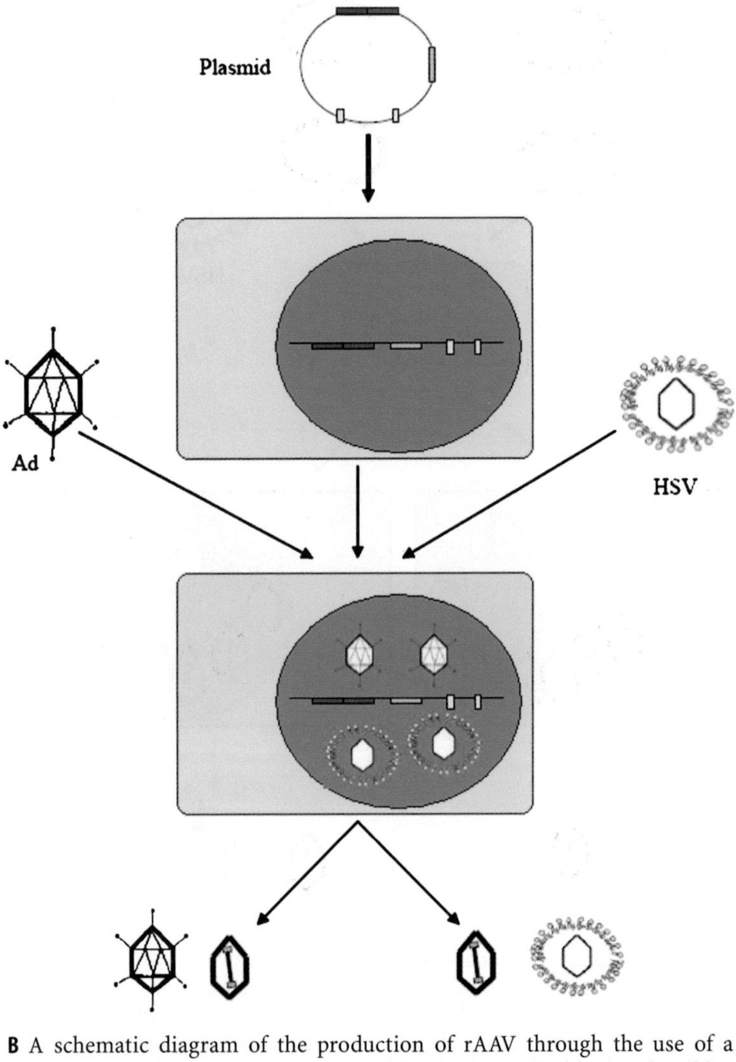

Fig. 2 **B** A schematic diagram of the production of rAAV through the use of a stable producer cell line. A plasmid containing rep (*blue box*), cap (*red box*), selectable marker (*green box*) and the transgene of interest flanked by ITRs (*light blue*) are transfected into cells. Cells that contain integrated copies of the plasmid are selected using a suitable media and subsequently infected with Ad or HSV to produce rAAV

In a similar study, a single plasmid containing the Ad helper and the rep and cap genes (pDG) was developed [36]. In this construct, the rep p5 promoter was replaced with the MMTV-LTR (mouse mammary tumor virus long terminal repeat) promoter to decrease the rep 78/68 expression levels and in turn increase the production of the capsid proteins. Cotransfection of 293 cells with pDG and AAV vector plasmid was sufficient to produce rAAV at titers of approximately 50–500 infectious particles/cell or 5×10^3 to 5×10^4 viral particles/cell. Allen et al. were able to limit the number of Ad helper genes to just one, the E4orf6 gene, to produce rAAV by utilizing helper AAV plasmids containing the MT (mouse metallothionein) and CMV (cytomegalovirus) promoters in replacement of the p5 and p40 promoters respectively. Yields of approximately 10 000 rAAV particles/cell were achieved after quadruple transfection of plasmids MT-rep, CMV-cap, vector DNA and the E4orf6 into 293 cells. Other studies utilizing variations of the transfection method with Ad mini-plasmids yielded titers of approximately 14 000 rAAV particles/cell [37, 47, 48].

Zhang et al. utilized DISC (disabled single-cycle virus) HSV (PS1) containing a deletion of the glycoprotein H (gH). This virus is infectious only if propagated in a complementing cell line [49], similar to Ad mutants grown on complementing 293 cells. They were able to use the DISC-HSV as a helper for rAAV replication, and have simulated to some extent the amplification of the rep and cap genomes seen in wtAAV infection by incorporating both these and vector sequences in HSV amplicons. In this system, increased production of AAV rep and cap proteins translated to an improved recovery of rAAV. When transfecting the HSV amplicon into BHK cells followed by infection with DISC-HSV helper virus, approximately 28 000 DNase resistant rAAV particles/cell are produced [50]. All of the above-described systems have allowed researchers to routinely generate AAV vectors for preclinical and phase I clinical trials.

2.2
Stable cell lines for production of AAV

Packaging cell lines are currently being developed for the large-scale production of rAAV. Vincent et al. were the first to demonstrate that a cell line stably expressing rep and cap could be used to produce rAAV. This method involved transfection of the rAAV genome and titers were low [120].

rAAV producer cell lines are generated by transiently transfecting the desired plasmid or plasmids containing the vector DNA with or without the essential rep and cap genes. A selectable marker is also present on each plasmid. After the cells have been transfected and selected in suitable growth media, stable cell lines are screened for viral production by Ad infection. The clones that produce the most rAAV are used as producer cell lines. Thus, this method can be used to produce rAAV without the need to transfect plasmid

DNA into cells. Infecting these cell lines with Ad or HSV is sufficient to generate rAAV (Fig. 2B). A HeLa cell line was generated as described above and infected with wtAd to produce rAAV [39, 50, 51]. This method was shown to produce approximately 40 000 particles/cell. Although this approach yields high titers of rAAV, a disadvantage of this approach is that new producer cell lines must be generated for each transgene construct with the risk of Ad contamination.

Many variations of this approach have also been tested. Gao et al. infected a HeLa cell line (B50) that has the rep-cap genes with endogenous promoters stably integrated. rAAV was produced in a two-step process [52]. The B50 cells were infected with an Ad defective in E2b to induce rep and cap expression and provide helper functions, followed by a hybrid virus in which the AAV vector is cloned in the E1 region of a replication-defective Ad. This method results in a 100-fold amplification and rescue of the AAV genome with a rAAV yield of 55 000 particles/cell that is wtAAV-free. The advantage of this method is that stable cell line selection is not necessary when producing rAAV with varying transgenes. In a similar study, but lacking a producer cell line, Zhang et al. generated rAAV using only Ad [53]. All components required for rAAV production were delivered to 293 cells by three Ad vectors. They estimated that 5.2×10^4 to 3×10^5 DNase resistant particles per cell are produced from this approach. A potential limitation of this approach is the possibility that some of the Ad may contaminate the rAAV preparations leading to the possibility of generating replicative rAAV in host cells.

The development of a stable cell line has not been limited to the use of Ad helper only. A similar production system for HSV-1 carrying the AAV rep and cap genes (d27.1-rc) has been described by Conway et al. [54]. This rHSV-1, although replication defective, expresses the HSV-1 early genes required for rAAV replication and packaging. 293 cells harboring an integrated copy of vector DNA are infected with the rHSV-1 to produce rAAV with the transgene of interest. rAAV vector yields were reported to be greater than 10^4 particles/cell.

Continued development of stable cell lines including integrated inducible systems to produce rAAV have been tested [55, 56]. Qiao et al. developed the "dual splicing switch" where an intron and three polyadenylation sequences are inserted into the protein coding region of the rep gene and flanked on either end by LoxP sites. The intron and polyadenylation sequences disrupt all four Rep genes because they contain transcription termination sequences. As a result, all of the Rep transcripts are prematurely terminated and the genes inactivated making this system viable in 293 cells expressing the E1 Ad gene. Removal of the terminator sequences by Cre protein reactivates the transcription of all four Rep proteins. The aforementioned cassette (derived from pXX2) along with the vector DNA cassette are stably integrated in 293 cells. Since 293 cells already stably express E1A and E1B genes, an adenovirus with E1A and E1B deleted would provide sufficient helper functions for rAAV production. Large-scale vector productions (20 × 5-cm plates) with GFP and minidystrophin cell lines yielded 6.8×10^{13} and 1.24×10^{14} vector genome particles per mL respectively. These yields are 30–60 fold higher in titer over the previously described triple transfection method with pXX2 and pXX6 [38].

2.3 The baculovirus system

The use of the baculovirus system to produce high titers of rAAV2 has been demonstrated [40, 57]. The recombinant baculovirus derived from the *Autographa californica* nuclear polyhedrosis virus, widely used for large-scale production of heterologous proteins in cultured insect cells, was used in this study. rAAV were produced by coinfecting three recombinant baculoviruses (a Rep-baculovirus, a VP-baculovirus, and an AAV ITR vector baculovirus) into suspension Sf9 cells. The yields from this system were compared to those obtained by transfection of 293 cells [36]. 4×10^{8} cells were used for rAAV production by infecting 200 ml of Sf9 cells or transfecting twenty 15-cm plates of 293 cells. PCR results from this study illustrated that 1.8×10^{13} viral genomes (vg) were produced in Sf9 cells compared with 2×10^{12} vg by transfection of 293 cells corresponding to approximately 5000 vg/293 cell and 50 000 vg/Sf 9 cell. Along with the baculovirus system, the HSV- and Ad-infected packaging cell lines represent the most promising for GMP scale production.

In summary, numerous successful methods producing high titers of rAAV have been developed. Each method has its advantages and disadvantages depending on its application. More importantly, as better methods of producing rAAV continue to be developed, the generation of purification methods that can be used to obtain high concentrations of uncontaminated vector becomes essential. The next section details recent advancements in vector purification.

2.4 Vector purification

The method used for the purification of rAAV particles is another critical step for the assessment of vectors in preclinical and clinical trials. It is essential to have purification methods that are suitable for purifying high titers of AAV from cell homogenate free of contaminating cellular proteins and/or helper virus. The traditional method used to purify AAV from infected or transfected cells has been density equilibrium gradients (cesium chloride, CsCl). Because cesium atoms are heavy, concentrated solutions of CsCl can form density gradients after only a few hours of ultracentrifugation. The buoyant density of a molecule is defined as the concentration of CsCl, in g/cm^3, at that exact point in the density gradient at which the molecule floats [58].

The density of AAV packaging a wild-type size genome is approximately 1.41 g/cm^3. This method allows for the physical separation of full particles (AAV packaging a genome) from empty particles based on their differences in density. Contents in the CsCl gradient are traditionally collected from the bottom of the gradient by dripping. Methods such as dot blot analysis or quantitative PCR, used to assess the presence of the viral genomes, are then carried out on each fraction of the gradient to identify the fraction(s) that contain AAV. There are many disadvantages to using CsCl gradients to purify AAV. Two or three rounds of CsCl centrifugation must be carried out to get purified AAV. Dialysis of CsCl fractions containing AAV against a physiological buffer is necessary prior to in vivo analysis because CsCl can exert toxic affects on animals in the study. This process can take up to two weeks from start to finish. It has also been shown that, even after CsCl purification, 1% of the input infectious Ad and/or Ad proteins can still be found as contaminates. Based on these limitations, a protocol that could reduce the preparation time without sacrificing purity and quality of AAV preparations was needed.

Research efforts in this area resulted in an immunoaffinity column to purify AAV2 particles [36]. The monoclonal antibody A20 used in the affinity columns, which recognizes assembled AAV2 capsids but not unassembled capsid proteins, has been characterized previously [59]. In this study, a known amount of infectious AAV2 in the cell homogenate was passed over the affinity column resulting in 70% recovery of the starting virus. The A20 affinity column was also used to purify a wtAAV2 stock that was generated using Ad. Dot blot hybridization of the starting material and the eluted virus from the column using a probe specific for the Ad genome showed that contaminating Ad could be completely removed from the AAV2 stock. The purity of the eluted AAV2 particles was determined to be 80% by utilizing polyacrylamide gels followed by silver staining. Affinity column purification allows for fast and easy preparation of concentrated and purified AAV2 stocks in a single step.

Zolotukhin et al. developed a novel way of purifying rAAV from crude lysates that involved the use of iodixanol gradients and heparin column purification [60]. The density of macromolecules in iodixanol is different from that in CsCl. Purified rAAV was found to have a density of 1.266 g/cm^3 in a continuous iodixanol gradient, which is equivalent to a 50% solution of iodixanol. To purify rAAV utilizing iodixanol, step gradients of 15%, 25%, 40%, and 60% were generated and the crude lysate was placed on top of the 15% step. After an hour of centrifugation, the majority of the rAAV bands were within the 40% density step. Approximately 75–80% of the rAAV in the crude lysate is recovered in the iodixanol fraction. For further purification after isolation of rAAV from iodixanol gradients, heparinized columns were used to determine if rAAV is capable of binding these columns. Heparan sulfate proteoglycan (HSPG) was recently identified to be a cell surface receptor for AAV2 [61]. HPLC chromatography utilizing UNO-S1 heparin columns was capable of yielding a rAAV product that is greater than 99% pure based on polyacrylamide gel electrophoresis and silver staining. Utilizing the calcium phosphate transfection protocol with pDG and vector plasmids, yields of 2×10^{10} infectious particles per ten 15 cm plates or approximately 3.1×10^{8} 293 cells can be acquired. When coinfecting with Ad to produce rAAV, the authors found the presence of less than 1% (0.0017%) of the starting Ad contaminating the heparin purified rAAV. A five-fold higher recovery of rAAV particles and a greater than 100-fold increase in infectivity was obtained using this method of purification when compared to the traditional CsCl gradient purification. The main disadvantage to using this system of purification is that AAV must have the ability to bind HSPG (currently only AAV2 and AAV3).

At this point, the only purification method that can be utilized to purify all of the new and novel AAV serotypes is the CsCl gradient method. Recently, methods have been developed that have the ability to purify all AAV serotypes by utilizing ion exchange chromatography [62, 63]. Gao et al. utilized cation and anion chromatography to purify rAAV in their B50/hybrid cell line production method. They utilized POROS 20 HE as the first column (cation) and the POROS 50 PI (anion) as the second column. The POROS 20 HE column was used to bind AAV from the cell lysate. This column alone was able to remove a majority of the infectious Ad used to produce rAAV, but not to the levels needed for clinical use. Therefore, the incorporation of a second column (anion exchange resin) was used for the purpose of binding the remaining Ad and cellular contaminates while allowing rAAV to flow through. Particle to infectious particle ratios were compared to other methods of purification and these ratios were between 4 and 40 for the two-column method. Based on these numbers, this was much lower than the B50/hybrid-produced rAAV purified by CsCl (150–200) and the 293 triple transfection method purified by CsCl (400–1000) in their hands. Kaludov et al. developed a similar method in order to purify rAAV serotypes 2, 4 and 5 using an HPLC anion exchange method. Conditions such as pH of buffers and concentration of the detergent octylglucopyranoside (OGP) were critical when purifying each serotype. The authors found that the particles were purified more than 60 000-fold from the crude lysate using this two-step procedure. Freeze-thaw lysate was solubilized with OGP and passed over the PI anion exchange resin column, enriching the virus over 200-fold. The PI eluate still contained a number of contaminating proteins. The eluate was then further purified and concentrated using a high molecular weight retention filter. The end-product consisted of over 90% AAV particles. Particle to infectious particle ratios were reported to be similar to those achieved by affinity column chromatography. Most importantly, this method was found to be amenable to scaling up. Using a 90 mL PI column, 10^{13} rAAV particles can be purified in three hours.

In summary, numerous methods have been developed for the production and purification (Fig. 3) of AAV that are focused on acquiring high vector yields free from contaminating cellular and helper virus proteins and are amenable to scaling up for human clinical trials. Needless to say, as better

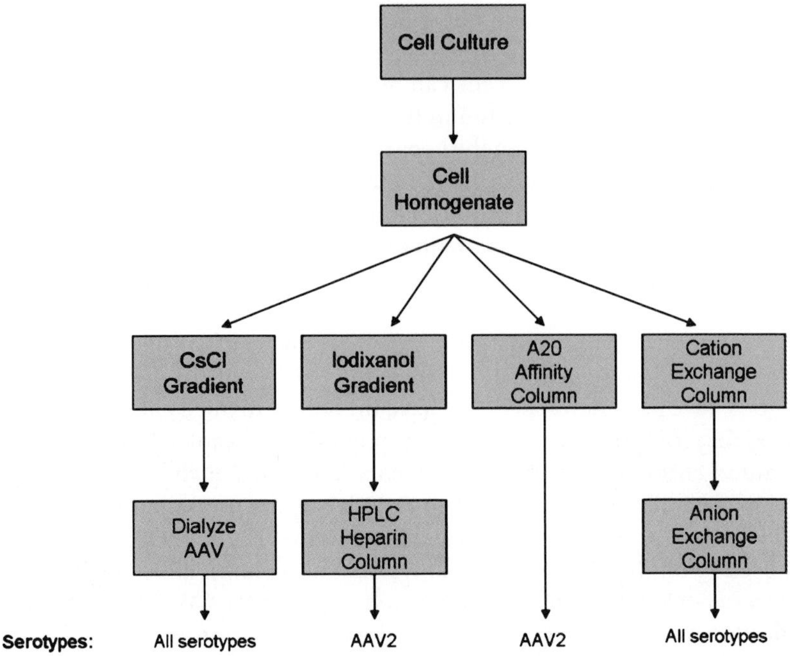

Fig. 3 A schematic representation of various methods (described in this chapter) used to purify rAAV from cell homogenate. The *bottom* of the schematic depicts the serotypes that can be purified using each method

methods are developed for generating high yields of rAAV, better methods of purification must be generated to acquire them in pure form. In addition, the biotechnology industry is building upon these numerous methods to produce rAAV by scaling up production utilizing large bioreactors containing liters of media and suspension cells, resulting in log increases of vector yield over the conventional small-scale methods described above.

3 Vector development

3.1 AAV2 Crystal structure

The structure of AAV2 has been determined at 3-Å resolution by X-ray crystallography [64], as shown in Fig. 4A. The AAV virion is composed of 60 capsid subunits arranged with $T = 1$ icosahedral symmetry. AAV contains three capsid proteins (Vp1, Vp2, and Vp3) in proportions of 1 : 1 : 8 that share overlapping sequences differing only at their *N* termini. Vp2 is 137 amino acids shorter than

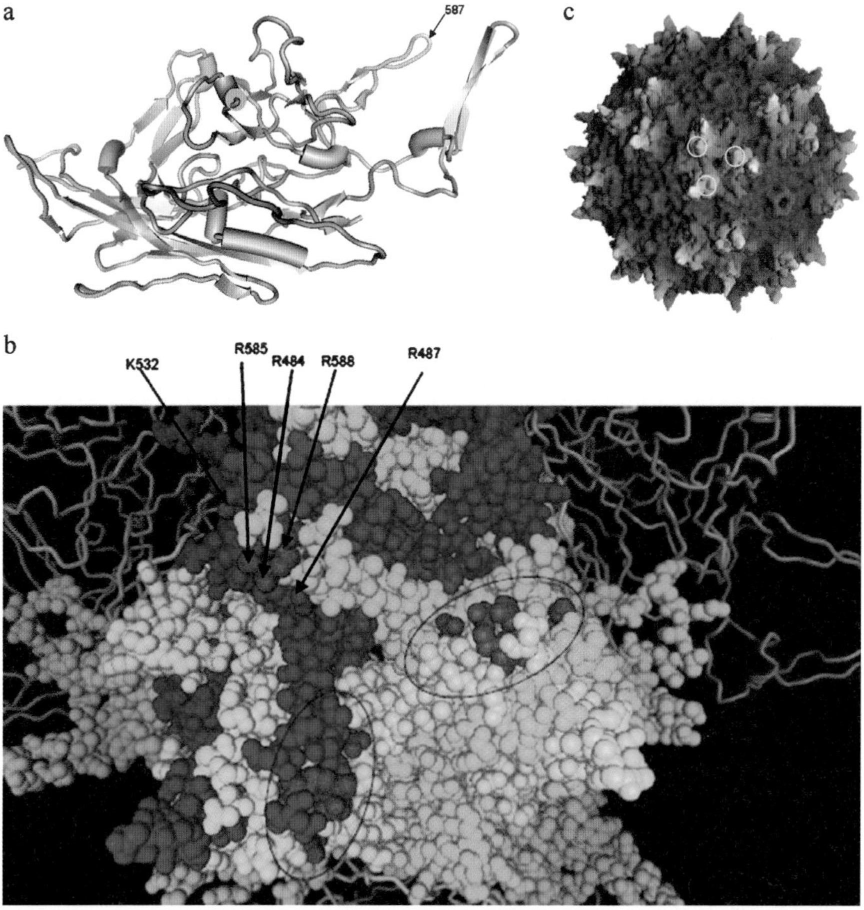

Fig. 4 Ribbon, space-filling, and cryo-EM diagrams of AAV2. **a** The crystal structure of the carboxy-terminal 519 amino acids of the AAV2 monomer subunit. N-terminal regions of Vp2 and Vp1 were not detected in the crystal structure. The amino acid (587) pointed to by an *arrow* represents the region of the monomer subunit where the majority of the successful retargeting ligands have been inserted. **b** A space-filling model of the three-fold axis of symmetry. Amino acids highlighted in *red* represent the amino acids necessary for heparan sulfate proteoglycan binding. R585 and R588 are supplied from one subunit and R484, R487, and K532 are supplied by a second subunit. Regions that are *circled* represent the other peaks that make up the three-fold spikes. **c** Depicts a cryo-EM image of the surface topology of AAV. The *yellow circles* identify the locations of the HSPG binding domain and the region where targeting ligands have been inserted. The authors would like to thank Dr. Michael Chapman for generously providing this image

Vp1 and is the product of an alternative start codon, while Vp3 is 65 residues shorter than Vp2. The 533 residue Vp3 is only visualized in the crystal structure, along with the corresponding regions of Vp1 and Vp2 occupying equal positions in the capsid, but not the unique *N*-termini of Vp1 and Vp2. It is possible to produce Vp3-only viruses [7] that can successfully package and protect its genome and bind and internalize into cells. However, these viruses are not infectious (meaning they are incapable of trafficking to the nucleus) potentially due to the lack of the phospholipase domain [15] present in the unique *N*-terminus of Vp1 and/or an NLS sequence.

The basic subunit structure of AAV is comprised of a jelly-roll β-barrel motif that comprises two antiparallel β-sheets. Long loop insertions between the strands comprise about 60% of the structure. These interstrand loops contain β-ribbons and other additional secondary structures that constitute the surface features of AAV. These, in turn, govern interactions with antibodies and cellular receptors. The most prominent feature of the surface topology of AAV is the three-fold proximal peaks clustering around each three-fold axis of symmetry, as shown in Fig. 4C. Loops from neighboring subunits interact to form the peaks. Located at the center of each peak is a subloop that is part of the GH loop of one subunit that is inserted between two other subloops from the GH loop of an adjacent three-fold symmetry subunit producing the three-fold peaks [64].

The crystal structure of AAV2 will provide the knowledge needed to analyze the existing mutagenesis data and to direct future retargeting modifications of AAV2 and the other serotypes. In the past, genetic modification of AAV2 virions has depended on random mutagenesis [65], alanine-scanning mutagenesis [66], and site-directed insertional mutagenesis [66–68]. All of these strategies utilized the crystal structure of canine parvovirus [69] (CPV) along with the amino acid alignments of AAV2 with CPV as a guide for mutagenesis. These strategies have determined surface-displayed domains and regions that interfere with primary receptor binding (HSPG). The genetic data acquired from epitope mapping and mutagenesis studies has now been integrated with the crystal structure to form a rational picture that links function with structure.

Prior to the AAV2 crystal structure being solved, HSPG was determined to be the primary receptor for AAV2 [61]. The disaccharides that make up heparan and heparan sulfate can be modified by N sulfation as well as 2-O and 6-O sulfation to give an overall negative charge at physiological pH. Therefore, heparan sulfate can interact with numerous proteins, primarily by electrostatic attraction between the sulfate groups and clusters of positively charged amino acids [70, 71]. Alanine-scanning mutagenesis conducted by Wu et al. identified the amino acid domain of 585–588 as essential for HSPG binding [66]. Amino acids R585 and R588, depicted in Fig. 4B, located on one of the peaks at the three-fold axis of symmetry, were thought to be the only amino acids necessary for HSPG interaction until two recent studies [72, 73] elucidated three more basic amino acids essential for this interaction. It was possible to narrow down the amino acids to target mutagenesis utilizing the crystal structure of CPV and later that of AAV2 with amino acid alignments of AAV serotypes 1–5. Both studies identified the basic amino acids of K532, R484, R487, R585, and R588 to be essential for HSPG interaction at the cell surface. K532, R487, R484 were found to play minor roles in HSPG binding while R585 and R588 were identified to be the most important. Figure 4B illustrates that all five amino acids are located in close proximity to one another, lining the canyon of the three-fold axis of symmetry.

3.2 Retargeting of rAAV2 to specific cell types

rAAV vectors possess a number of attractive features that include a relatively low pathogenicity, ability to transduce both dividing and nondividing cells, and long-term transgene expression in vitro and in vivo. rAAV vectors have been used to attain expression of a large number of genes in a variety of cells and organs including muscle, brain, eye, liver, and lung [1–4]. However, its broad host range is a disadvantage for in vivo gene therapy because it does not allow for tissue-specific transduction. Thus, efforts have been ongoing to target the well-characterized AAV2 based vectors to specific receptors.

At least two different strategies/methods are possible to achieve alternate receptor targeting of AAV [74]. The authors of this review [74] refer to these two different strategies as indirect and direct targeting. Indirect targeting is described as the interaction between the rAAV vector and the target cell, that is mediated by an associated molecule which is bound to the viral capsid and interacts with a cell surface receptor. Direct targeting involves the insertion of a ligand into the capsid structure that mediates cell-specific targeting of the vector.

When utilizing indirect targeting, knowledge of the three-dimensional structure of the virus is not necessary when high-affinity viral surface binding molecules (such as antibodies) are available. The bispecific molecules must bind to cell-specific receptors to allow uptake and proper intracellular viral processing. The success of this strategy is dependent upon the stability of the interaction of the virus with the bispecific molecule.

Targeting AAV2 using a bispecific antibody mediating the interaction between the virus and target cell was first described by Bartlett et al. (1999) [75]. The two monoclonal antibodies used to produce the bispecific molecule were AP-2 and A20. Heterodimers were generated by chemically cross-linking the Fab′ arms, one with specificity toward $\alpha_{IIb}\beta_3$ integrin (AP-2 antibody) and the other with specificity toward the intact capsid (A20 antibody). Fibrinogen is the ligand for $\alpha_{IIb}\beta_3$ integrin and it becomes internalized via endocytosis. Thus, AAV2 was expected to become internalized via receptor-mediated endocytosis similarly to wtAAV2 infection when targeted to this integrin. The targeted vector was able to transduce MO7e (human megakaryocytic leukemia cell line) and DAMI (human megakaryoblastoid cell line) cells that have been characterized as nonpermissive to wtAAV2 infection. The targeted AAV2 vector was able to transduce these cells 70-fold more efficiently. A subsequent study utilized a high-affinity biotin-avidin interaction as a molecular bridge to cross-link targeting ligands (EGF and FGF1α) genetically fused to core-streptavidin [76]. This strategy was chosen because the avidin-biotin complex represents the highest affinity interaction between a ligand and a protein known in nature. Conjugation of the bispecific targeting protein to the capsid was accomplished by biotinylating purified rAAV2. The modified vectors, targeted to epidermal growth factor receptor (EGFR) or fibroblast growth factor 1α receptor (FGFR1α), resulted in an increase ($>$ 100-fold) in transduction efficiency of EGFR positive SKOV3.ip1 (human ovarian cancer cell line) cells. The issue that remains to be resolved is the ability of these viruses to target similar cell types in vivo and to assess the stability of the virus-bispecific antibody complexes.

Four important constraints are required for the generation of targeting vectors via modifications of the capsid. The first is to identify the site of insertion on the capsid to place the targeting ligand to ensure that DNA packaging is unaffected and the ligand is exposed on the surface of the virus. Utilizing the crystal structure of AAV2, it is now possible to make educated guesses at identifying the regions of the capsid to insert targeted ligands. The second constraint involves modifying the vector so that it can no longer bind to its native cellular receptor(s), becoming dependent upon the targeting ligand for infection. The HSPG binding domain has been mapped to the GH loops located at the three-fold axis of symmetry on AAV2 [72, 73]. Therefore, it would be possible to insert a targeting ligand into this region to ablate the vectors' ability to bind HSPG and consequently retarget the vector to a specific cell type. The third constraint is the choice of the targeting peptide. It is complicated to determine the secondary structure of the ligand after insertion into the capsid. Thus, choosing a small structure-independent ligand is preferable to avoid destabilization of the capsid. The fourth constraint is that the ligand should target the vector to a receptor for internalization in a way that allows efficient transport of the virus and release of the viral DNA in the cell nucleus.

The first demonstration of a genetic modification leading to retargeting of AAV2, prior to the solving of the crystal structure, was described by Girod et al. [77]. They identified six sites on the capsid via a sequence alignment of AAV2 with CPV that were expected to be exposed on the surface of the capsid without disrupting functions essential to the viral life cycle. The ligand inserted into these six sites was the 14 amino acid L14 peptide that contains an RGD motif of the laminin fragment P1, which is the target for several integrin receptors. The L14 insertion into amino acid region 587 (depicted in Fig. 4A) was successful for retargeting AAV2 to infect mouse melanoma and rat schwannoma cell lines, which are incapable of binding to and infection by wtAAV2. In another study, amino acid region 587 was also used for insertion of an endothelial-specific peptide (SIGYPLP) that generated an AAV2 vector capable of infecting human umbilical vein endothelial cells and human saphenous vein endothelial cells [78]. Grifman et al. also utilized the insertion site at amino acid 587 to place the Myc epitope and the CD13 ligand (NGRAHA) [68]. Again, successful retargeting of AAV2 was achieved by gaining the ability to transduce Kaposi's sarcoma and rhabdomyosarcoma cell lines. These studies identified that insertion of a ligand into the amino acid 587 region inhibited AAV2 from binding its native receptor HSPG and confirmed that infection of the target cells is HSPG-independent. The peptide ligands that were chosen for these studies were identified by phage display prior to insertion into the capsid. Other noteworthy studies that were successful at retargeting AAV to cell lines of interest are detailed in references [79–84].

A study conducted by Wu et al. generated 93 mutants at 59 different positions on the AAV2 capsid to characterize functional domains on the AAV2 capsid [66]. Through their mutagenesis, regions were identified on the surface of the capsid that could possibly accept the insertion of a ligand. The regions identified were amino acids 266, 328, 447, 522, 553, 591, 664 in Vp3 as well as 34 in Vp1 and the *N*-terminus of Vp2 (138). The unique discovery in this study was that placing a serpin ligand at positions 34 and 138 of Vp1 and Vp2 respectively enhanced the ability of AAV2 to infect IB3 cells. This study and another study by Warrington et al. illustrated that insertions, as large as GFP, into the *N*-terminus of Vp2 are exposed on the surface of the capsid [16]. It is probable that targeting ligands placed in this region will enable AAV to target specific cell lines of interest. Mutations abrogating the HSPG binding domain at the three-fold axis of symmetry may be necessary to increase the specificity of the retargeting effort when positioning ligands at the *N*-terminus of Vp2.

Recently, a novel approach was developed that aided in the selection process of ligands to retarget AAV2 to specific cell types [85]. A random AAV2 peptide library was generated in which each particle displays a random 7-mer peptide at the capsid surface at amino acid position 588. This library was produced in a three-step process that ensured encoding of the displayed peptides by the packaged DNA. The diversity of the produced library of plasmids was 1.1×10^8 clones per library. Transfection of the library into producer cells would produce chimeric capsids displaying more than a single type of peptide insert and the packaged capsid gene would not encode the capsid mutants displayed on the viral surface. This would impede the selection process and the identification of selected capsid mutants. To resolve this, the library containing mutant capsid genes was packaged into capsids composed partially of wt capsid protein by cotransfecting wt rep-cap plasmid (pXX2) along with the AAV plasmid library. These particles were referred to as "AAV library transfer shuttles". Shuttle vectors were purified by heparin columns at a level of about 60% of wild-type AAV2. 293T cells were then infected with an MOI of 1 and 5 per cell. Infecting with a low MOI with the shuttle vectors led to the production of randomized AAV capsid library particles that contained the corresponding cap gene. The selection process then begins by infecting a cell line(s) of interest with the randomized AAV capsid library and Ad to amplify internalized AAV library clones. In this study, the cell line of interest was coronary endothelial cells. Amplified AAV from these cells were harvested and subjected to two more rounds of selection to enrich with AAV particles that successfully infect coronary endothelial cells. DNA was isolated from the cell lysate and AAV to PCR amplify the region containing the ligand to identify the sequence(s) necessary to infect the cell line of interest. A similar study was conducted to achieve AAV2 vectors targeted to cardiac tissue [86]. In vivo studies were carried out in the mouse via intravenous injection to assess luciferase activity in the heart. Transgene expression was five-fold higher in the hearts of animals injected with the targeted vector when compared to animals injected with the wt capsid rAAV2. This novel vector targeting strategy is advantageous over other approaches, such as phage display, in that selection happens within the context of viral capsid. Once the library has been generated, it is possible to screen numerous cell lines of interest to identify library clones that are efficient at targeting them.

It will be important to the field of gene therapy to develop targeting schemes to generate AAV2 vectors that could potentially be injected systemically, lacking its native tropism and having the ability to target the cell type(s) of interest. The aforementioned studies have established the groundwork for this concept. To further aid in targeting, tissue-specific promoters will need to be utilized to add another level of safety. It has also been shown that AAV-mediated transduction can be negatively impacted after internalization into the cell by proteasome degradation [87] and second strand synthesis [88]. Therefore, a better understanding of intracellular processing of AAV targeted vectors will be necessary, because these vectors may be transported into cellular compartments from which they will never escape or are processed in ways that prevent nuclear processing or gene expression.

3.3 Clades of AAV

Despite the longevity of transgene expression achieved by AAV2, its application has been limited by low levels of transgene expression in some tissues. This has been attributed to blocks at the level of cell binding and entry and intracellular processing. Progress in circumventing these issues has been made through the development of vectors based on other previously described serotypes that enter the cells through the use of receptors distinct from those utilized by AAV2. Intracellular processing and kinetics of uncoating may also be different for each serotype [89, 90]. AAV gene therapists have redirected their focus by integrating the other serotypes in preclinical studies using animal models to characterize the tropisms of each serotype. Each serotype has been shown to be advantageous over another based on their ability to transduce similar or different tissues more efficiently and/or evade the immune system.

Through work conducted by Gao et al., new serotypes of AAV have been identified (AAV 7, 8 and 9) along with a total of 108 new and unique isolates from 55 human and 53 nonhuman primates [91–93]. Table 1 illustrates capsid amino acid sequence diversity among AAV serotypes 1 through 9. The goal of their studies was to isolate new AAV serotypes with advantageous characteristics such as evasion of the human immune system (isolating AAV from nonhuman primates) and unique tropism. These new isolates were identified using PCR screening techniques on 259 human samples of 18 different tissue types derived from 250 individuals. Nonhuman primate data was gathered from similar tissues isolated from rhesus monkeys, cynomolgus and pigtailed macaques, baboons, and chimpanzees. Sequence variations in the isolated AAVs were identified as being located in hypervariable regions (HVRs) on the surface of the virus. Nine of the HVRs were mapped to the existing crystal structure of AAV2 in the Vp3 region. The three-fold proximal peaks, identified as the location of HSPG binding for AAV2, were found to contain the majority of the variability. Utilizing sequence alignments, they were able to generate a phylogenetic tree describing the evolution/homologous recombination phenomenon and to place the 108 isolates into nine clades. It is believed that the mechanism for the detected evolution seems to be a high rate of homologous recombination between different parental viruses infecting the same cell. The result is extensive swapping of HVRs of the capsid gene sequence, resulting in a variety of hybrids that could potentially have different tropisms and serology. This can be described as nature's system of "direct targeting" to generate tissue-specific AAV.

One of the main advantages to utilizing serotypes derived from nonhuman primates is their ability to evade the human immune system. This would allow the use of AAV serotype vectors in patients carrying neutralizing antibodies

Table 1 Capsid amino acid homology among AAV serotypes 1 through 9. AAV4 and AAV5 share the least homology with the other serotypes

Serotype	AAV1	AAV2	AAV3	AAV4	AAV5	AAV6	AAV7	AAV8	AAV9
AAV1	100	83.3	86.8	63	57.9	99.2	84.8	83.9	82.2
AAV2	83.3	100	87.9	59.9	57.3	83.4	82.4	82.8	81.7
AAV3	86.8	87.9	100	62.8	58.4	87	84.8	85.7	83.6
AAV4	63	59.9	62.8	100	52.5	62.7	63.4	63.1	62.2
AAV5	57.9	57.3	58.4	52.5	100	58	57.7	57.6	56.7
AAV6	99.2	83.4	87	62.7	58	100	85	84	82.1
AAV7	84.8	82.4	84.8	63.4	57.7	85	100	87.8	81.8
AAV8	83.9	82.8	85.7	63.1	57.6	84	87.8	100	85.3
AAV9	82.2	81.7	83.6	62.2	56.7	82.1	81.8	85.3	100

to various AAV serotypes, either from acquired infection with the virus, or from previous treatment with serotype vectors. AAV serotypes may be distinct enough from each other to allow repeated patient treatment with two or more different members of the AAV family. The majority of the serotypes (AAV1, 2, 3, 5, 6, and 9) that have been characterized to date have been detected and isolated from human tissue. It has been shown in studies that neutralizing antibodies to some of these serotypes have been found in the human population. For example, Xiao et al. and Hildinger et al. screened 77 and 85 normal human subjects, respectively, for the presence of neutralizing antibodies (nAb) to AAV1 and 2. They concluded that approximately 20% had nAbs to AAV1 and 27% to AAV2 [94, 95]. nAbs have also been detected for AAV5 in the human population [96, 97]. Numerous studies carried out on rAAV vectors in animals and nonhuman primates have established that both humoral and cell-mediated immune responses can limit sustained transgene expression [4, 98, 99]. An interesting stipulation to these conclusions is that the cellular immune response to rAAV appears to be dependent upon the route of administration [100].

As described, it was recognized by researchers that efficient and controlled gene transfer to target cells and evasion of the host immune system will be crucial to the success and advancement of the gene therapy field. It is plausible that the development of targeted tissue-specific vectors from serotypes other than type 2 will be generated for therapies that depend on readministration of vector. This will allow for continued treatment of patients by evading neutralizing antibodies produced from previous treatments. It is apparent that the challenges for targeted gene transfer have been acknowledged and an established groundwork for addressing them has been recognized. With the ability to PCR from tissue unique AAV capsid sequences and the advent of the targeting field, an exciting future of AAV vector development for gene therapy is on the horizon.

4 Preclinical and clinical studies

4.1 Preclinical studies

The availability of high-titer, highly purified rAAV preparations has allowed for testing of rAAV gene delivery in many in vivo settings, including numerous rodent and nonhuman primate models. However, these studies support the need to develop alternative AAV serotypes as vectors since they possess advantageous characteristics over rAAV2 vectors. The next few paragraphs will give a brief overview of the preclinical data using rAAV serotypes as vectors followed by human clinical applications that have occurred or are ongoing.

Preclinical studies utilizing AAV serotypes 1 through 9 have lead to the identification of serotypes with advantageous characteristics over AAV2, such as tissue tropism and efficiency of transduction. The use of alternative AAV serotypes for transduction of the central nervous system [1, 101], skeletal muscle [92, 93, 95, 102, 103], liver [7, 95, 104, 105], eye [5, 7], pancreas [106–108], heart [109], and lung [110–113] have been met with success and are described in detail in their respective references.

The development of several disease models in mice and larger animals such as Parkinson's, Alzheimer's, hemophilia, diabetes, obesity, alpha-1-antitrypsin deficiency, Canavan disease, cystic fibrosis, lysosomal storage diseases, Duchenne muscular dystrophy, and so on, have the potential to lead to new breakthroughs in rAAV vector application. Initial preclinical studies in some of the above disease models have progressed to the clinic for human therapy applications.

4.2 Clinical trials

To date, 25 Phase I and II human clinical trials have been set up using AAV as a vector. The majority of the human trials have been associated with cystic fibrosis (7) and the rest are as follows: hemophilia B, prostate and melanoma cancers, Canavan disease, amyotrophic lateral sclerosis (ALS), alpha-1-antitrypsin, Alzheimer's, Parkinson's, muscular dystrophy, rheumatoid arthritis, and HIV vaccines [121].

The first rAAV human gene therapy clinical trials began in 1995 under the direction of Dr. Flotte and Johns Hopkins University on cystic fibrosis [114]. The disease results from mutations in the gene for the cystic fibrosis transmembrane conductance regulator (CFTR), a cAMP-activated Cl^- channel in the apical membrane of most secretory epithelia. This study utilized rAAV-CFTR vectors for administration to the respiratory tract and the vector was found to be well-tolerated in human patients. There was a trend to indicate that the vector was biologically active for the correction of the chloride channel [115, 116]. To date, there have been four phase I and three phase II clinical trials that have utilized rAAV-CFTR for treatment of CFTR in human patients. Phase II clinical trials found that the vector was well-tolerated with multiple doses utilizing different routes of administration (maxillary sinus and aerosol) [117, 118].

A phase I clinical trial was recently finished on hemophilia B [119]. Hemophilia B is an X-linked bleeding disorder (coagulopathy) caused by mutations leading to loss of function of coagulation factor IX. A dose-escalation study on adult men with severe hemophilia B (Factor IX $< 1\%$) with rAAV ranging from 2×10^{11} vector genomes (vg)/kg to 1.8×10^{12} vg/kg injected at multiple intramuscular sites showed no evidence of toxicity up to 40 months after injection. Southern blot and immunohistochemical analysis of muscle biopsies taken two and ten months after injection confirmed gene transfer into muscle. Circulating levels of Factor IX were detected to be less than 2% and most less than 1%. It was concluded from this study that more extensive transduction of muscle will be required to reach above 1% Factor IX circulating levels and the doses of rAAV tested were safe. Another phase I clinical trial of hemophilia B carried out by Avigen involved direct infusion of rAAV-Factor IX into the liver at three increasing dose levels. Subjects one through four received the low doses of rAAV and all of them tolerated the procedure well with no side effects. These subjects were unable to sustain Factor IX levels above 1%. One subject receiving the higher dose achieved circulating levels of Factor IX in excess of 10% of normal for about four weeks and in week four experienced an elevation in the levels of two liver enzymes leading to a rapid decline in Factor IX levels. This lead Avigen to suspend further enrollment until further data was gathered and confirmed by the FDA [122]. It is important to note that a liver response similar to this has never been observed in any of the preclinical animal studies.

Currently, phase I clinical trials are ongoing for alpha-1-antitrypsin deficiency (AAT) [118] and Canavan disease (Robert Wood Johnson Medical School in NJ and UNC-Chapel Hill Gene Therapy Center). The AAT clinical trial is utilizing rAAV2 to deliver the human AAT gene intramuscularly. AAT deficiency in humans can lead to lung diseases as serious as emphysema. Four cohorts of three subjects are being used with a dose escalation between them ranging from 2.1×10^{12} vg/patient to 7×10^{13} vg/patient. To date, rAAV2-AAT was found to be safe with all vector doses tested in the human patients. Testing of rAAV1-AAT in the phase I clinical trials is also expected to occur [124]. The Canavan clinical trial is being carried out on a group of ten patients using rAAV2 mediated delivery of human aspartoacylase gene (ASPA). Canavan disease is a childhood neurodegenerative disorder. This autosomal recessive disease is caused by mutations or deletions in the ASPA gene. An increase in the substrate molecule, N-acetyl-aspartate occurs with the lack of functional ASPA. This results in spongiform degeneration of the brain. In this ongoing study, 9×10^{11} rAAV2 encoding human ASPA gene are surgically delivered to six sites of the brain in each human patient. Preliminary immunological data suggests that there are no deleterious effects of AAV2 in the brain or other organs. Expansion into phase II and III trials will be essential for ongoing monitoring of patients and further assessment of safety issues.

The Neurologix company was approved by the FDA for phase I clinical trials for the treatment of Parkinson's disease in August of 2002. The first Phase I trial of this treatment began in August 2003. rAAV2-GAD (glutamic acid decarboxylase) was delivered stereotactically to the subthalamic nucleus of the brain where Parkinson's disease is extremely overactive [123]. The patients will receive low, medium and high doses of the gene treatment, and will undergo brain scans, neurologic and motor tests for a year.

Ceregene is planning to initiate additional gene therapy clinical trials during the coming year, including the FDA-approved Alzheimer's disease clinical trial beginning October 1, 2004 (rAAV2-NGF-nerve growth factor) and Parkinson's disease (rAAV2-GDNF-Glial cell line Derived Neurotrophic Factor), also expected to begin during 2004. A clinical trial in amyotrophic lateral sclerosis (ALS or Lou Gehrig's disease) is expected to be initiated in 2005 [125].

5
Conclusion

The intricate biology of AAV as a nonpathogenic human virus has lead to its development into a promising viral vector. As a better understanding of AAV biology becomes available, we will see a direct impact on vector development (such as the determination of AAV receptors and viral tropism). The long-term potential of this vector will only be realized when the action of this reagent in humans is understood completely.

References

1. Xiao X, Li J, McCown TJ, Samulski RJ (1997) Exp Neurol 144:113
2. Snyder RO, Miao CH, Patijn GA, Spratt SK, Danos O, Nagy D, Gown AM, Winther B, Meuse L, Cohen LK, Thompson AR, Kay MA (1997) Nat Genet 16:270
3. Flotte TR, Afione SA, Conrad C, McGrath SA, Solow R, Oka H, Zeitlin PL, Guggino WB, Carter BJ (1993) P Natl Acad Sci USA 90:10613
4. Fisher KJ, Jooss K, Alston J, Yang Y, Haecker SE, High K, Pathak R, Raper SE, Wilson JM (1997) Nat Med 3:306
5. Acland GM, Aguirre GD, Ray J, Zhang Q, Aleman TS, Cideciyan AV, Pearce-Kelling SE, Anand V, Zeng Y, Maguire AM, Jacobson SG, Hauswirth WW, Bennett J (2001) Nat Genet 28:92
6. Auricchio A (2003) Vision Res 43:913
7. Rabinowitz JE, Rolling F, Li C, Conrath H, Xiao W, Xiao X, Samulski RJ (2002) J Virol 76:791
8. Atchison RW, Casto BC, Hammon WM (1965) Science 149:754
9. Buller RM, Janik JE, Sebring ED, Rose JA (1981) J Virol 40:241
10. Srivastava A, Lusby EW, Berns KI (1983) J Virol 45:555
11. Lusby E, Fife KH, Berns KI (1980) J Virol 34:402
12. King JA, Dubielzig R, Grimm D, Kleinschmidt JA (2001) Embo J 20:3282
13. Chejanovsky N, Carter BJ (1989) Virology 173:120
14. Girod A, Wobus CE, Zadori Z, Ried M, Leike K, Tijssen P, Kleinschmidt JA, Hallek M (2002) J Gen Virol 83:973
15. Zadori Z, Szelei J, Lacoste MC, Li Y, Gariepy S, Raymond P, Allaire M, Nabi IR, Tijssen P (2001) Dev Cell 1:291
16. Warrington KH Jr, Gorbatyuk OS, Harrison JK, Opie SR, Zolotukhin S, Muzyczka N (2004) J Virol 78:6595

17. Muzyczka N (1992) Curr Top Microbiol Immunol 158:97
18. Berns KI, Linden RM (1995) Bioessays 17:237
19. Samulski RJ, Zhu X, Xiao X, Brook JD, Housman DE, Epstein N, Hunter LA (1991) Embo J 10:3941
20. Kotin RM, Berns KI (1989) Virology 170:460
21. Cheung AK, Hoggan MD, Hauswirth WW, Berns KI (1980) J Virol 33:739
22. Laughlin CA, Cardellichio CB, Coon HC (1986) J Virol 60:515
23. McLaughlin SK, Collis P, Hermonat PL, Muzyczka N (1988) J Virol 62:1963
24. Chang LS, Shi Y, Shenk T (1989) J Virol 63:3479
25. Samulski RJ, Shenk T (1988) J Virol 62:206
26. Janik JE, Huston MM, Rose JA (1981) P Natl Acad Sci USA 78:1925
27. Laughlin CA, Jones N, Carter BJ (1982) J Virol 41:868
28. Richardson WD, Westphal H (1984) J Virol 51:404
29. Carter BJ, Antoni BA, Klessig DF (1992) Virology 191:473
30. Chang LS, Shenk T (1990) J Virol 64:2103
31. Weindler FW, Heilbronn R (1991) J Virol 65:2476
32. Samulski RJ, Chang LS, Shenk T (1989) J Virol 63:3822
33. Vincent KA, Piraino ST, Wadsworth SC (1997) J Virol 71:1897
34. Ogasawara Y, Urabe M, Ozawa K (1998) Microbiol Immunol 42:177
35. Li J, Samulski RJ, Xiao X (1997) J Virol 71:5236
36. Grimm D, Kern A, Rittner K, Kleinschmidt JA (1998) Hum Gene Ther 9:2745
37. Collaco RF, Cao X, Trempe JP (1999) Gene 238:397
38. Xiao X, Li J, Samulski RJ (1998) J Virol 72:2224
39. Clark KR, Voulgaropoulou F, Fraley DM, Johnson PR (1995) Hum Gene Ther 6:1329
40. Ruffing M, Zentgraf H, Kleinschmidt JA (1992) J Virol 66:6922
41. Chen C, Okayama H (1987) Mol Cell Biol 7:2745
42. Ferrari FK, Samulski T, Shenk T, Samulski RJ (1996) J Virol 70:3227
43. McCarty DM, Christensen M, Muzyczka N (1991) J Virol 65:2936
44. Pereira DJ, McCarty DM, Muzyczka N (1997) J Virol 71:1079
45. Pereira DJ, Muzyczka N (1997) J Virol 71:4300
46. Pereira DJ, Muzyczka N (1997) J Virol 71:1747
47. Matsushita T, Elliger S, Elliger C, Podsakoff G, Villarreal L, Kurtzman GJ, Iwaki Y, Colosi P (1998) Gene Ther 5:938
48. Salvetti A, Oreve S, Chadeuf G, Favre D, Cherel Y, Champion-Arnaud P, David-Ameline J, Moullier P (1998) Hum Gene Ther 9:695
49. Zhang X, De Alwis M, Hart SL, Fitzke FW, Inglis SC, Boursnell ME, Levinsky RJ, Kinnon C, Ali RR, Thrasher AJ (1999) Hum Gene Ther 10:2527
50. Clark KR (2002) Kidney Int 61 Suppl 1:9
51. Clark KR, Liu X, McGrath JP, Johnson PR (1999) Hum Gene Ther 10:1031
52. Gao GP, Qu G, Faust LZ, Engdahl RK, Xiao W, Hughes JV, Zoltick PW, Wilson JM (1998) Hum Gene Ther 9:2353
53. Zhang X, Li CY (2001) Mol Ther 3:787
54. Conway JE, Rhys CM, Zolotukhin I, Zolotukhin S, Muzyczka N, Hayward GS, Byrne BJ (1999) Gene Ther 6:986
55. Qiao C, Wang B, Zhu X, Li J, Xiao X (2002) J Virol 76:13015
56. Inoue N, Russell DW (1998) J Virol 72:7024
57. Urabe M, Ding C, Kotin RM (2002) Hum Gene Ther 13:1935
58. Sambrook J, Russell DW (eds)(2001) Cesium chloride and cesium chloride equilibrium density gradients. In: Molecular cloning: A laboratory manual, 3rd edn. Cold Spring Harbor Laboratory Press, New York, pp 1.154–155

59. Wistuba A, Kern A, Weger S, Grimm D, Kleinschmidt JA (1997) J Virol 71:1341
60. Zolotukhin S, Byrne BJ, Mason E, Zolotukhin I, Potter M, Chesnut K, Summerford C, Samulski RJ, Muzyczka N (1999) Gene Ther 6:973
61. Summerford C, Samulski RJ (1998) J Virol 72:1438
62. Gao G, Qu G, Burnham MS, Huang J, Chirmule N, Joshi B, Yu QC, Marsh JA, Conceicao CM, Wilson JM (2000) Hum Gene Ther 11:2079
63. Kaludov N, Handelman B, Chiorini JA (2002) Hum Gene Ther 13:1235
64. Xie Q, Bu W, Bhatia S, Hare J, Somasundaram T, Azzi A, Chapman MS (2002) P Natl Acad Sci USA 99:10405
65. Rabinowitz JE, Xiao W, Samulski RJ (1999) Virology 265:274
66. Wu P, Xiao W, Conlon T, Hughes J, Agbandje-McKenna M, Ferkol T, Flotte T, Muzyczka N (2000) J Virol 74:8635
67. Girod A, Ried M, Wobus C, Lahm H, Leike K, Kleinschmidt J, Deleage G, Hallek M (1999) Nat Med 5:1052
68. Grifman M, Trepel M, Speece P, Gilbert LB, Arap W, Pasqualini R, Weitzman MD (2001) Mol Ther 3:964
69. Tsao J, Chapman MS, Agbandje M, Keller W, Smith K, Wu H, Luo M, Smith TJ, Rossmann MG, Compans RW et al. (1991) Science 251:1456
70. Hileman RE, Fromm JR, Weiler JM, Linhardt RJ (1998) Bioessays 20:156
71. Mulloy B, Linhardt RJ (2001) Curr Opin Struct Biol 11:623
72. Opie SR, Warrington KH Jr, Agbandje-McKenna M, Zolotukhin S, Muzyczka N (2003) J Virol 77:6995
73. Kern A, Schmidt K, Leder C, Muller OJ, Wobus CE, Bettinger K, Von der Lieth CW, King JA, Kleinschmidt JA (2003) J Virol 77:11072
74. Buning H, Ried MU, Perabo L, Gerner FM, Huttner NA, Enssle J, Hallek M (2003) Gene Ther 10:1142
75. Bartlett JS, Kleinschmidt J, Boucher RC, Samulski RJ (1999) Nat Biotechnol 17:181
76. Ponnazhagan S, Mahendra G, Kumar S, Thompson JA, Castillas M Jr (2002) J Virol 76:12900
77. Girod A, Ried M, Wobus C, Lahm H, Leike K, Kleinschmidt J, Deleage G, Hallek M (1999) Nat Med 5:1438
78. Nicklin SA, Buening H, Dishart KL, de Alwis M, Girod A, Hacker U, Thrasher AJ, Ali RR, Hallek M, Baker AH (2001) Mol Ther 4:174
79. Shi W, Bartlett JS (2003) Mol Ther 7:515
80. Work LM, Nicklin SA, Brain NJ, Dishart KL, Von Seggern DJ, Hallek M, Buning H, Baker AH (2004) Mol Ther 9:198
81. White SJ, Nicklin SA, Buning H, Brosnan MJ, Leike K, Papadakis ED, Hallek M, Baker AH (2004) Circulation 109:513
82. Shi W, Arnold GS, Bartlett JS (2001) Hum Gene Ther 12:1697
83. Loiler SA, Conlon TJ, Song S, Tang Q, Warrington KH, Agarwal A, Kapturczak M, Li C, Ricordi C, Atkinson MA, Muzyczka N, Flotte TR (2003) Gene Ther 10:1551
84. Nicklin SA, Dishart KL, Buening H, Reynolds PN, Hallek M, Nemerow GR, von Seggern DJ, Baker AH (2003) Cancer Lett 201:165
85. Muller OJ, Kaul F, Weitzman MD, Pasqualini R, Arap W, Kleinschmidt JA, Trepel M (2003) Nat Biotechnol 21:1040
86. Perabo L, Buning H, Kofler DM, Ried MU, Girod A, Wendtner CM, Enssle J, Hallek M (2003) Mol Ther 8:151
87. Duan D, Yue Y, Yan Z, Yang J, Engelhardt JF (2000) J Clin Invest 105:1573
88. Qing K, Hansen J, Weigel-Kelley KA, Tan M, Zhou S, Srivastava A (2001) J Virol 75:8968

89. Thomas CE, Storm TA, Huang Z, Kay MA (2004) J Virol 78:3110
90. Bantel-Schaal U, Hub B, Kartenbeck J (2002) J Virol 76:2340
91. Gao G, Alvira MR, Somanathan S, Lu Y, Vandenberghe LH, Rux JJ, Calcedo R, Sanmiguel J, Abbas Z, Wilson JM (2003) P Natl Acad Sci USA 100:6081
92. Gao G, Vandenberghe LH, Alvira MR, Lu Y, Calcedo R, Zhou X, Wilson JM (2004) J Virol 78:6381
93. Gao GP, Alvira MR, Wang L, Calcedo R, Johnston J, Wilson JM (2002) P Natl Acad Sci USA 99:11854
94. Hildinger M, Auricchio A, Gao G, Wang L, Chirmule N, Wilson JM (2001) J Virol 75:6199
95. Xiao W, Chirmule N, Berta SC, McCullough B, Gao G, Wilson JM (1999) J Virol 73:3994
96. Erles K, Sebokova P, Schlehofer JR (1999) J Med Virol 59:406
97. Georg-Fries B, Biederlack S, Wolf J, zur Hausen H (1984) Virology 134:64
98. Xiao X, Li J, Samulski RJ (1996) J Virol 70:8098
99. Hernandez YJ, Wang J, Kearns WG, Loiler S, Poirier A, Flotte TR (1999) J Virol 73:8549
100. Brockstedt DG, Podsakoff GM, Fong L, Kurtzman G, Mueller-Ruchholtz W, Engleman EG (1999) Clin Immunol 92:67
101. Kaspar BK, Vissel B, Bengoechea T, Crone S, Randolph-Moore L, Muller R, Brandon EP, Schaffer D, Verma IM, Lee KF, Heinemann SF, Gage FH (2002) P Natl Acad Sci USA 99:2320
102. Chao H, Monahan PE, Liu Y, Samulski RJ, Walsh CE (2001) Mol Ther 4:217
103. Chao H, Liu Y, Rabinowitz J, Li C, Samulski RJ, Walsh CE (2000) Mol Ther 2:619
104. Grimm D, Zhou S, Nakai H, Thomas CE, Storm TA, Fuess S, Matsushita T, Allen J, Surosky R, Lochrie M, Meuse L, McClelland A, Colosi P, Kay MA (2003) Blood 102:2412
105. Mingozzi F, Schuttrumpf J, Arruda VR, Liu Y, Liu YL, High KA, Xiao W, Herzog RW (2002) J Virol 76:10497
106. Flotte T, Agarwal A, Wang J, Song S, Fenjves ES, Inverardi L, Chesnut K, Afione S, Loiler S, Wasserfall C, Kapturczak M, Ellis T, Nick H, Atkinson M (2001) Diabetes 50:515
107. Prasad KM, Yang Z, Bleich D, Nadler JL (2000) Gene Ther 7:1553
108. Wang AY, Peng PD, Ehrhardt A, Storm TA, Kay MA (2004) Hum Gene Ther 15:405
109. Du L, Kido M, Lee DV, Rabinowitz JE, Samulski RJ, Jamieson SW, Weitzman MD, Thistlethwaite PA (2004) Mol Ther 10:604
110. Beck SE, Jones LA, Chesnut K, Walsh SM, Reynolds TC, Carter BJ, Askin FB, Flotte TR, Guggino WB (1999) J Virol 73:9446
111. Halbert CL, Rutledge EA, Allen JM, Russell DW, Miller AD (2000) J Virol 74:1524
112. Halbert CL, Allen JM, Miller AD (2001) J Virol 75:6615
113. Zabner J, Seiler M, Walters R, Kotin RM, Fulgeras W, Davidson BL, Chiorini JA (2000) J Virol 74:3852
114. Flotte T, Carter B, Conrad C, Guggino W, Reynolds T, Rosenstein B, Taylor G, Walden S, Wetzel R (1996) Hum Gene Ther 7:1145
115. Wagner JA, Reynolds T, Moran ML, Moss RB, Wine JJ, Flotte TR, Gardner P (1998) Lancet 351:1702
116. Wagner JA, Messner AH, Moran ML, Daifuku R, Kouyama K, Desch JK, Manley S, Norbash AM, Conrad CK, Friborg S, Reynolds T, Guggino WB, Moss RB, Carter BJ, Wine JJ, Flotte TR, Gardner P (1999) Laryngoscope 109:266
117. Wagner JA, Nepomuceno IB, Messner AH, Moran ML, Batson EP, Dimiceli S, Brown BW, Desch JK, Norbash AM, Conrad CK, Guggino WB, Flotte TR,

Wine JJ, Carter BJ, Reynolds TC, Moss RB, Gardner P (2002) Hum Gene Ther 13:1349
118. Flotte TR, Brantly ML, Spencer LT, Byrne BJ, Spencer CT, Baker DJ, Humphries M (2004) Hum Gene Ther 15:93
119. Manno CS, Chew AJ, Hutchison S, Larson PJ, Herzog RW, Arruda VR, Tai SJ, Ragni MV, Thompson A, Ozelo M, Couto LB, Leonard DG, Johnson FA, McClelland A, Scallan C, Skarsgard E, Flake AW, Kay MA, High KA, Glader B (2003) Blood 101:2963
120. Vincent KA, Moore GK, Haigwood NL (1990) Vaccine 90:353
121. Edelstein M (2005) The Journal of Gene Medicine clinical trial website. Wiley, New York (see http://www.wiley.co.uk/wileychi/genmed/clinical, last accessed 19th September 2005)
122. Avigen, Inc. (2005) Website. Avigen Inc., Alameda, CA (see http://www.avigen.com, last accessed 19th September 2005)
123. Neurologix, Inc. (2005) Website. Neurologix, Inc., Fort Lee, NJ (see http://www.neurologix.net, last accessed 19th September 2005)
124. Flotte TR (2004) Preclinical and phase I/II clinical trials of recombinant adeno-associated virus (rAAV)-alpha-1 antitrypsin vectors. In: Xth Parvovirus Workshop, 8–12 September 2004, St. Petersburg, FL
125. Ceregene, Inc. (2005) Website. Ceregene, Inc., San Diego, CA (see http://www.ceregene.com, last accessed 19th September 2005)

Adv Biochem Engin/Biotechnol (2005) 99: 147–167
DOI 10.1007/10_006

Published online: 8 November 2005

Advanced Targeting Strategies for Murine Retroviral and Adeno-associated Viral Vectors

Julie H. Yu · David V. Schaffer (✉)

Department of Chemical Engineering and Helen Wills Neuroscience Institute, University of California, 201 Gilman Hall, Berkeley, CA 94720, USA
Schaffer@cchem.berkeley.edu

Abstract Targeted gene delivery involves broadening viral tropism to infect previously nonpermissive cells, replacing viral tropism to infect a target cell exclusively, or stealthing the vector against nonspecific interactions with host cells and proteins. These approaches offer the potential advantages of enhanced therapeutic effects, reduced side effects, lowered dosages, and enhanced therapeutic economics. This review will discuss a variety of targeting strategies, both genetic and nongenetic, for re-engineering the tropism of two representative enveloped and nonenveloped viruses, murine retrovirus and adeno-associated virus. Basic advances in understanding the structural biology and virology of the parent viruses have aided rational design efforts to engineer novel properties into the viral attachment proteins. Furthermore, even in the absence of basic, mechanistic knowledge of viral function, high-throughput library and directed evolution approaches can yield significant improvements in vector function. These two complementary strategies offer the potential to gain enhanced molecular control over vector properties and overcome challenges in generating high titer, stealthy, retargeted vectors.

Keywords Adeno-associated virus · Lentivirus · Retrovirus · Targeting · Viral vector

1 Introduction

The two major classes of gene delivery vehicles, viral and synthetic, have in many senses complementary advantages and disadvantages. Synthetic vehicles have the significant benefit that their chemical compositions and properties can readily be varied and controlled, a capability that makes them very flexible particles to engineer. However, their delivery efficiencies are generally not yet high. In contrast, over evolutionary timescales, viruses have acquired numerous strategies to overcome gene transfer barriers and can therefore deliver their genetic cargo with high efficiencies. However, a number of their delivery properties are not yet optimized for human therapeutic use, for the simple reason that nature did not evolve them explicitly for this purpose. Therefore, it is desirable to engineer novel properties into viral vectors, but this goal can be challenging because they are very complex and intricate entities.

Targeted gene expression is an attractive goal for gene delivery systems. It can be achieved by transductional targeting (the delivery of genes to specific cells) as well as transcriptional targeting (the use of promoters that mediate gene expression only in targeted cell types). This review focuses on the former approach, and readers are referred to several recent reviews for discussion of tissue-specific promoters [1, 2]. There are three potential goals of transductional targeting. First, the tropism or delivery range of vectors must sometimes be broadened to allow them to infect cells ordinarily resistant to transduction with the virus. Second, targeted delivery to *only* a specific cell type can be advantageous, since gene products that are therapeutic in some cellular settings can have side effects in others. This goal requires engineering a vector to reduce or eliminate its natural delivery properties and *replace* these properties with a novel, desired tropism. Finally, a vector can be enhanced to improve its stealth; in other words to reduce potentially undesirable interactions with cells or proteins including components of the immune system. It should be noted that these latter two goals, replacing tropism and vector stealth, have the advantage of potentially reducing the vector dosage needed for an application, which can lessen side effects as well as enhance the economics of gene medicines.

There has been success in targeted delivery by a number of viral vehicles, including the enveloped alphavirus, herpesvirus [3], retrovirus, and lentivirus, as well as the nonenveloped viruses adenovirus [4] and adeno-associated virus (AAV). This review will focus on one promising vector class from each category, retrovirus/lentivirus and AAV.

Three general strategies for engineering novel properties into viral vectors have been developed (Fig. 1), and all essentially strive to achieve a high level of control over the molecular properties of these vehicles. The first is to enhance vector properties through nongenetic approaches. These include using

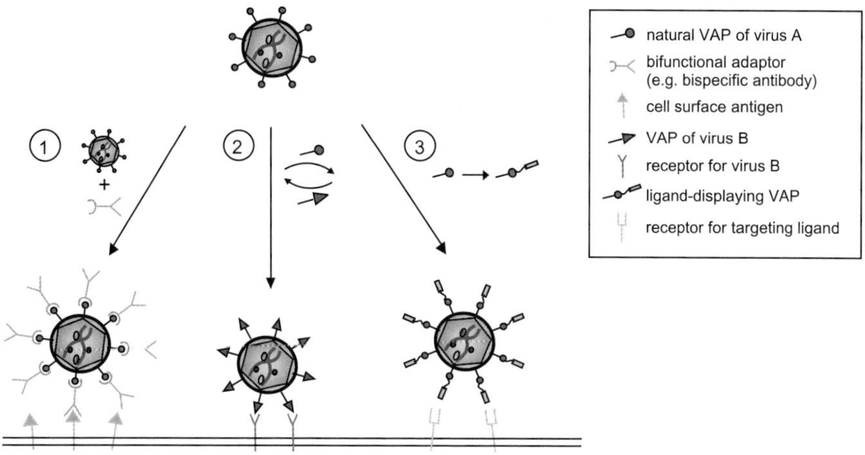

Fig. 1 Three approaches to targeting through the viral attachment protein (VAP). These include: 1) bridging with an adaptor molecule (nongenetic), 2) pseudotyping with a compatible viral capsid or envelope protein, and 3) genetic engineering of the VAP to insert new target cell specificity

bispecific antibodies, chemically cross-linking ligands to the viral surface to enhance binding to specific cell types, and grafting of polymers that resist protein adsorption to reduce interactions with the immune system. These latter two methods essentially attempt to merge the benefits of viral vectors with a major advantage of synthetic systems, the precise chemical control of vector properties. A second approach is to replace the viral attachment protein (VAP) of a vector with the corresponding VAP of another compatible virus to alter its tropism, known as pseudotyping. Although we will discuss numerous examples of these first two approaches, this review focuses mainly on the third promising strategy, the genetic engineering of viral attachment proteins for targeted gene delivery. If successful, this approach offers the highest potential for precise control over vector properties.

2 Retroviral and Lentiviral Vector Targeting

The *Retroviridae* are a family of enveloped viruses with a diploid, positive-stranded RNA genome. Retroviral vectors are very promising vehicles for delivering therapeutic genes to cells because they offer the advantage of stably integrating their genomic information into the chromosomes of their host's DNA. In addition, their simple gene composition has allowed vectors to be engineered to contain all viral functions needed to enter the cell but none of the viral gene sequences. Relatively recent work with this vector resulted in the apparently permanent cure of children with SCID-X1 in the first successful gene therapy clinical trial [5]. Though two of nine children treated later experienced a severe adverse effect from the therapy [6], therapies based on retroviral vectors still comprise 28% of the clinical trials in progress today [7]. The results of the SCID-X1 trial further emphasize the need to develop safe and regulatable gene delivery vectors.

The majority of retroviral vectors in clinical trials are based on murine leukemia virus (MLV), a type-C simple retrovirus that is nonpathogenic to humans [8]. MLV genomes consist of three genes, *gag*, *pol*, and *env*, which encode all of the necessary proteins for the retrovirus to complete its life cycle. The *env* gene produces two protein subunits, transmembrane (TM) and surface (SU), which are cleaved from the same precursor and associate to form the Env protein, the VAP for MLV. This protein is directed into the endoplasmic reticulum where it is glycosylated and folded. Correctly folded proteins associate into homotrimers and are processed through the Golgi apparatus before export to the cellular membrane. As the virus assembles near the cell membrane, the envelope proteins concentrate at the site of budding through a mechanism that is not yet well understood.

During cellular entry, the envelope glycoproteins of the retrovirus mediate its attachment to the cell surface and subsequent fusion and insertion of the capsid into the cytoplasm. Since the envelope protein is the primary molecule that comes into contact with the cell surface, the majority of targeting efforts have been directed towards its modification and enhancement. The three strategies for engineering vector specificity discussed above (nongenetic modification, pseudotyping, and genetic engineering) have been applied to retroviruses. Success has been limited primarily by the fact that the binding of the natural retroviral envelope protein to its cell surface receptor triggers the fusogenic activity that mediates viral fusion and cell entry. This intimate coupling of binding and fusion makes it difficult to re-engineer the binding specificity without significantly compromising fusion, and thereby reducing viral titer. Therefore, engineering viral envelope proteins for novel specificity while fully retaining vector packaging and infection functionalities remains an important goal. Extensive investigations of different MLV envelope proteins have provided information on tolerable insertion sites and important regions for functionality [9–11], and these studies have provided a foundation for attempting the envelope protein modifications discussed below.

One major disadvantage of using simple retroviruses as gene therapy vectors is their inability to infect post-mitotic cells, a desired tissue target for many therapies. Much work has therefore been conducted to engineer complex retroviruses, lentiviruses and foamy viruses, which are able to transduce nondividing cells, to serve as gene therapy vehicles [12, 13]. Like all enveloped viruses, both of these classes of retroviruses also attach to their target cell through an envelope glycoprotein. One of the earliest strategies for targeting consisted of packaging retroviral and lentiviral vectors with the viral attachment proteins of other enveloped proteins, or pseudotyping [14, 15]. While this approach can successfully swap vector tropism with that of other viruses, in many cases it is not useful if there are no envelope proteins available to target delivery specifically to a desired cell population. Still, the success of pseudotyping has shown that retroviral envelope proteins are very modular, and foreign glycoproteins can be efficiently incorporated into fully infectious retroviruses and lentiviruses. Because of this feature, glycoproteins that are engineered for a desired function in one type of virus can often be readily interchanged and utilized in other enveloped viral vectors.

2.1 Direct Targeting with Retroviral Glycoproteins by Genetic and Nongenetic Approaches

Investigations in direct targeting began with the simple idea that if a vector could be engineered to attach to a cell through tissue-specific cell surface molecules, infection would predominantly occur in that tissue type. Strategies to promote such specific binding have included the genetic incorporation of ligands such as growth factors [16–18], peptides [19, 20], and single-chain antibodies [21–24]. In addition, nongenetic bifunctional adaptor molecules such as biotin-streptavidin or bispecific antibodies that bridge interactions between the virus and cell have been explored [25–29]. Furthermore, there is also a class of matrix-targeting vectors that incorporate collagen-binding domains that can direct the vector to extracellular matrix exposed during metastatic cancer [30–32]. While targeting matrix may concentrate vectors in the region of interest, it does not eliminate the possibility of infecting bystander cells. Much of this early work in direct targeting with retroviral vectors has been well reviewed [33, 34]. Some of this work demonstrated that enhanced binding of viruses to a desired cell type could be achieved; however, infection was also inefficient because the mutant envelope proteins fail to induce fusion [35]. The envelope protein has binding and fusion activities that are coupled in a complex and not fully elucidated mechanism; therefore, it has been difficult to re-engineer binding activity to a novel receptor target but maintain the same, efficient level of fusion [36]. In addition to inefficient cell entry, modified envelopes often reduce the packaging efficiency of the vectors, which again results in low overall titers. Limited success at targeting has been achieved by coexpressing the modified targeting envelope with the wild-type envelope in an attempt to improve fusion activity [16, 18, 21, 24]. The wild-type protein serves as an escort to the targeting protein to provide the means for fusion. However, this can only *broaden* rather than *replace* viral tropism since the wild-type proteins are free to interact with native receptors on nontargeted cells. Ideally, coexpression of a binding-defective envelope that can still trigger fusion would lead to more stringent targeting. This idea of separating the mechanisms of binding and fusion is further explored below in the discussion of pseudotyping with pH-dependent glycoproteins. It is now clear that specific attention needs to be invested to ensure that the fusion mechanism of the virus is not impaired by envelope modifications.

Previous work identified a strategy of receptor co-operation whereby retroviruses expressing two different receptor binding domains linked by optimized proline-rich spacers can only infect cells expressing both receptors [37]. Martin et al. have applied this strategy by using single-chain antibodies that recognize high molecular weight melanoma-associated antigen (HMWMAA) and carcinoembryonic antigen (CEA) to target tumor cells [38]. The spacer length separating these dual targeting antibodies was optimized such that binding by both receptors was necessary to induce infection. It is proposed that binding of the first (targeting) receptor induces a conformational change in the proline spacer that allows binding of the second (viral/entry) receptor to occur. These vectors showed significant improvement in infection of tumors over previous strategies based on vectors that coexpressed the same scFv antibodies and wild-type Env [39–41]. Though the specificity and infection levels of these vectors are encouraging, low packaging efficiencies must be addressed before the strategy can be used clinically. An earlier version of the HMWMAA-targeting vector was the first to show selective transduction of targeted cells and reduced transduction of nontargeted cells in vivo [42]. However, these vectors transduced tissue at 10% of the efficiency of control vectors expressing wild-type Env. Future in vivo work will determine whether the combination of receptor cooperation and established binding strategies can increase transduction efficiency.

A nongenetic approach, the conjugation of targeting proteins to a virus after it has been packaged using receptor-ligand bridges, has been explored by Young and colleagues [43]. Earlier work showed enhanced infection of subgroup A avian leucosis viruses (ALV-A) when $EGFR^+$ target cells had been incubated with a bridge consisting of the soluble form of the ALVA receptor fused to EGF [44]. Another version of the system using ALV-B and a similar fusion protein had the improved feature of successfully allowing the virus, rather than the cells, to be preloaded with the targeting molecule [43]. This work also showed that the targeting virions could be produced directly from packaging cells by coexpression of the fusion protein. In addition, this group has had success using vascular endothelial growth factor (VEGF) and heregulin to target cells, thus establishing a broader applicability of this strategy [45, 46]. These bridge proteins, termed guided adaptors for targeted entry (GATEs), allow the native envelope-receptor interactions to be preserved since they are attached to the viral envelope protein rather than genetically incorporated into them. However, the stability of the protein-protein interaction between the virus and the bridge molecule is crucial to the success of this strategy in vivo. In addition, until the mechanism of binding-triggered fusion is better described, it is unclear whether fusion can be triggered by the preloading and thus pose a toxicity threat to producer cells. It is important to note that the success of using EGF as a targeting molecule in this study may be due to the novel two-step cell entry mechanism of ALV that is still not completely understood, but seems to require both pH-independent binding and pH-dependent fusion events that are relatively uncoupled [47, 48].

One approach that combines a nongenetic bridging strategy that relies on post-packaging modifications as well as a genetically modified envelope protein utilizes antibody-antigen interactions to direct viruses to cells [27, 28]. To achieve this, envelope proteins are engineered to express the antibody binding domain of protein A. Prior to cellular infection, these viruses are incubated with monoclonal antibodies specific for the desired target cell. This modular system allows for several different cell types to be targeted without the need to re-engineer the vector. Preferential infection in the presence of targeting antibody with these modified viruses was first shown with vectors pseudotyped with the envelope protein of Sindbis virus, an alphavirus that utilizes pH-dependent fusion [29]. More recently, antibody-mediated targeting has been shown to be successful in replication-competent retroviruses with modified ecotropic and amphotropic Env proteins [49]. These envelopes incorporated the same protein A IgG-binding domain and were complexed with anti-human epidermal growth factor-like receptor-2 (HER2) antibodies to target HER2, a receptor overexpressed on 30% of breast cancer cells. Similar to previous direct targeting experiments with MLV envelope, viruses harboring only the chimeric envelope were able to bind to but not infect target cells. Infection was achieved by coexpressing wild-type envelope with the HER2 targeting envelopes. Significant enhancement of infection was seen on murine NIH3T3 cells engineered to overexpress HER2, but not on human mammary carcinoma cell lines despite evidence of enhanced binding in the presence of the anti-HER2 antibody. This discrepancy between the infection of engineered target cells and the natural target cells with the same receptor shows that cellular factors other than receptor expression must be explored in the development of targeting strategies. This difference underscores the importance of the need for more in vivo work in the development of targeting strategies.

An alternative pseudotyping method that utilizes incorporation of viral receptor proteins rather than viral envelope proteins was first shown in lentiviral vectors by Endres et al. [50]. This strategy takes advantage of the fact that virally infected cells express viral glycoproteins on their surface. Vectors that express the cellular receptor for the viral envelope protein are targeted to infected cells by exploiting the virus-cell binding in a reverse-directional manner. Endres et al. were able to selectively target HIV and SIV infected cell lines with lentiviral vectors pseudotyped with HIV and SIV receptor CD4 and coreceptor CXCR4 or CCR5 [50]. This idea has been further extended to create MLV vectors that are reverse-targeted to cells expressing RSV and MLV glycoproteins [51] and HIV-infected cells [52] by pseudotyping with the corresponding receptor(s). This approach has specific potential for anti-HIV therapies. Successful HIV infection requires the binding of a coreceptor subsequent to attachment to the viral receptor CD4. Bittner et al. have shown that MLV and lentiviral vectors pseudotyped with a hybrid CD4/CXCR4 receptor can successfully transduce HIV Env-expressing cells [53].

Recent evidence of nonspecific, receptor-independent adsorption of retroviruses to cells suggests that earlier results that apparently demonstrated enhanced binding of targeting vectors may warrant re-examination [54]. Pizzato et al. attempted to target MLV vectors to ovarian cancer cells by fusing a single-chain antibody directed against the α folate receptor, which is overexpressed on ovarian cancer cells, to amphotropic and ecotropic MLV Env [55]. FACS analysis using fluorescently labeled anti-Env antibodies to detect the association of viruses with cells indicated that virions with ligand-incorporated glycoproteins had enhanced binding to target cells. Similar to previous targeting studies, this enhanced binding did not result in enhanced infection. However, further analysis by immunofluorescence microscopy showed that virions with wild-type envelope, ligand-incorporated envelope, or no envelope at all associated equally well with the cell surface. Furthermore, this group has also shown via confocal microscopy that a fluorescence increase detected by FACS, which could be interpreted as increased viral binding, is actually due to interactions between cells and a contaminant in the viral stock, soluble vesicles containing the SU protein [54]. Both results have major implications on other direct targeting studies that have reported enhanced binding based on FACS analysis but limited infectivity of target cells.

These results highlight several major issues that must be considered in the design of targeting retroviral vectors. First, if particles associate equally well with cells regardless of the presence of a targeting ligand, the inclusion of this ligand could be inconsequential unless the binding affinity and kinetics of this association are such that they favor actual infection. Further investigation of the mechanism of nonspecific retroviral binding may be required to aid the design of better ligands. Secondly, these results reiterate the fact that a productive ligand-receptor interaction does not necessarily catalyze viral fusion. The targeting vector must have a specific strategy designed to trigger viral entry. Depending on the results of these two issues, it may be more desirable to focus on mechanisms to trigger cell-specific fusion rather than cell-specific binding. Lastly, widespread nonspecific adsorption would increase the required therapeutic dosage and adversely affect the therapy's economics; therefore, it would be worth exploring ways to reduce nonspecific binding.

2.2 Direct Targeting via Genetic Engineering of pH-dependent Glycoproteins

Fusion of enveloped viruses with their target cells can be triggered by two mechanisms. Most retroviruses undergo pH-independent binding where fusion is triggered by interaction with the cellular receptor upon binding. Other enveloped viruses employ a pH-dependent fusion mechanism that occurs inside the cells. Upon binding, these viruses undergo receptor-mediated endocytosis, and the subsequent reduction in endosomal pH triggers fusion between the virus and the endosome. Although retroviral binding and cell entry are intimately associated events in wild-type virions, work with recombinant HIV has shown that it is possible to decouple these two mechanisms and still generate fully infectious particles in vitro [56]. In general, however, because it has been difficult to engineer binding without compromising the fusion trigger of envelope proteins with pH-independent triggers, glycoproteins with uncoupled cell surface binding and pH-dependent fusion activities offer strong potential for engineering novel binding specificities.

Influenza hemagglutinin (HA) protein is an extensively studied fusion protein that undergoes a conformational change under acidic pH. Retroviruses were shown to specifically bind to target cells when pseudotyped with HA proteins from fowl plague virus fused to four different targeting ligands, including EGF, an anti-human MHC class I scFv, an anti-HMWMAA scFv, and an IgG Fc-binding polypeptide of protein A [57]. This binding could be abolished by the addition of neutralizing antibodies to the targeting ligand, thus confirming the mechanism of viral attachment. Some chimeric viruses showed greater selection for infecting appropriate target cells, but this effect was masked somewhat by the basal level of infection through the natural HA receptor, sialic acid. This approach therefore involved broadening rather than swapping viral tropism. A later improvement to this strategy was made by Lin et al. by coexpressing a fusion-defective, ligand-containing Moloney MLV Env protein with a mutant HA protein defective in its ability to bind sialic acid [58]. By completely separating the mechanisms of binding and fusion, these vectors showed a 10-fold increase in titer of targeted cell types over the control cell type. This effect was eliminated in the presence of competing soluble targeting ligand. In addition, this increase was found only in vectors coexpressing both envelope types and was not seen when mixing vectors expressing only one type. This indicates that the functions of binding and fusion cannot operate in *trans* across separate virions and suggests that fusion of these chimeras must be triggered by the internalized virus after binding.

The G glycoprotein from vesicular stomatitis virus (VSV-G) is commonly used to pseudotype retroviral and lentiviral vectors due to its broad tropism and the capacity of the pseudotyped particles to be concentrated by ultracentrifugation [15]. VSV-G pseudotyped viruses are internalized by the cell, and fusion between the virus and endosome occurs at approximately pH 6 [8]. Though VSV-G pseudotyped lentiviral vectors can be inactivated by human serum [59], recent evidence shows that this may be prevented by PEGylation of the vector, thus strengthening the opportunities for the use of this glycoprotein in vivo [60]. Attempts at targeting modifications to VSV-G have been limited, however, due to the lack of a three-dimensional crystal structure and incomplete understanding of its fusion mechanism. Recently, MLV-based and HIV-1-based vectors pseudotyped with a modified VSV-G protein expressing a collagen binding domain of von Willebrand factor were shown to have increased attachment on a collagen-matrix while retaining their infectivity [61]. However, the ligand modifications resulted in a temperature-sensitive defect in the intracellular trafficking of the protein that could be restored if the viruses were packaged at the permissive temperature of 30 °C. Nonetheless, this work demonstrates the first successful modification to VSV-G that still allows for functional pseudotyping.

2.3 Inverse Targeting by Receptor Sequestration and Proteolytic Cleavage

The lessons learned from early work in direct targeting led to a complementary approach that exploits the fact that retroviruses can remain bound to the surface of cells through a targeting ligand without being internalized by the cell. In inverse targeting, retroviral vectors with envelope proteins displaying high affinity ligands are blocked from infecting cells that express the cognate receptor due to the sequestration of the virus by that receptor. Cosset et al. showed that retroviral vectors with amphotrophic envelope proteins displaying EGF were able to bind to cells that expressed EGFR [17]. However, unlike the pH-dependent entry mechanisms described above for EGF-ALV Env vectors, fusion could not be triggered by the low pH environment that the bound virus may encounter during intracellular EGFR trafficking. Therefore, despite viral attachment, successful infection did not occur on EGFR$^+$ cells that do not display the virus' natural receptor, Ram-1. The modified vectors were, however, able to infect EGFR$^-$, RAM+ cells, confirming infectivity of the vectors. Infection of Ram-1$^+$ cells with the modified vector was slightly reduced compared to wild-type, which is most likely attributed to steric effects of the incorporated ligand. This group has shown analogous results with virus that displays stem cell factor (SCF) [62]. This inverse targeting strategy offers benefits over initial direct targeting approaches because it allows the vector to exploit the virus' natural fusion pathway for cell entry. However, since this method relies on the absence of a specific receptor on the targeted cells, it is limited in scope and application.

The phenomenon of vector sequestration has led to another targeting strategy whose specificity is based upon cell surface proteases rather than receptors. Here, envelope proteins display a high-affinity ligand tethered to their N-terminus by a peptide containing a protease cleavage site. The target cells must express the proper receptor to enable the anchoring of the virus to the cell surface. If the cell also expresses the appropriate protease, the high-affinity ligand is cleaved, and the virus is allowed to enter the cell through its natural receptor. This was first demonstrated by fusing EGF to amphotrophic MLV envelope via a factor Xa protease recognition site [63]. The modified vectors were able to bind EGF receptors on human cells in vitro, but did not proceed with gene transfer until they were cleaved by factor Xa protease. Comparable results were found using similarly displayed EGF on spleen necrosis viruses (SNV), an avian retrovirus [64]. Additional work has shown parallel results for vectors fused to protease-cleavable insulin-like growth factor (IGF-I), albeit to a much lesser extent, suggesting that the success of this method may rely on the specific ligand chosen [65]. When used to pseudotype a lentiviral vector, the factor Xa-targeting, EGF-displaying envelope proteins were shown to be effective in altering the biodistribution of the vector in vivo [66]. Here, vectors that had EGF incorporated into their envelope had a lower infectivity in the liver (EGF-rich cells) than vectors with wild-type envelope. Infectivity levels were restored by competing with soluble EGF as well as by introducing factor Xa protease. In vivo studies using matrix-metalloproteinase (MMP) cleavable retroviruses have shown promising results targeting MMP-rich and carcinoembryonic antigen (CEA) expressing tumor xenografts in nude mice [67, 68].

Despite these promising results, there are two shortcomings that need improvement. Nonspecific infection of bystander cells can arise from the incomplete masking of the envelope's native binding activity, and large dosages of vector may be required to overcome the high levels of vector sequestration by nontargeted, $EGFR^+$ cells. A strategy that addresses both problems uses larger trimeric leucine zipper peptides or the trimeric C-terminal domain of CD40 ligand to sterically block infection of cells [69]. These peptides replace EGF in the previous strategy and are again linked to the amphotrophic envelope protein by a protease cleavage site. Infectivity of the modified vectors was low compared to a control and was restored upon addition of factor Xa protease. However, the incorporation rate of these bulky chimeric envelope proteins into virions during production was reduced and may present an obstacle when trying to achieve high packaging titers. Unless the packaging efficiency of these vectors can be improved, the benefit of sterically blocking infection of nontarget cells may not present as much of an advantage as the enhanced binding efficiency provided by a small targeting ligand.

2.4 Directed Evolution Methods for Retroviral Targeting

Though the above strategies have shown that vector retargeting is possible, they also illustrate the difficulty associated with predicting the modifications needed to impart new functionality while retaining the vector's ability to package and infect at high titer. One approach that circumvents the need for complex rational design is directed evolution. Directed evolution emulates the process of natural evolution by generating large libraries of mutants or variants and screening these libraries for improved function, an approach that has previously been applied with great success to enzyme and antibody engineering [70–72].

To explore the possibility of altering the tropism of a vector through directed evolution of the envelope gene, the DNA of a family of MLV *env* genes was shuffled and resulted in a clone that had an entirely new tropism not present in any of the parents [73]. As an additional example, phage display libraries have been used to search for optimal peptide sequences for cell specific binding [74, 75]. Since selection for binding activity is conducted in the context of a phage coat protein, however, it does not ensure that the binding specificity will be maintained when imported into a viral vector. In contrast, the identification of tolerable insertion sites in retroviruses has allowed the creation of retroviral display libraries that can be used to identify and select polypeptides with specific interactions in mammalian cells [9, 76]. After screening libraries for infectious mutants with desirable targeting features, the responsible modifications can be analyzed by sequencing the "successful" viral genomes. In the first application of this approach to retroviral targeting, the screening of a random display FeLV Env library resulted in the identification of a mutant with specificity for D17 canine osteocarcinoma cells, a tropism distinct from any FeLV subgroup [77]. The direct screening of random-display libraries also enables the discovery of mutants with altered tropism via unknown receptors. Bupp et al. were successful in identifying a mutant FeLV envelope protein that preferentially targets 143B cells and 293T cells via a novel, unidentified receptor [78].

The approach of using library and directed evolution in a mammalian cell context offers several major benefits. First, since the mutants are being screened directly for infectivity, vectors that have desired binding properties but unsuccessful entry properties, such as those in the early direct targeting literature, are not selected. Second, this approach greatly enhances the diversity of cell types that can be targeted since it can be applied to cells that are not yet well characterized. The studies utilizing random display libraries mentioned above each focused on identifying optimal virus-receptor interactions, but the strategy has also been shown to be successful in identifying optimized protease-activated retroviruses by screening libraries of protease cleavable peptides [79].

In summary, vectors with enhanced function must still be improved for high titer and stability during purification and storage before they will be able to be used clinically. However, the promising results of both rational and directed evolution approaches to vector targeting have laid the foundation for addressing these and other aspects of vector optimization.

3 Targeting Adeno-associated Viral Vectors

AAV is a nonenveloped virus with a 4.7 kb single-stranded DNA genome, and it belongs to the family *Parvoviridae* and genus *Dependovirus* [80, 81]. Its genome contains only two open reading frames flanked by short inverted terminal repeats (ITRs); however, the economical use of alternative splicing and start codons allows AAV to express seven partially-overlapping proteins from this short sequence. The first gene *rep* encodes four proteins necessary for protein replication (Rep78, Rep68, Rep52, and Rep40), and the second gene *cap* uses alternative splicing and start codons to express three structural proteins, VP1–3. VP3 is the shortest, whereas VP1 and VP2 include the entire VP3 sequence plus additional N-terminal extensions. Sixty subunits of VP1–3 self-assemble in an approximate stoichiometric ratio of 1 : 1 : 20 to generate the viral capsid. There are a number of known, relatively highly homologous, primate AAV serotypes [82, 83]. However, AAV2, which was first converted into a recombinant vector by Muzyczka et al. and Samulski et al. in the 1980s, is the best characterized and has received the most attention as a gene delivery vehicle [84–86]. The biology, highly promising gene delivery properties, and clinical application of AAV are discussed in more detail in another article in this volume (Grieger and Samulski).

The three general strategies for viral vector targeting have also been applied to AAV. First, as a nongenetic targeting approach, Bartlett et al. used a bispecific antibody to mediate the interaction between the AAV vector and a specific cell surface receptor expressed on human megakaryocytes [87]. The resulting vectors could transduce certain megakaryocyte cell lines at levels 70-fold above background, and the targeting was both selective and restrictive in that the endogenous tropism of the modified vectors was significantly reduced. In a second example, biotin was recently chemically crosslinked to the AAV2 capsid, and streptavidin was used as a bridge for the attachment of targeting ligands. Specifically, streptavidin genetically fused to the ligands epidermal growth factor or fibroblast growth factor-1 mediated a greater than 100-fold increase in gene delivery to cell lines ordinarily resistant to AAV gene delivery [88]. Finally, chemical approaches can be used to reduce the interaction of AAV with components of the immune system. We have recently crosslinked polyethylene glycol (PEG) to the surface of AAV2 to shield the virus from neutralization by serum antibodies [89]. In the future, PEG coating may be combined with targeting ligands as a chemical approach to replace AAV targeting specificity.

The delivery specificity of AAV2 can also be altered by vector pseudotyping, since different AAV serotypes bind to distinct receptors on cell surfaces [85, 90–93]. In particular, AAV2 binds to heparan sulfate as its primary receptor and FGFR and an integrin as secondary receptors [94–96]. It has also been shown that AAV3 also binds heparan sulfate, AAV4 and AAV5 bind to specific sialic acid linkages [97], and AAV5 binds to the PDGF receptor [98], whereas the receptors of other serotypes are not yet known. These serotypes are capable of packaging recombinant AAV2 genomes composed of a reporter gene flanked by the short AAV2 ITRs [99]. Consistent with the distinct cell binding properties of the parent serotypes, the different pseudotyped vectors possess distinct gene delivery properties. For example, in the brain AAV2 selectively transduces neurons, AAV5 transduces both neurons and astrocytes, and AAV4 preferentially delivers genes to ependymal cells [93]. Different gene delivery properties have also been reported for the different serotypes in the retina, where AAV2 transduces phororeceptors, AAV5 capsid directs gene delivery to both retinal pigment epithelium (RPE) and photoreceptors, and AAV1 delivery is restricted to the RPE [100].

3.1
Targeting via Genetic Engineering of the AAV Capsid

As with retroviral and lentiviral vectors, pseudotyping offers new vector tropism options. However, if no natural capsid that offers precisely the desired gene delivery properties is available, it would be desirable to engineer the AAV capsid for "custom" targeted gene delivery. In contrast to retrovirus, to date only direct targeting approaches have been applied for AAV. In addition, at the time that such work was initiated, the crystal structure for AAV2 was not available, making targeting via genetic engineering efforts challenging.

In the first AAV targeting effort, Yang et al. genetically fused a single chain antibody directed against the CD34 antigen, a marker of hematopoietic stem cells, to the N-terminus of VP2 [101]. This large insertion significantly increased selective gene delivery to a $CD34^+$ cell line, although the viral titers were low. This important study represented the first successful targeted AAV gene delivery. However, it also demonstrated that the use of smaller, less disruptive targeting moieties, as well as insertion into optimal locations on the capsid, may be required to enhance viral packaging and targeting efficacy. These constraints likely result from the need for the viral proteins to undergo a complex self-assembly process to generate the capsid, a process readily interrupted by insertional mutagenesis.

The next successful effort involved the incorporation of small targeting peptides into the capsid. By superimposing the AAV2 capsid sequence onto the known structure of canine parvovirus (CPV), Girod et al. chose six sites putatively on the viral surface for the insertion of a ligand containing the integrin-targeting amino acid sequence RGD. The insertion sites were predicted to lie on various surface loops of the VP structure, and one mutant infected cell lines that were resistant to infection by wild-type AAV2, but with a modest titer [102]. This successful insertion occurred at amino acid 587, based upon numbering of VP1, within the surface accessible loop 4 of the capsid protein.

Subsequent efforts involved widespread insertional mutagenesis to probe the capsid for sites exposed to the surface and tolerant of peptide insertions. One earlier study of basic AAV biology explored the effects of small peptide insertions on viral packaging and infectivity [103]. More recent studies have comprehensively scanned the capsid to search for insertion sites for targeting peptides. Rabinowitz et al. employed linker insertional mutagenesis into several dozen sites of the vector and found several regions amenable to the insertion of small peptides [104].

In addition, Wu et al. generated 93 mutants by epitope tag or ligand insertion, or alanine scanning mutagenesis. This work identified numerous locations where insertion disabled the virus, as well as located regions on VP1 and VP2 where the insertion of a serpin receptor peptide ligand successfully altered viral tropism [105]. Interestingly, insertion of their targeting peptide within loop 4 was not tolerated, indicating that the optimal insertion site could be dependent on the specific identity of the targeting peptide. Furthermore, mutants defective in heparin binding were identified, which could aid future efforts to fully replace viral tropism. In another study, Shi et al. generated capsid mutants with peptide insertions at 38 locations, and the addition of the peptide fragment of luteinizing hormone to VP1 and VP2 enabled selective targeting to an ovarian cancer cell line [106]. Shi and Bartlett followed up on this work by inserting RGD peptides into several permissive sites they had previously identified. The resulting vectors exhibited significantly increased transduction to cell lines that express integrins but low levels of heparan sulfate [107].

While these comprehensive insertional mutagenesis efforts were in progress, Grifman again employed a comparison of the AAV and CPV to identify accessible regions on the viral capsid for the insertion of the tumor-targeting NGRAHA peptide that binds to CD13, a vascular receptor and regulator of angiogenesis [108]. Insertion of this motif in loop 4 of the capsid led to a 20-fold increase in delivery to several tumor cell lines. Finally, Baker used phage display to identify peptides that bound venous endothelial cells and inserted them into position 587 of the AAV2 capsid, the site previously identified by Girod et al. [102]. Several of the resulting vectors exhibited reduced accumulation in the liver and enhanced transduction of the vena cava in vivo [109].

In a modular approach that combined genetic and nongenetic targeting approaches, Ried et al. generated AAV2 vectors with capsids displaying immunoglobulin-binding domains borrowed from protein A [110], elegant work similar to that conducted with retrovirus [27, 28]. These vectors, when coupled to antibodies against receptors such as CD29, CD117, or CXCR4, specifically transduced human hematopoietic cell lines. Furthermore, this approach is highly modular in that it readily allows different antibodies to be employed.

These insertional mutagenesis and targeting studies identified numerous regions of VP1–3 exposed to the viral surface, findings that were further confirmed and substantiated by the solution of the AAV2 crystal structure [111]. Furthermore, as compared to the previously determined structures for CPV and insect parvovirus, the AAV2 crystal structure revealed the presence of prominent peaks on the viral surface at the three-fold axis of symmetry, and a cluster of positively charged residues on the side of this peak likely mediates binding to the receptor heparan sulfate. The AAV2 structure, as well as the recently elucidated AAV5 structure [112], will likely aid further efforts to rationally design novel capsid properties.

3.2
Directed Evolution and Library Methods for AAV Targeting

Prior targeting work involved insertion of a defined targeting sequence into numerous locations on the viral surface. In contrast, a recent elegant library approach inserted random peptides into a single location on the viral surface to identify novel targeting sequences for specific cellular targets [113]. Specifically, a random seven amino acid peptide sequence was inserted into position 588 at the peak of the three-fold axis peak, again a previously identified insertion site [102]. In an approach analogous to phage display, the resulting library was selected for the ability to infect human primary coronary endothelial cells, cells nonpermissive for AAV infection, and variants with a consensus sequence emerged and were able to infect cells at levels as high as 630-fold higher than wild-type virus. The molecular mechanisms of AAV gene delivery are being progressively elucidated [114–118], knowledge that will enable further rational design. Until the viral structure-function relationships are fully elucidated, however, such library approaches offer a high-throughput means to solve problems in targeted delivery.

We have recently applied a directed evolution approach in order to address the problem of antibody neutralization of AAV. It is clear that there is significant sequence and functional diversity in the AAV capsid [82, 83, 93, 99], and the capsid is therefore potentially reasonably plastic and tolerant of point mutation. We have therefore applied a directed evolution approach involving random mutagenesis of the capsid and selection for variants that are not neutralized by serum containing anti-AAV antibodies. This approach has led to the generation of viral variants with mutations that render them largely resistant to antibodies that neutralize wild-type virus, and analysis of their sequences could yield further insights into the mechanism of viral cell entry (Maheshri et al., submitted). Furthermore, this approach of evolving novel "custom serotypes" can be applied to a variety of challenges in AAV gene delivery.

4
Summary

Significant progress has been made in the molecular engineering of viral attachment proteins for the three goals of broadening tropism, replacing tropism, and stealthing vectors to reduce unwanted molecular interactions. In efforts to fulfill these three targeting goals, the different design challenges and constraints of enveloped and nonenveloped viruses have been addressed.

Envelope proteins can be highly challenging to engineer since in many cases receptor binding directly triggers viral fusion; however, these proteins can typically tolerate relatively large genetic insertions since structurally they need only assemble into small multimers such as trimers. In contrast, capsid proteins must multimerize into highly organized structures, and even small peptides can readily disable viral function if not inserted into the correct sites. However, the binding and endosome disruption activities of these proteins are not typically intimately coupled, potentially making it somewhat easier to re-engineer binding specificities.

For both enveloped and nonenveloped vectors, there are three strategies for developing targeting vectors (Fig. 1), and each has characteristic advantages. Nongenetic targeting approaches offer the potential for highly modular systems, where a single vector product can be adapted with antibody or chemical prosthetics for specific applications. Future work must be conducted to fully explore the potential of this approach. In addition, pseudotyping with viral attachment proteins from related viruses offers a selection of viral tropism options, and it is possible that one of these options could match the precise needs of a given gene delivery application. Finally, genetic engineering offers the potential for rapid targeting or stealthing without the need for an extra prosthetic. With pseudotyping and genetic engineering, however, different clinical applications will likely require the generation and large-scale production of different targeted vector variants. For genetic engineering strategies, rational design methods can be effective when a significant amount of structural information about the viral attachment protein is known. In parallel, library and directed evolution methods have the potential to yield useful vector products as well as yield new insights into viral structure and function, particularly for viruses for which little structural information is known. Major progress has been made in pursuing these strategies for targeting vectors, and future work will reveal whether they can solve the challenges that remain, particularly the generation of high titer vector variants with fully replaced tropism.

Acknowledgements The authors wish to thank James Koerber for a critical reading of this manuscript. J.Y. is supported by a Graduate Research Fellowship from the Whitaker Foundation. D.S. is supported by EB003007 and generous funding from the ALS Association.

References

1. Robson T, Hirst DG (2003) J Biomed Biotechnol 2003:110–137
2. Romano G (2004) Drug News Perspect 17:85–90
3. Burton EA, Bai Q, Goins WF, Glorioso JC (2001) Adv Drug Deliv Rev 53:155–170
4. Wickham TJ (2003) Nat Med 9:135–139
5. Cavazzana-Calvo M, Hacein-Bey S, de Saint Basile G, Gross F, Yvon E et al. (2000) Science 288:669–672

6. Hacein-Bey-Abina S, Von Kalle C, Schmidt M, McCormack MP, Wulffraat N et al. (2003) Science 302:415–419
7. Edelstein M (2004) The Journal of Gene Medicine website for gene therapy clinical trials worldwide. Wiley, Chichester, UK (see http://www.wiley.co.uk/genmed/clinical/, last accessed 20th September 2005)
8. Fields BN, Knipe DM, Howley PM, Griffin DE (2001) Fields Virology. Lippincott Williams & Wilkins, Philadelphia, PA
9. Kayman SC, Park H, Saxon M, Pinter A (1999) J Virol 73:1802–1808
10. Wu BW, Lu J, Gallaher TK, Anderson WF, Cannon PM (2000) Virology 269:7–17
11. Rothenberg SM, Olsen MN, Laurent LC, Crowley RA, Brown PO (2001) J Virol 75:11851–11862
12. Naldini L, Blomer U, Gallay P, Ory D, Mulligan R et al. (1996) Science 272:263–267
13. Russell DW, Miller AD (1996) J Virol 70:217–222
14. Wilson C, Reitz MS, Okayama H, Eiden MV (1989) J Virol 63:2374–2378
15. Burns JC, Friedmann T, Driever W, Burrascano M, Yee JK (1993) P Natl Acad Sci USA 90:8033–8037
16. Kasahara N, Dozy AM, Kan YW (1994) Science 266:1373–1376
17. Cosset FL, Morling FJ, Takeuchi Y, Weiss RA, Collins MK et al. (1995) J Virol 69:6314–6322
18. Nguyen TH, Pages JC, Farge D, Briand P, Weber A (1998) Hum Gene Ther 9:2469–2479
19. Valsesia-Wittmann S, Drynda A, Deleage G, Aumailley M, Heard JM et al. (1994) J Virol 68:4609–4619
20. Gollan TJ, Green MR (2002) J Virol 76:3558–3563
21. Somia NV, Zoppe M, Verma IM (1995) P Natl Acad Sci USA 92:7570–7574
22. Chu TH, Dornburg R (1995) J Virol 69:2659–2663
23. Chu TH, Dornburg R (1997) J Virol 71:720–725
24. Martin F, Kupsch J, Takeuchi Y, Russell SJ, Cosset FL et al. (1998) Hum Gene Ther 9:737–746
25. Roux P, Jeanteur P, Piechaczyk M (1989) P Natl Acad Sci USA 86:9079–9083
26. Etienne-Julan M, Roux P, Carillo S, Jeanteur P, Piechaczyk M (1992) J Gen Virol 73 (Pt 12):3251–3255
27. Ohno K, Meruelo D (1997) Biochem Mol Med 62:123–127
28. Ohno K, Sawai K, Iijima Y, Levin B, Meruelo D (1997) Nat Biotechnol 15:763–767
29. Morizono K, Bristol G, Xie YM, Kung SK, Chen IS (2001) J Virol 75:8016–8020
30. Hall FL, Gordon EM, Wu L, Zhu NL, Skotzko MJ et al. (1997) Hum Gene Ther 8:2183–2192
31. Hall FL, Liu L, Zhu NL, Stapfer M, Anderson WF et al. (2000) Hum Gene Ther 11:983-993
32. Gordon EM, Chen ZH, Liu L, Whitley M, Wei D et al. (2001) Hum Gene Ther 12:193–204
33. Russell SJ, Cosset FL (1999) J Gene Med 1:300–311
34. Lavillette D, Russell SJ, Cosset FL (2001) Curr Opin Biotechnol 12:461–466
35. Zhao Y, Zhu L, Lee S, Li L, Chang E et al. (1999) Natl Acad Sci USA 96:4005–4010
36. Barnett AL, Davey RA, Cunningham JM (2001) P Natl Acad Sci USA 98:4113–4118
37. Valsesia-Wittmann S, Morling FJ, Hatziioannou T, Russell SJ, Cosset FL (1997) Embo J 16:1214–1223
38. Martin F, Chowdhury S, Neil SJ, Chester KA, Cosset FL et al. (2003) J Virol 77:2753–2756
39. Martin F, Neil S, Kupsch J, Maurice M, Cosset FL et al. (1999) J Virol 73:6923–6929

40. Konishi H, Ochiya T, Chester KA, Begent RH, Muto T et al. (1998) Hum Gene Ther 9:235–248
41. Khare PD, Shao-Xi L, Kuroki M, Hirose Y, Arakawa F et al. (2001) Cancer Res 61:3705
42. Martin F, Chowdhury S, Neil S, Phillipps N, Collins MK (2002) Mol Ther 5:269–274
43. Boerger AL, Snitkovsky S, Young JA (1999) P Natl Acad Sci USA 96:9867–9872
44. Snitkovsky S, Young JA (1998) P Natl Acad Sci USA 95:7063–7068
45. Snitkovsky S, Niederman TM, Mulligan RC, Young JA (2001) J Virol 75:1571–1575
46. Snitkovsky S, Young JA (2002) Virology 292:150–155
47. Mothes W, Boerger AL, Narayan S, Cunningham JM, Young JA (2000) Cell 103:679–689
48. Melikyan GB, Barnard RJ, Markosyan RM, Young JA, Cohen FS (2004) J Virol 78:3753–3762
49. Tai CK, Logg CR, Park JM, Anderson WF, Press MF et al. (2003) Hum Gene Ther 14:789–802
50. Endres MJ, Jaffer S, Haggarty B, Turner JD, Doranz BJ et al. (1997) Science 278:1462–1464
51. Balliet JW, Bates P (1998) J Virol 72:671–676
52. Somia NV, Miyoshi H, Schmitt MJ, Verma IM (2000) J Virol 74:4420–4424
53. Bittner A, Mitnacht-Kraus R, Schnierle BS (2002) J Virol Methods 104:83–92
54. Pizzato M, Marlow SA, Blair ED, Takeuchi Y (1999) J Virol 73:8599–8611
55. Pizzato M, Blair ED, Fling M, Kopf J, Tomassetti A et al. (2001) Gene Ther 8:1088–1096
56. Sharma S, Miyanohara A, Friedmann T (2000) J Virol 74:10790–10795
57. Hatziioannou T, Delahaye E, Martin F, Russell SJ, Cosset FL (1999) Hum Gene Ther 10:1533–1544
58. Lin AH, Kasahara N, Wu W, Stripecke R, Empig CL et al. (2001) Hum Gene Ther 12:323–332
59. DePolo NJ, Reed JD, Sheridan PL, Townsend K, Sauter SL et al. (2000) Mol Ther 2:218–222
60. Croyle MA, Callahan SM, Auricchio A, Schumer G, Linse KD et al. (2004) J Virol 78:912–921
61. Guibinga GH, Hall FL, Gordon EM, Ruoslahti E, Friedmann T (2004) Mol Ther 9:76–84
62. Fielding AK, Maurice M, Morling FJ, Cosset FL, Russell SJ (1998) Blood 91:1802–1809
63. Nilson BH, Morling FJ, Cosset FL, Russell SJ (1996) Gene Ther 3:280–286
64. Merten CA, Engelstaedter M, Buchholz CJ, Cichutek K (2003) Virology 305:106–114
65. Chadwick MP, Morling FJ, Cosset FL, Russell SJ (1999) J Mol Biol 285:485–494
66. Peng KW, Pham L, Ye H, Zufferey R, Trono D et al. (2001) Gene Ther 8:1456–1463
67. Peng KW, Vile R, Cosset FL, Russell SJ (1999) Gene Ther 6:1552–1557
68. Chowdhury S, Chester KA, Bridgewater J, Collins MK, Martin F (2004) Mol Ther 9:8592
69. Morling FJ, Peng KW, Cosset FL, Russell SJ (1997) Virology 234:51–61
70. Stemmer WP (1994) Nature 370:389–391
71. Boder ET, Wittrup KD (1997) Nat Biotechnol 15:553–557
72. Arnold FH, Volkov AA (1999) Curr Opin Chem Biol 3:54–59
73. Soong NW, Nomura L, Pekrun K, Reed M, Sheppard L et al. (2000) Nat Genet 25:436–439
74. Barry MA, Dower WJ, Johnston SA (1996) Nat Med 2:299–305

75. Engelstadter M, Bobkova M, Baier M, Stitz J, Holtkamp N et al. (2000) Hum Gene Ther 11:293–303
76. Buchholz CJ, Peng KW, Morling FJ, Zhang J, Cosset FL et al. (1998) Nat Biotechnol 16:951–954
77. Bupp K, Roth MJ (2002) Mol Ther 5:329–335
78. Bupp K, Roth MJ (2003) Hum Gene Ther 14:1557–1564
79. Schneider RM, Medvedovska Y, Hartl I, Voelker B, Chadwick MP et al. (2003) Gene Ther 10:1370–1380
80. Xiao X, Li J, McCown TJ, Samulski RJ (1997) Exp Neurol 144:113–124
81. Bueler H (1999) Biol Chem 380:613–622
82. Gao GP, Alvira MR, Wang L, Calcedo R, Johnston J et al. (2002) P Natl Acad Sci USA 99:11854–11859
83. Gao G, Alvira MR, Somanathan S, Lu Y, Vandenberghe LH et al. (2003) P Natl Acad Sci USA 100:6081–6086
84. Rutledge EA, Halbert CL, Russell DW (1998) J Virol 72:309–319
85. Gao GP, Alvira MR, Wang L, Calcedo R, Johnston J et al. (2002) P Natl Acad Sci USA 99:11854–11859
86. Samulski RJ, Chang LS, Shenk T (1989) J Virol 63:3822–3828
87. Bartlett JS, Kleinschmidt J, Boucher RC, Samulski RJ (1999) Nat Biotechnol 17:181–186
88. Ponnazhagan S, Mahendra G, Kumar S, Thompson JA, Castillas M Jr (2002) J Virol 76:12900–12907
89. Lee G, Maheshri N, Kaspar B, Schaffer DV (2005) Biotechnol Bioeng 92:24–34
90. Chiorini JA, Yang L, Liu Y, Safer B, Kotin RM (1997) J Virol 71:6823–6833
91. Chiorini JA, Kim F, Yang L, Kotin RM (1999) J Virol 73:1309–1319
92. Chirmule N, Propert K, Magosin S, Qian Y, Qian R et al. (1999) Gene Ther 6:1574–1583
93. Davidson BL, Stein CS, Heth JA, Martins I, Kotin RM et al. (2000) P Natl Acad Sci USA 97:3428–3432
94. Summerford C, Samulski RJ (1998) J Virol 72:1438–1445
95. Summerford C, Bartlett JS, Samulski RJ (1999) Nat Med 5:78–82
96. Qing K, Mah C, Hansen J, Zhou S, Dwarki V et al. (1999) Nat Med 5:71–77
97. Walters RW, Yi SM, Keshavjee S, Brown KE, Welsh MJ et al. (2001) J Biol Chem 276:20610–20616
98. Di Pasquale G, Davidson BL, Stein CS, Martins I, Scudiero D et al. (2003) Nat Med 9:1306–1312
99. Rabinowitz JE, Rolling F, Li C, Conrath H, Xiao W et al. (2002) J Virol 76:791–801
100. Auricchio A, Kobinger G, Anand V, Hildinger M, O'Connor E et al. (2001) Hum Mol Genet 10:3075–3081
101. Yang Q, Mamounas M, Yu G, Kennedy S, Leaker B et al. (1998) Hum Gene Ther 9:1929–1937
102. Girod A, Ried M, Wobus C, Lahm H, Leike K et al. (1999) Nat Med 5:1052–1056
103. Hermonat PL, Labow MA, Wright R, Berns KI, Muzyczka N (1984) J Virol 51:329–339
104. Rabinowitz JE, Xiao W, Samulski RJ (1999) Virology 265:274–285
105. Wu P, Xiao W, Conlon T, Hughes J, Agbandje-McKenna M et al. (2000) J Virol 74:8635–8647
106. Shi W, Arnold GS, Bartlett JS (2001) Hum Gene Ther 12:1697–1711
107. Shi W, Bartlett JS (2003) Mol Ther 7:515–525
108. Grifman M, Trepel M, Speece P, Gilbert LB, Arap W et al. (2001) Mol Ther 3:964–975

109. White SJ, Nicklin SA, Buning H, Brosnan MJ, Leike K et al. (2004) Circulation 109:513–519
110. Ried MU, Girod A, Leike K, Buning H, Hallek M (2002) J Virol 76:4559–4566
111. Xie Q, Bu W, Bhatia S, Hare J, Somasundaram T et al. (2002) P Natl Acad Sci USA 99:10405–10410
112. Walters RW, Agbandje-McKenna M, Bowman VD, Moninger TO, Olson NH et al. (2004) J Virol 78:3361–3371
113. Muller OJ, Kaul F, Weitzman MD, Pasqualini R, Arap W et al. (2003) Nat Biotechnol 21:1040–1046
114. Bartlett JS, Wilcher R, Samulski RJ (2000) J Virol 74:2777–2785
115. Duan D, Li Q, Kao AW, Yue Y, Pessin JE et al. (1999) J Virol 73:10371–10376
116. Hansen J, Qing K, Srivastava A (2001) J Virol 75:4080–4090
117. Qing K, Hansen J, Weigel-Kelley KA, Tan M, Zhou S et al. (2001) J Virol 75:8968–8976
118. Hansen J, Qing K, Srivastava A (2001) Mol Ther 4:289–296

Adv Biochem Engin/Biotechnol (2005) 99: 169–191
DOI 10.1007/10_007

Published online: 1 November 2005

Lentiviral Vectors

Nils Loewen (✉) · Eric M. Poeschla (✉)

Molecular Medicine Program, Mayo Clinic College of Medicine, Guggenheim 18,
Rochester, MN 55905, USA
N-loewen@northwestern.edu, poeschla.eric@mayo.edu

Abstract We review the use of lentiviral vectors in current human gene therapy applications that involve genetic modification of nondividing tissues with integrated transgenes. Safety issues, including insertional mutagenesis and replication-competent retroviruses, are discussed. Innate cellular defenses against retroviruses and their implications for human gene therapy with different lentiviral vectors are also addressed.

1 Retroviral vectors in gene therapy

A capacity for permanent genetic modification is the key advantage of retroviral vectors in the treatment of chronic diseases and genetic defects. Integration into the host genome is a requisite part of the retroviral life cycle. No other viral vector type allows reliable integration, although this property has now been mimicked with interesting results by some nonviral gene therapy vectors [1,2]. Lentiretroviruses and type C retroviruses have different characteristics, which are reflected in how they are used in gene therapy. Lentiviruses possess complex genomes and are found in primates (SIV, HIV), ungulates (EIAV, CAEV, BIV, Visna) and felidae (FIV). This review will compare and contrast features of these different retroviral vectors, evaluate selected recent preclinical and clinical advances, highlight some aspects that concern safety, and point out the potential relevance for gene therapy of recent basic research findings about lentiviral restriction factors.

Lentiviruses generally display a long latency for disease causation and infect nondividing host cells such as macrophages. To gain access to chromatin, type C retroviruses require breakdown of the nuclear envelope during mitosis [3, 4], and therefore type C vectors cannot be used to transduce nondividing cells. In contrast, lentiviral preintegration complexes are imported into nondividing cell nuclei [5], which allows transduction of terminally differentiated and nondividing cells such as neurons [6–8]. Besides this central advantage, other less well-documented features may add to the attractiveness of lentiviral vectors. Hypermethylation and other mechanisms that result in silencing may interfere with long-term type C murine leukemia virus (MLV)-based retroviral vector function [9]. Incorporation of chromatin insulator elements to reduce this problem has met with variable success [10–14]. While certain modifications can increase resistance to MLV vector silencing [15–17], lentiviral vectors have been shown to achieve long-term expression in different kinds of stem cells [18–20] and may be less prone to silencing in general. Another potential advantage is that the action of the Rev gene restricts transfer vector RNA splicing, which may facilitate the transfer of sequences that contain cryptic splice acceptors [21–23].

Early lentiviral vectors were derived from HIV-1 [6, 24, 25]. Nonprimate lentivirus-based vector systems were then devised, beginning with feline immunodeficiency virus (FIV) [7, 26, 27]. Systems were soon developed from equine infectious anemia virus (EIAV) [8, 28], and later from the other ungulate lentiviruses: caprine arthritis encephalitis virus (CAEV) [29], bovine immunodeficiency virus (BIV) [30, 31], and Visna virus [32]. It is not yet clear whether nonprimate lentiviral vectors will provide biosafety advantages in human gene therapy. The parental viruses do not propagate in or cause disease in humans. On the other hand, more is known about HIV-1 and over 20 antiretroviral drugs now exist that could be used to treat an HIV-1-derived replication-competent retrovirus (RCR) [33]. Human cell restriction of nonprimate retroviruses, as discussed below, may impact on the use of different vector systems in complex ways as well. Restriction might be viewed as either desirable or undesirable depending on the particular gene therapy setting. Cellular restriction factors like TRIM5alpha [34] can be saturated and overcome by excess virion particles, as seen with the typically high inputs used in many gene therapy applications, and they would provide an innate impediment to RCR propagation.

In recent years, over 263 clinical trials with type C retroviral vectors [35] and one trial with lentiviral vectors have been initiated [36]. Almost all of the trials target diseases that have a poor prognosis with conventional treatment (45% solid cancers, 39% hematological malignancies, 28% HIV, 28% monogenetic (SCID, ADA, and so on)), with the exception of some innovative approaches for chronic conditions (5.9%) (such as rheumatoid arthritis, Alzheimer's disease, erythropoietin in hemodialysis patients, superficial corneal opacities, degenerative joint disease) [35]. Retroviral and adenoviral vectors account for 27% and 26%, respectively, of all gene therapy vectors used, followed by naked DNA (15%) and lipofection (9%) [35]. Although 64% of all current gene therapy studies are phase 1 studies assessing safety, first therapeutic successes have been reported. The most prominent is the treatment of a fatal form of severe combined immunodeficiency-X1 (SCID-X1), an X-linked hereditary disorder characterized by an early block in the development of T and natural killer (NK) cells due to mutations in the γc cytokine receptor subunit. Autologous cytokine-stimulated hematopoietic stem cells were transduced ex vivo with a type-C murine leukemia virus (MLV)-based retroviral vector expressing the receptor subunit and infused [37]. T and NK cells counts and function were restored to normal levels in nine out of ten children. As discussed further below, this clear-cut success in reversing severe immune deficiency was offset by the delayed appearance of T cell malignancy in two children, apparently due to insertional mutagenesis [38]. The ongoing lentiviral vector trial employs HIV vector VRX496, which expresses a 937-base antisense sequence against the HIV-1 envelope gene [36, 39, 40].

Promising results with lentiviral vectors have been achieved in animal models. More recent successes include correction of β-thalassaemia [22, 23, 41] and sickle cell disease [42] in mouse models after transduction of hematopoietic stem cells, with improvement observed in important hematological parameters. Pawliuk et al. transduced progenitor cells with lentiviral vectors expressing a beta-globin gene variant that prevents hemoglobin S polymerization. Erythroid-specific, long-term expression was seen, with a decrease in erythrocyte sickling and correction of anemia, red cell morphology and splenomegaly [42]. Sadelain and colleagues were able to express human beta-globin under the transcriptional control of segments of its locus control region [23]. Previous efforts to treat thalassaemia by gene therapy have often been complicated by insufficient globin transgene expression, position effects and transcriptional silencing. In long-term recipients of unselected transduced bone marrow cells, tetramers of murine and human beta-globin molecules contributed to 13% of the total hemoglobin in mature red cells of normal mice. Beta-thalassaemia mice that received transduced bone marrow cells had mature red cells with 17–24% of total hemoglobin consisting of two murine alpha-globin and two human beta-globin molecules. Galimi et al. showed that Fanconi anemia, an inherited cancer susceptibility syndrome caused by mutations in a DNA repair pathway, can be corrected with lentiviral vectors by ex vivo transduction of the normal gene into quiescent hematopoietic progenitor cells from Fanca(−/−) and Fancc(−/−) mice [43]. Efficiency was high enough that no cell purification or cytokine prestimulation was necessary. Resistance to DNA-damaging agents was fully restored, allowing for in vivo selection of transduced cells with nonablative doses of cyclophosphamide.

The nervous system has also been a principal focus for lentiviral vector-based gene therapy, and several recent studies are notable. Improvement of neurological function and prevention of nigrostriatal degeneration was recently observed in a primate model of Parkinson's disease [44]. In this study, rhesus monkeys received injections of lentiviral vectors expressing glial cell-derived neurotrophic factor (GDNF) into the striatum and substantia nigra and were followed for eight months. Vector treatment reversed neurological deficits in young animals with toxin-induced Parkinson's disease and augmented dopaminergic function in nonlesioned aged monkeys. Also using GDNF, Georgievska et al. demonstrated protection of dopaminergic neurons in a rat forebrain bundle axotomy model of Parkinson's disease [45]. Lentiviral vectors expressing arylsulfatase A have been shown to confer long-term protection from learning impairments in a mouse model of metachromatic leukodystrophy [46]. Durable, widespread expression of the transgene was seen after stereotactical injection of hippocampal fimbria. The same group achieved similar results by transplanting hematopoietic stem cells transduced ex vivo with the arylsulfatase A gene, which prevented the development of motor conduction impairment, learning and coordination deficits, and neuropathological abnormalities typical of the disease [47]. Building on earlier work demonstrating that rabies G glycoprotein can mediate retrograde axonal transport of vector particles [48], Azzouz et al. showed that rabies-G pseudotyping of an EIAV-based VEGF-expressing lentiviral vector prolonged survival in a mouse model of amyotrophic lateral sclerosis after a single injection into distal muscles [49]. Histopathological defects of mucopolysaccharidosis type VII in a mouse model were shown to regress after CNS injection with beta-glucoronidase expressing FIV vectors [50]. Moreover, when tested in mice with established impairments in spatial learning and memory, there was significant recovery of behavioral function [51].

In a study of gene therapy for muscular dystrophy, mesoangioblasts from juvenile alpha-sarcoglycan null mice, a model for limb-girdle muscular dystrophy, were genetically modified with lentiviral vectors to express alpha-sarcoglycan. When injected into the femoral artery, these modified vessel-associated stem cells corrected the dystrophic phenotype morphologically and functionally in most downstream muscles [52]. Success was attributed to widespread distribution of the cells through capillary networks. Kobinger et al. injected muscles of mice with muscular dystrophy with lentiviral vectors encoding the mini-dystrophin [53]. Restored dystrophin expression could be seen in muscle progenitor and satellite cells providing functional correction in skeletal muscles.

In the eye, Takahashi et al. rescued rd mice with retinitis pigmentosa by transducing the rod photoreceptor cGMP phosphodiesterase beta subunit (PDEbeta) [54]. Follow-up for six months confirmed expression and reversed pathology. Working in the anterior chamber of the eye, Loewen et al. achieved high-grade, stable GFP expression in a glaucoma-critical structure in vivo (the aqueous humor outflow tract) [55]. Extensive, targeted expression resulted after a single transcorneal injection in cats, persisted for ten months, and could be monitored noninvasively over time in living animals, suggesting an approach to glaucoma gene therapy.

Spermatogenesis could be restored in lentiviral vector-injected testes of Sl/Sl(d) mutant mice lacking a transmembrane ligand on Sertoli cells [56]. Sperm collected from recipient testes were able to generate normal pups after intracytoplasmic sperm injection, while none of the offspring carried the transgene, suggesting that lentiviral vectors might be used for gene therapy of male infertility without the risk of germ-line transmission.

Systemic (intravenous) administration of lentiviral vectors has generally resulted in the most expression in liver, spleen and bone marrow, with lesser transduction in other organs [57, 58]. Other vectors, such as AAV, have also resulted in high levels of gene transfer after intravenous delivery [59].

2 Retroviral mutagenesis

Retroviral vectors are mutagenic by definition because the host genome sequence is irreversibly altered. HIV-1 and MLV both integrate preferentially into transcriptionally active regions during productive infection [60–62] (reviewed in [63]). However, the linear placements of their proviruses appear to differ with respect to functional elements within the cellular transcription units. HIV-1 integration favors regions downstream of promoters [60–62]. In contrast, MLV integrations show a predilection for promoter regions, favoring the 2.5 kb surrounding the transcriptional start site [61].

MLV vectors were recently used to treat severe combined immunodeficiency-X1 (SCID-X1) [37]. This trial illustrates both the therapeutic potential and the risks associated with retroviral vectors. SCID-X1 is an X-linked hereditary disorder characterized by mutations in the γc cytokine receptor subunit, which produce an early block in the development of T and natural killer (NK) cells. Children with this disorder are profoundly immunocompromised and require bone marrow transplantation to survive. In the trial, cycling hematopoietic stem cells were transduced ex vivo with a murine leukemia viral vector encoding the receptor subunit and returned intravenously. T and NK cell numbers and function were restored to levels comparable to age-matched controls in nine out of ten children [37, 64], providing the first substantial, unambiguous evidence for efficacy of gene therapy in humans. Success was greatly facilitated in this trial by the marked proliferative advantage conferred to the gene-corrected γc-expressing lymphocyte progenitor cells compared to unmodified host cells. However, two patients developed a T cell malignancy during follow-up [38, 65]. Both participants were found to have vector integration in proximity to the LMO2 proto-oncogene promoter, leading to aberrant transcription and expression of LMO2, a transcription factor and central regulator of hematopoiesis [66]. There is evidence, however, that it is the combination of γc transduction and LMO2 insertion that produced the high frequency of leukemogenesis observed in this trial [67]; retroviral vectors without co-operatively oncogenic transgenes like the γc cytokine receptor subunit therefore probably present much less risk. Several other adjustments in method are also possible to reduce risk. In addition to the transduction of fewer cells initially in order to reduce the probability of undesirable integration sites, use of self-inactivating vectors (SIN) represents one strategy that could help reduce the probability of such adverse events in the future. SIN vectors have deletions in the 3' LTR (of the U3 element), which may reduce the chance of aberrant transcription of host genes in proximity to an integrated vector [68–71]. The problem is not completely solved however, since internal promoter/enhancers in SIN vectors may also have potential to alter host cell oncogene transcription, as can the vector insertion itself. Coexpression of a suicide gene in transduced cells is a strategy that could be used to ablate transduced clones [72, 73] should unforeseen pathogenic properties develop in vivo. It is not yet known how the aforementioned differences in integration site preferences between MLV and lentiviral vectors will impact on the risk of insertional mutagenesis.

Other investigators have proposed inducible vectors that modulate expression when a regulatory small molecule is withdrawn or added [74–77]. In an attempt to partially reverse transgene insertion, vectors have also been flanked with cre-lox cassettes that allow excision with cre-recombinase [78–81]. However, multiple copies of cre-lox retroviruses in a given cell can result in translocation when distant lox sites react with each other and cause cell death from chromosomal aberration [81]. Chromatin insulators may also be instrumental for isolating the genetic apparatus of the integrated vector from surrounding chromatin, which could reduce vector silencing by blocking encroachment by heterochomatin, and in turn can reduce untoward effects of vector sequences on host gene expression [10–14].

In specialized circumstances, integrase-mutant vectors might be an alternative that could be used to reduce risks associated with insertional mutagenesis [82–84]. In nondividing cells, internal transgene expression from unintegrated lentiviral DNA can be abundant and persistent, which can be demonstrated with specific catalytic center integrase mutants [84]. Vargas et al. reported an integrase-defective lentiviral vector in which use of the simian virus 40 (SV40) origin of replication (oriT) to prolong episome maintenance permitted long-term expression. In the presence of the SV40 large T antigen (TAg), episomal DNA expression persisted until the experimental endpoint of 60 days [82]. Similarly, Lu et al. were able to achieve transgene expression in primary macrophages for several weeks and also in Jurkat cells using this approach [83]. In contrast, incorporation into vectors of the Epstein-Barr virus (EBV) origin (oriP) in EBV nuclear antigen 1-expressing cells did not lead to extended unintegrated DNA expression [83]. Both of the latter groups showed that the vector remained episomal and transcriptionally active specifically in cells expressing the SV40 large T antigen, providing a novel system for cell-specific persistent gene expression. However, caveats are that coexpression of the T antigen is generally undesirable, the utility is likely to be highly cell-type-specific, and the approach dispenses with the main advantage of a retroviral vector, namely permanent integration. The approach has greatest potential for situations, e.g., gene repair strategies, where transient expression in non-dividing cells is the goal.

3 Replication competent retroviruses

MLV vectors have been used in thousands of animal studies and have been administered to more than 1200 patients to date [35, 85]. Early studies of rhesus macaques transduced with MLV vectors highlighted risks associated with RCRs when MLV RCRs induced rapidly fatal T-cell lymphomas in three out of ten monkeys [86]. Homologous recombination between producer cells and transfer vector resulted in incorporation of LTR and encapsidation elements to form a replication-competent retroviral genome [87]. Chronic retroviral infection of bone marrow stem cells with an MLV of altered host range appeared to have disrupted critical growth control genes, resulting in stepwise cell transformation and clonal tumors [88]. The Center for Biologics Evaluation and Research of the Food and Drug Administration subsequently established guidelines for testing for RCRs in retroviral vector-based gene therapy products with recommendations that 5% of each vector lot and 10^8 cells or 1% of vector producer and target cells be tested for RCR [89].

First-generation assays to detect RCR in vector supernatants scored changes in morphology (syncytial foci) [90] but were replaced with more sensitive assays based on rescue of markers after prolonged culture of a permissive cell line (RCR infection allows packaging of beta-galactosidase encoding vector RNA) [91–94]. Protocols to detect replication-competent lentiviruses (RCL) have adapted these techniques but may use different markers (such as p24 capsid protein concentration for HIV vectors) [95] or PCR-based methods [96]. Assays require three weeks of culture and are capable of detecting one RCL in a background of 2.5×10^8 transducing units of vector in a single test culture [95] or 10–100 copies of target sequence, respectively [96].

Construction of systems in which viral protein encoding sequences are separated from those sequences needed for encapsidation and reverse transcription has made generation of RCRs by recombination into one single viral genome less likely. Split gag/pol packaging systems [97] and Rev-independent humanized gag/pol expression cassettes [98, 99] may provide additional safety. However, it is not clear that the removal from lentiviral vector systems of a requirement for Rev for viral structural gene expression would provide a net safety advantage [100]. RCR has been observed when generation was either deliberately facilitated [101] or when these species arose secondary to recombination [102–106]. Identification of RNA packaging signals in nonprimate lentiviruses has been initiated [107–109] and will facilitate minimization of homologous recombination. Mobilization of integrated lentiviral vectors can occur by infection with wild type virus [110]. There is no significant nucleotide sequence homology between lentiviruses and human endogenous retroviruses.

4 Immunogenicity and toxicity

Lentiviral vector preparation involves a variety of reagents that can contribute to the heterogeneity of the final product. Serum and cellular proteins and fragments are sources of potentially immunogenic material. Production by transient transfection can lead to carryover of input plasmid DNA; this can be reduced with DNAse or benzoase treatment [111]. Vector-producing cells not only generate viral proteins and packageable RNAs, but the transgene-encoded protein is readily expressed as well. Transfer of the preformed protein to target cells can lead to pseudotransduction [112, 113]. Use of tissue-specific promoters in retro- and lentiviral vectors for transcriptional targeting can limit expression to the desired cell type [114–122]. It is difficult to remove virus-like capsid structures, microvesicles [123] and membranous particles [124] generated during vector production because of their similar physical and chemical properties. Material that is co-pelleted during vector concentration can be immunogenic [125–127]. Purification methods that have been proposed for selected applications have included sucrose cushion separation [128], sedimentation gradient banding [129–131] or column purification [126, 127]. Column or gradient purifications achieve better separation but have some disadvantages. Exchange chromatography has the advantage that larger volumes can be processed without repeated loading [126, 127]. Recoveries can be low but recent improvements with rates between 22–68% and 78–143 fold concentrations have been reported [126, 127]. In comparison, sucrose gradient purification is cumbersome and only permits the handling of small volumes [128]. Iodixanol (velocity gradient) purification has also been applied to lentiviral purification but presents similar problems for scale-up [132].

Stable packaging cell lines obviate transient introduction of DNA, facilitate formal characterization of biological reagents and allow a more standardized vector production for commercial application. Generation of retroviral producer cell lines that generate VSV-G pseudotyped particles has been difficult because VSV-G associated cytotoxicity can result in negative selection. Titers of up to 5–6 $\times$ 10^6 TU/ml of VSV-G-pseudotyped vectors have been reported using inducible systems [133–136]. Other envelopes have been used to pseudotype lentiviral vectors, including A-MLV (broad tropism, titer 10^5 TU/ml) [137], Mokola and rabies (neurotropism, titer 10^6 TU/ml) [48, 138], Ebola (some airway tropism, titer 10^5 TU/ml) [139], HTLV-I (extended tropism of HIV vectors, titer 10^3 TU/ml) [140], influenza HA (airway, 10^2 TU/ml) [139], RSV (airway, 10^3 TU/ml) [139], Marburg (airway, titer not determined) [141] and RD114 (hematopoietic, 10^5 TU/ml) [142]. Pseudotyping of lentiviral vectors with RD114, a feline endogenous retrovirus, has been reported to improve the complement stability of vectors [143, 144]. Ross River virus glycoproteins were efficiently incorporated into FIV vectors, generating unconcentrated vector titers of up to 5 $\times$ 10^5 TU/ml [145]. Baculovirus GP64 may produce similar stability during ultracentrifugation, while eliminating toxicity in packaging cells [146, 147], but tropism is restricted in many cell types and species, in particular in cells of hematopoietic origin [147]. Ikeda et al. reported continuous high-titer HIV-1 vector production by cells in which HIV-1 Gag-Pol is stably expressed [148]. Prior attempts to make stable HIV-1 packaging cells by transfection of plasmids encoding HIV-1 Gag-Pol were limited by low levels of p24 antigen, probably due to the cytotoxicity of HIV-1 protease. This could now be overcome by expressing HIV-1 Gag-Pol from an MLV vector. When MLV 4070A or RD114 envelopes were used, production of 10^7 TU/ml could be achieved for three months.

5
Lentiviral vectors and innate cellular immunity

Host defense mechanisms against viral pathogens can act before cell contact, at cell entry or during uncoating, transcription and translation. While more complex adaptive immune systems requiring participation of complex multicellular regulatory networks are found in vertebrates, forms of innate immunity that utilize autonomous cellular defenses targeted to particular classes of pathogens developed in early single cell eukaryotes and exist today in all four eukaryote kingdoms. Recently discovered retroviral restriction factors [149–159] limit retroviral infection by interfering with viral capsid transit at a post-entry step.

As discussed above, retroviruses pose a risk of mutagenesis due to the alteration of the genetic code permanently through insertion. The human genome is riddled with full length and partial retroviruses, highlighting the significance of exposure and defense: while only 1.2% of the genome codes for about 20 000 to 25 000 expressed genes [39], as much as 8% of the human genome sequence consists of HERVs and related elements [160–162].

Retroelements as a whole contribute to a remarkable 45% of the genome sequence [161–163]. Considering the potential consequences of retroviral insertion, it is not surprising that effective mechanisms have evolved to limit infection or mitigate its effects. As discussed below, recent results suggest that remnants of ancient retroviral infection provide protection against retroviruses today [164].

6 Genome defense

Although adaptive immunity (such as antibodies, antigen-specific cytotoxic T lymphocytes) are critical to antiviral defense, cells have evolved other strategies to block viruses even before the adaptive responses can be generated. Transcriptional silencing by condensation, methylation, binding of methyl-CpG-binding protein 2 to methylated promoters or recruitment of histone deacetylases is a common mechanism used to control retroelements in vertebrates [165, 166]. Interferons constitute another nonspecific antiviral intracellular mechanism against viruses that can be found in vertebrates, invertebrates and insects (reviewed in [167]). Species-specific post-entry restriction factors are a third form of innate cellular antiviral defenses that confer resistance to retroviruses and consequently to retroviral vectors in gene therapy. Retroviruses and derived vectors display characteristic patterns of species- and cell type-dependent efficacy. These relative restrictions to infection localize to the post-entry phase of the viral life cycle, before or synchronous with reverse transcription, and so persist even when receptor-specific entry is circumvented by pseudotyping. The restriction factors Lv1 [150, 151, 168], Ref1 [150–152, 168] and Fv1 (Friend virus susceptibility 1) [151–159, 168] all target, with high specificity, the incoming viral capsid of HIV-1 and SIV in primate cells, N-tropic murine leukemia virus (MLV) and EIAV in human cells, and N- or B-tropic MLV in murine cells, respectively. The human, rhesus and African green monkey (AGM) TRIM5alpha genes have now been identified to encode Ref1 and Lv1 antiretroviral activities [34, 168–171]. HIV-1 has evolved an interaction between its capsid protein and the host protein cyclophilin A to counteract restriction [172].

Fv1-mediated restriction maps to one amino acid of the viral capsid (position 110), while determinants for Lv1 and Ref1 appear to be more complex, but are similarly determined by the capsid [150, 152, 153, 173–175]. Interestingly, there is no sequence similarity between the endogenous Fv1 and MLV Gag. Fv1 instead shares 60% of the nucleotide sequence of the mouse endogenous retrovirus MERV-L and the human endogenous retrovirus HERV-L [156, 157]. In analogy to Lv1 and Ref1, expression of Fv1 in nonmurine cells imparts a restricting phenotype [158]. TRIM5alpha is a prerequisite for Fv1 to establish a restricting phenotype [168]. The inhibition mediated by these three restriction factors displays kinetics consistent with competitive inhibition, and can be abrogated in a saturable manner using a high multiplicity of infection. Saturation can be accomplished with particles of the same retrovirus or a different retrovirus, as long as it is also restricted and is capable of entry into the target cell [151]. Therefore, virus-like particles (genomeless particles produced by expression of Gag/Pol alone, as well as a suitable envelope) will also suffice. For instance, HIV-1 does not replicate in African Green Monkey (AGM) cells, yet preincubation or co-infection with HIV-1 allows 100-fold more efficient infection with N-MLV, which is normally restricted [151].

Another recently discovered mechanism for intrinsic cellular immunity [176] is the viral genome editing enzyme APOBEC3G [177] (and to some extent other APOBEC proteins). APOBEC3G is counteracted by lentiviral Vif proteins, which target it for destruction in the cellular proteasome. Countering APOBEC proteins appears to be the sole reason for the evolution of Vif proteins by lentiviruses. Importantly, APOBEC3G associates with the viral genome in the cell in which the virus is assembled rather than in the target cell. During reverse transcription, it acts to edit the minus strand DNA, resulting in cytidine deamination and hence plus strand G to A mutations that are highly deleterious or lethal to viral genetic stability [178, 179]. Since lentiviral vectors can be prepared in cells that lack APOBEC3G expression, this restriction does not hinder gene therapy with lentiviral vectors, and Vif can be eliminated from the system. The elimination of Vif can be considered an additional safeguard against the propagation of RCR.

7 Ref1 restriction in human gene therapy

Restriction factors decrease the transduction efficiency of EIAV-based lentiviral vectors [151]. Data are lacking for other ungulate virus vectors, but it is probable that these will also be restricted by Ref1/TRIM5alpha. For example, FIV is restricted by both rhesus and human TRIM5alpha [181]. Restriction may not necessarily be a detriment in the gene therapy setting since high vector inputs are typically used and these (as well as added VLPs) may allow saturation. Situations where vectors are diluted and individual cells encounter less capsid, for instance when administered intravenously, will make restriction more overt. Effective transduction of human tissues has been seen with nonprimate lentiviral vectors, for example [180] and [182]. Since a principal concern with all retroviral vector systems is the generation and propagation of RCR, the presence of saturable restriction also provides an innate defense against systemic propagation of nonprimate lentivirus RCRs. Thus, restriction has the potential to be an advantage or a disadvantage, depending on the particular application, retroviral vector, and gene therapy setting. Better understanding of lentiviral restriction may permit the engineering of vector capsids that escape restriction factors, similar to certain MLV capsid mutants [155]. Conversely, engineering of restriction factors might enable their use in gene therapy for HIV disease [183].

8 Conclusion

Lentiviral vectors allow genetic modification of nondividing tissues with permanently integrated transgenes. They have the most potential in the treatment of inborn errors of metabolism and chronic diseases in which prolonged or life-long therapeutic gene expression is needed. Successful correction of genetic defects has been demonstrated in various animal studies. Further study of the propensity of these vectors to cause adverse effects through insertional mutagenesis is needed. Innate post-entry restrictions in target cells need to be taken into account for future applications of lentiviral vectors to gene therapy.

Acknowledgements We thank Yasuhiro Ikeda for critically reviewing the manuscript. We apologize to those authors whose works were not cited in this selective review of more recent developments due to space limitations.

References

1. Groth AC, Olivares EC, Thyagarajan B, Calos MP (2000) A phage integrase directs efficient site-specific integration in human cells. P Natl Acad Sci USA 97(11):5995–6000
2. Ortiz-Urda S, Thyagarajan B, Keene DR, Lin Q, Fang M, Calos MP, Khavari PA (2002) Stable nonviral genetic correction of inherited human skin disease. Nat Med 8(10):1166–1170
3. Miller DG, Adam MA, Miller AD (1990) Gene transfer by retrovirus vectors occurs only in cells that are actively replicating at the time of infection. Mol Cell Biol 10(8):4239–4242
4. Lewis PF, Emerman M (1994) Passage through mitosis is required for oncoretroviruses but not for the human immunodeficiency virus. J Virol 68(1):510–516
5. Fouchier RA, Malim MH (1999) Nuclear import of human immunodeficiency virus type-1 preintegration complexes. Adv Virus Res 52:275–299
6. Naldini L, Bloemer U, Gallay P, Ory D, Mulligan R, Gage FH, Verma IM, Trono D (1996) In vivo gene delivery and stable transduction of nondividing cells by a lentiviral vector. Science 272(5259):263–267
7. Poeschla E, Wong-Staal F, Looney D (1998) Efficient transduction of nondividing cells by feline immunodeficiency virus lentiviral vectors. Nat Med 4(3):354–357

8. Mitrophanous K, Yoon S, Rohll J, Patil D, Wilkes F, Kim V, Kingsman S, Kingsman A, Mazarakis N (1999) Stable gene transfer to the nervous system using a non-primate lentiviral vector. Gene Ther 6(11):1808–1818
9. Cherry SR, Biniszkiewicz D, van Parijs L, Baltimore D, Jaenisch R (2000) Retroviral expression in embryonic stem cells and hematopoietic stem cells. Mol Cell Biol 20(20):7419–7426
10. Rivella S, Callegari JA, May C, Tan CW, Sadelain M (2000) The cHS4 insulator increases the probability of retroviral expression at random chromosomal integration sites. J Virol 74(10):4679–4687
11. Yannaki E, Tubb J, Aker M, Stamatoyannopoulos G, Emery DW (2002) Topological constraints governing the use of the chicken HS4 chromatin insulator in oncoretrovirus vectors. Mol Ther 5(5):589–598
12. Emery DW, Yannaki E, Tubb J, Stamatoyannopoulos G (2000) A chromatin insulator protects retrovirus vectors from chromosomal position effects. P Natl Acad Sci USA 97(16):9150–9155
13. Hejnar J, Hajkova P, Plachy J, Elleder D, Stepanets V, Svoboda J (2001) CpG island protects Rous sarcoma virus-derived vectors integrated into nonpermissive cells from DNA methylation and transcriptional suppression. P Natl Acad Sci USA 98(2):565–569
14. Jakobsson J, Rosenqvist N, Thompson L, Barraud P, Lundberg C (2004) Dynamics of transgene expression in a neural stem cell line transduced with lentiviral vectors incorporating the cHS4 insulator. Exp Cell Res 298(2):611–623
15. Robbins PB, Yu XJ, Skelton DM, Pepper KA, Wasserman RM, Zhu L, Kohn DB (1997) Increased probability of expression from modified retroviral vectors in embryonal stem cells and embryonal carcinoma cells. J Virol 71(12):9466–9474
16. Prasad Alur RK, Foley B, Parente MK, Tobin DK, Heuer GG, Avadhani AN, Pongubala J, Wolfe JH (2002) Modification of multiple transcriptional regulatory elements in a Moloney murine leukemia virus gene transfer vector circumvents silencing in fibroblast grafts and increases levels of expression of the transferred enzyme. Gene Ther 9(17):1146–1154
17. Haas DL, Lutzko C, Logan AC, Cho GJ, Skelton D, Jin Yu X, Pepper KA, Kohn DB (2003) The Moloney murine leukemia virus repressor binding site represses expression in murine and human hematopoietic stem cells. J Virol 77(17):9439–9450
18. Consiglio A, Gritti A, Dolcetta D, Follenzi A, Bordignon C, Gage FH, Vescovi AL, Naldini L (2004) Robust in vivo gene transfer into adult mammalian neural stem cells by lentiviral vectors. P Natl Acad Sci USA 101(41):14835–14840
19. Pfeifer A, Ikawa M, Dayn Y, Verma IM (2002) Transgenesis by lentiviral vectors: lack of gene silencing in mammalian embryonic stem cells and preimplantation embryos. P Natl Acad Sci USA 99(4):2140–2145
20. Gropp M, Itsykson P, Singer O, Ben-Hur T, Reinhartz E, Galun E, Reubinoff BE (2003) Stable genetic modification of human embryonic stem cells by lentiviral vectors. Mol Ther 7(2):281–287
21. Malim MH, Hauber J, Le SY, Maizel JV, Cullen BR (1989) The HIV-1 rev transactivator acts through a structured target sequence to activate nuclear export of unspliced viral mRNA. Nature 338(6212):254–257
22. Puthenveetil G, Scholes J, Carbonell D, Xia P, Qureshi N, Zeng L, Li S, Yu Y, Hiti AL, Yee JK, Malik P (2004) Successful correction of the human beta-thalassemia major phenotype using a lentiviral vector. Blood 104(12):3445–3453

23. May C, Rivella S, Callegari J, Heller G, Gaensler KM, Luzzatto L, Sadelain M (2000) Therapeutic haemoglobin synthesis in beta-thalassaemic mice expressing lentivirus-encoded human beta-globin. Nature 406(6791):82–86
24. Parolin C, Sodroski J (1995) A defective HIV-1 vector for gene transfer to human lymphocytes. J Mol Med 73(6):279–288
25. Poznansky M, Lever A, Bergeron L, Haseltine W, Sodroski J (1991) Gene transfer into human lymphocytes by a defective human immunodeficiency virus type 1 vector. J Virol 65(1):532–536
26. Johnston JC, Gasmi M, Lim LE, Elder JH, Yee JK, Jolly DJ, Campbell KP, Davidson BL, Sauter SL (1999) Minimum requirements for efficient transduction of dividing and nondividing cells by feline immunodeficiency virus vectors. J Virol 73(6):4991–5000
27. Curran MA, Kaiser SM, Achacoso PL, Nolan GP (2000) Efficient transduction of nondividing cells by optimized feline immunodeficiency virus vectors. Mol Ther 1(1):31–38
28. Olsen JC (1998) Gene transfer vectors derived from equine infectious anemia virus. Gene Ther 5(11):1481–1487
29. Mselli-Lakhal L, Favier C, Da Silva Teixeira MF, Chettab K, Legras C, Ronfort C, Verdier G, Mornex JF, Chebloune Y (1998) Defective RNA packaging is responsible for low transduction efficiency of CAEV-based vectors. Arch Virol 143(4):681–695
30. Berkowitz R, Ilves H, Lin WY, Eckert K, Coward A, Tamaki S, Veres G, Plavec I (2001) Construction and molecular analysis of gene transfer systems derived from bovine immunodeficiency virus. J Virol 75(7):3371–3382
31. Metharom P, Takyar S, Xia HH, Ellem KA, Macmillan J, Shepherd RW, Wilcox GE, Wei MQ (2000) Novel bovine lentiviral vectors based on Jembrana disease virus. J Gene Med 2(3):176–185
32. Berkowitz RD, Ilves H, Plavec I, Veres G (2001) Gene transfer systems derived from Visna virus: analysis of virus production and infectivity. Virology 279(1):116–129
33. Poeschla EM (2003) Nonprimate lentiviral vectors. Curr Opin Mol Ther 5(5):529–540
34. Stremlau M, Owens CM, Perron MJ, Kiessling M, Autissier P, Sodroski J (2004) The cytoplasmic body component TRIM5alpha restricts HIV-1 infection in Old World monkeys. Nature 427(6977):848–853
35. Edelstein M (2004) Journal of Gene Medicine website on gene therapy trials worldwide. Wiley, Chichester (see http://www.wiley.com/legacy/wileychi/genmed/clinical/, last accessed 21st September 2005)
36. MacGregor RR (2001) Clinical protocol A phase 1 open-label clinical trial of the safety and tolerability of single escalating doses of autologous CD4 T cells transduced with VRX496 in HIV-positive subjects. Hum Gene Ther 12(16):2028–2029
37. Cavazzana-Calvo M, Hacein-Bey S, de Saint Basile G, Gross F, Yvon E, Nusbaum P, Selz F, Hue C, Certain S, Casanova JL, Bousso P, Deist FL, Fischer A (2000) Gene therapy of human severe combined immunodeficiency (SCID)-X1 disease. Science 288(5466):669–672
38. Hacein-Bey-Abina S, von Kalle C, Schmidt M, Le Deist F, Wulffraat N, McIntyre E, Radford I, Villeval JL, Fraser CC, Cavazzana-Calvo M, Fischer A (2003) A serious adverse event after successful gene therapy for X-linked severe combined immunodeficiency. N Engl J Med 348(3):255–256
39. Humeau LM, Binder GK, Lu X, Slepushkin V, Merling R, Echeagaray P, Pereira M, Slepushkina T, Barnett S, Dropulic LK, Carroll R, Levine BL, June CH, Dropulic B (2004) Efficient lentiviral vector-mediated control of HIV-1 replication in CD4 lymphocytes from diverse HIV+ infected patients grouped according to CD4 count and viral load. Mol Ther 9(6):902–913

40. Manilla P, Rebello T, Afable C, Lu X, Slepushkin V, Humeau LM, Schonely K, Ni Y, Binder GK, Levine BL, MacGregor RR, June CH, Dropulic B (2005) Regulatory considerations for novel gene therapy products: a review of the process leading to the first clinical lentiviral vector. Hum Gene Ther 16(1):17–25
41. Moreau-Gaudry F, Xia P, Jiang G, Perelman NP, Bauer G, Ellis J, Surinya KH, Mavilio F, Shen CK, Malik P (2001) High-level erythroid-specific gene expression in primary human and murine hematopoietic cells with self-inactivating lentiviral vectors. Blood 98(9):2664–2672
42. Pawliuk R, Westerman KA, Fabry ME, Payen E, Tighe R, Bouhassira EE, Acharya SA, Ellis J, London IM, Eaves CJ, Humphries RK, Beuzard Y, Nagel RL, Leboulch P (2001) Correction of sickle cell disease in transgenic mouse models by gene therapy. Science 294(5550):2368–2371
43. Galimi F, Noll M, Kanazawa Y, Lax T, Chen C, Grompe M, Verma IM (2002) Gene therapy of Fanconi anemia: preclinical efficacy using lentiviral vectors. Blood 100(8):2732–2736
44. Kordower JH et al. (2000) Neurodegeneration prevented by lentiviral vector delivery of GDNF in primate models of Parkinson's disease. Science 290(5492):767–773
45. Georgievska B, Kirik D, Rosenblad C, Lundberg C, Bjorklund A (2002) Neuroprotection in the rat Parkinson model by intrastriatal GDNF gene transfer using a lentiviral vector. Neuroreport 13(1):75–82
46. Consiglio A, Quattrini A, Martino S, Bensadoun JC, Dolcetta D, Trojani A, Benaglia G, Marchesini S, Cestari V, Oliverio A, Bordignon C, Naldini L (2001) In vivo gene therapy of metachromatic leukodystrophy by lentiviral vectors: correction of neuropathology and protection against learning impairments in affected mice. Nat Med 7(3):310–316
47. Biffi A, De Palma M, Quattrini A, Del Carro U, Amadio S, Visigalli I, Sessa M, Fasano S, Brambilla R, Marchesini S, Bordignon C, Naldini L (2004) Correction of metachromatic leukodystrophy in the mouse model by transplantation of genetically modified hematopoietic stem cells. J Clin Invest 113(8):1118–1129
48. Mazarakis ND, Azzouz M, Rohll JB, Ellard FM, Wilkes FJ, Olsen AL, Carter EE, Barber RD, Baban DF, Kingsman SM, Kingsman AJ, O'Malley K, Mitrophanous KA (2001) Rabies virus glycoprotein pseudotyping of lentiviral vectors enables retrograde axonal transport and access to the nervous system after peripheral delivery. Hum Mol Genet 10(19):2109–2121
49. Azzouz M, Ralph GS, Storkebaum E, Walmsley LE, Mitrophanous KA, Kingsman SM, Carmeliet P, Mazarakis ND (2004) VEGF delivery with retrogradely transported lentivector prolongs survival in a mouse ALS model. Nature 429(6990):413–417
50. Stein CS, Kang Y, Sauter SL, Townsend K, Staber P, Derksen TA, Martins I, Qian J, Davidson BL, McCray PB Jr (2001) In vivo treatment of hemophilia A and mucopolysaccharidosis type VII using nonprimate lentiviral vectors. Mol Ther 3(6):850–856
51. Brooks AI, Stein CS, Hughes SM, Heth J, McCray PM, Sauter SL Jr, Johnston JC, Cory-Slechta DA, Federoff HJ, Davidson BL (2002) Functional correction of established central nervous system deficits in an animal model of lysosomal storage disease with feline immunodeficiency virus-based vectors. P Natl Acad Sci USA 99(9):6216–6221
52. Sampaolesi M, Torrente Y, Innocenzi A, Tonlorenzi R, D'Antona G, Pellegrino MA, Barresi R, Bresolin N, De Angelis MG, Campbell KP, Bottinelli R, Cossu G (2003) Cell therapy of alpha-sarcoglycan null dystrophic mice through intra-arterial delivery of mesoangioblasts. Science 301(5632):487–492

53. Kobinger GP, Louboutin JP, Barton ER, Sweeney HL, Wilson JM (2003) Correction of the dystrophic phenotype by in vivo targeting of muscle progenitor cells. Hum Gene Ther 14(15):1441–1449
54. Takahashi M, Miyoshi H, Verma IM, Gage FH (1999) Rescue from photoreceptor degeneration in the rd mouse by human immunodeficiency virus vector-mediated gene transfer. J Virol 73(9):7812–7816
55. Loewen N, Fautsch MP, Teo WL, Bahler CK, Johnson DH, Poeschla EM (2004) Long-term, targeted genetic modification of the aqueous humor outflow tract coupled with noninvasive imaging of gene expression in vivo. Invest Ophthalmol Vis Sci 45(9):3091–3098
56. Ikawa M, Tergaonkar V, Ogura A, Ogonuki N, Inoue K, Verma IM (2002) Restoration of spermatogenesis by lentiviral gene transfer: offspring from infertile mice. P Natl Acad Sci USA 99(11):7524–7529
57. Pan D, Gunther R, Duan W, Wendell S, Kaemmerer W, Kafri T, Verma IM, Whitley CB (2002) Biodistribution and toxicity studies of VSVG-pseudotyped lentiviral vector after intravenous administration in mice with the observation of in vivo transduction of bone marrow. Mol Ther 6(1):19–29
58. Peng KW, Pham L, Ye H, Zufferey R, Trono D, Cosset FL, Russell SJ (2001) Organ distribution of gene expression after intravenous infusion of targeted and untargeted lentiviral vectors. Gene Ther 8(19):1456–1463
59. Wang Z, Zhu T, Qiao C, Zhou L, Wang B, Zhang J, Chen C, Li J, Xiao X (2005) Adeno-associated virus serotype 8 efficiently delivers genes to muscle and heart. Nat Biotechnol 23(3):321–328
60. Schroder AR, Shinn P, Chen H, Berry C, Ecker JR, Bushman F (2002) HIV-1 integration in the human genome favors active genes and local hotspots. Cell 110(4):521–529
61. Wu X, Li Y, Crise B, Burgess SM (2003) Transcription start regions in the human genome are favored targets for MLV integration. Science 300:1749–1751
62. Han Y, Lassen K, Monie D, Sedaghat AR, Shimoji S, Liu X, Pierson TC, Margolick JB, Siliciano RF, Siliciano JD (2004) Resting CD4+ T cells from human immunodeficiency virus type 1 (HIV-1)-infected individuals carry integrated HIV-1 genomes within actively transcribed host genes. J Virol 78(12):6122–6133
63. Trono D (2003) Virology: Picking the right spot. Science 300(5626):1670–1671
64. Hacein-Bey-Abina S et al. (2002) Sustained correction of X-linked severe combined immunodeficiency by ex vivo gene therapy. N Engl J Med 346(16):1185–1193
65. Hacein-Bey-Abina S et al. (2003) LMO2-associated clonal T cell proliferation in two patients after gene therapy for SCID-X1. Science 302(5644):415–419
66. Stocking C, Bergholz U, Friel J, Klingler K, Wagener T, Starke C, Kitamura T, Miyajima A, Ostertag W (1993) Distinct classes of factor-independent mutants can be isolated after retroviral mutagenesis of a human myeloid stem cell line. Growth Factors 8(3):197–209
67. Dave UP, Jenkins NA, Copeland NG (2004) Gene therapy insertional mutagenesis insights. Science 303(5656):333
68. Yu SF, von Ruden T, Kantoff PW, Garber C, Seiberg M, Ruther U, Anderson WF, Wagner EF, Gilboa E (1986) Self-inactivating retroviral vectors designed for transfer of whole genes into mammalian cells. P Natl Acad Sci USA 83(10):3194–3198
69. Olson P, Nelson S, Dornburg R (1994) Improved self-inactivating retroviral vectors derived from spleen necrosis virus. J Virol 68(11):7060–7066
70. Miyoshi H, Blomer U, Takahashi M, Gage FH, Verma IM (1998) Development of a self-inactivating lentivirus vector. J Virol 72(10):8150–8157

71. Iwakuma T, Cui Y, Chang LJ (1999) Self-inactivating lentiviral vectors with U3 and U5 modifications. Virology 261(1):120–132
72. Moolten FL (1986) Tumor chemosensitivity conferred by inserted herpes thymidine kinase genes: paradigm for a prospective cancer control strategy. Cancer Res 46(10):5276–5281
73. Obaru K, Fujii S, Matsushita S, Shimada T, Takatsuki K (1996) Gene therapy for adult T cell leukemia using human immunodeficiency virus vector carrying the thymidine kinase gene of herpes simplex virus type 1. Hum Gene Ther 7(18):2203–2208
74. Kafri T, van Praag H, Gage FH, Verma IM (2000) Lentiviral vectors: regulated gene expression. Mol Ther 1(6):516–521
75. Vigna E, Cavalieri S, Ailles L, Geuna M, Loew R, Bujard H, Naldini L (2002) Robust and efficient regulation of transgene expression in vivo by improved tetracycline-dependent lentiviral vectors. Mol Ther 5(3):252–261
76. Sirin O, Park F (2003) Regulating gene expression using self-inactivating lentiviral vectors containing the mifepristone-inducible system. Gene 323:67–77
77. Pollock R, Issner R, Zoller K, Natesan S, Rivera VM, Clackson T (2000) Delivery of a stringent dimerizer-regulated gene expression system in a single retroviral vector. P Natl Acad Sci USA 97(24):13221–13226
78. Silver DP, Livingston DM (2001) Self-excising retroviral vectors encoding the Cre recombinase overcome Cre-mediated cellular toxicity. Mol Cell 8(1):233–243
79. Chang LJ, Zaiss AK (2003) Self-inactivating lentiviral vectors and a sensitive Cre-loxP reporter system. Methods Mol Med 76:367–382
80. Ahmed BY, Chakravarthy S, Eggers R, Hermens WT, Zhang JY, Niclou SP, Levelt C, Sablitzky F, Anderson PN, Lieberman AR, Verhaagen J (2004) Efficient delivery of Cre-recombinase to neurons in vivo and stable transduction of neurons using adeno-associated and lentiviral vectors. BMC Neurosci 5:4
81. Pfeifer A, Brandon EP, Kootstra N, Gage FH, Verma IM (2001) Delivery of the Cre recombinase by a self-deleting lentiviral vector: efficient gene targeting in vivo. P Natl Acad Sci USA 98(20):11450–11455
82. Vargas J Jr, Gusella GL, Najfeld V, Klotman ME, Cara A (2004) Novel integrase-defective lentiviral episomal vectors for gene transfer. Hum Gene Ther 15(4):361–372
83. Lu R, Nakajima N, Hofmann W, Benkirane M, Jeang KT, Sodroski J, Engelman A (2004) Simian virus 40-based replication of catalytically inactive human immunodeficiency virus type 1 integrase mutants in nonpermissive T cells and monocyte-derived macrophages. J Virol 78(2):658–668
84. Saenz D, Loewen N, Peretz M, Whitwam T, Barraza R, Howell K, Holmes JH, Good M, Poeschla EM (2004) Unintegrated lentiviral DNA persistence and accessibility to expression in nondividing cells: analysis with class I integrase mutants. J Virol 78:2906–2920
85. Pages JC, Bru T (2004) Toolbox for retrovectorologists. J Gene Med6(Suppl 1):S67–S82
86. Donahue RE, Kessler SW, Bodine D, McDonagh K, Dunbar C, Goodman S, Agricola B, Byrne E, Raffeld M, Moen R et al. (1992) Helper virus induced T cell lymphoma in nonhuman primates after retroviral mediated gene transfer. J Exp Med 176(4):1125–1135
87. Purcell DF, Broscius CM, Vanin EF, Buckler CE, Nienhuis AW, Martin MA (1996) An array of murine leukemia virus-related elements is transmitted and expressed in a primate recipient of retroviral gene transfer. J Virol 70(2):887–897
88. Vanin EF, Kaloss M, Broscius C, Nienhuis AW (1994) Characterization of replication-competent retroviruses from nonhuman primates with virus-induced T-cell lym-

phomas and observations regarding the mechanism of oncogenesis. J Virol 68(7):4241–4250
89. Guidance for Industry, U.S. Department of Health and Human Services, Food and Drug Administration, Center for Biologics Evaluation and Research (CBER) (2001) Supplemental guidance on testing for replication-competent retrovirus in retroviral vector-based gene therapy products and during follow-up of patients in clinical trials using retroviral vectors. Hum Gene Ther 12(3):315–320
90. Haapala DK, Robey WG, Oroszlan SD, Tsai WP (1985) Isolation from cats of an endogenous type C virus with a novel envelope glycoprotein. J Virol 53(3):827–833
91. Danos O, Mulligan RC (1988) Safe and efficient generation of recombinant retroviruses with amphotropic and ecotropic host ranges. P Natl Acad Sci USA 85(17):6460–6464
92. Printz M, Reynolds J, Mento SJ, Jolly D, Kowal K, Sajjadi N (1995) Recombinant retroviral vector interferes with the detection of amphotropic replication competent retrovirus in standard culture assays. Gene Ther 2(2):143–150
93. Forestell SP, Dando JS, Bohnlein E, Rigg RJ (1996) Improved detection of replication-competent retrovirus. J Virol Methods 60(2):171–178
94. Reeves L, Duffy L, Koop S, Fyffe J, Cornetta K (2002) Detection of ecotropic replication-competent retroviruses: comparison of s(+)/l(–) and marker rescue assays. Hum Gene Ther 13(14):1783–1790
95. Escarpe P, Zayek N, Chin P, Borellini F, Zufferey R, Veres G, Kiermer V (2003) Development of a sensitive assay for detection of replication-competent recombinant lentivirus in large-scale HIV-based vector preparations. Mol Ther 8(2):332–341
96. Sastry L, Xu Y, Johnson T, Desai K, Rissing D, Marsh J, Cornetta K (2003) Certification assays for HIV-1-based vectors: frequent passage of gag sequences without evidence of replication-competent viruses. Mol Ther 8(5):830–839
97. Wu X, Wakefield JK, Liu H, Xiao H, Kralovics R, Prchal JT, Kappes JC (2000) Development of a novel trans-lentiviral vector that affords predictable safety. Mol Ther 2(1):47–55
98. Kotsopoulou E, Kim VN, Kingsman AJ, Kingsman SM, Mitrophanous KA (2000) A Rev-independent human immunodeficiency virus type 1 (HIV-1)-based vector that exploits a codon-optimized HIV-1 gag-pol gene. J Virol 74(10):4839–4852
99. Wagner R, Graf M, Bieler K, Wolf H, Grunwald T, Foley P, Uberla K (2000) Rev-independent expression of synthetic gag-pol genes of human immunodeficiency virus type 1 and simian immunodeficiency virus: implications for the safety of lentiviral vectors. Hum Gene Ther 11(17):2403–2413
100. Hope T (2002) Improving the post-transcriptional aspects of lentiviral vectors. Curr Top Microbiol Immunol 261:179–189
101. Segall HI, Yoo E, Sutton RE (2003) Characterization and detection of artificial replication-competent lentivirus of altered host range. Mol Ther 8(1):118–129
102. Otto E, Jones-Trower A, Vanin EF, Stambaugh K, Mueller SN, Anderson WF, McGarrity GJ (1994) Characterization of a replication-competent retrovirus resulting from recombination of packaging and vector sequences. Hum Gene Ther 5(5):567–575
103. Garrett E, Miller AR, Goldman JM, Apperley JF, Melo JV (2000) Characterization of recombination events leading to the production of an ecotropic replication-competent retrovirus in a GP+envAM12-derived producer cell line. Virology 266(1):170–179
104. Scarpa M, Cournoyer D, Muzny DM, Moore KA, Belmont JW, Caskey CT (1991) Characterization of recombinant helper retroviruses from Moloney-based vectors in ecotropic and amphotropic packaging cell lines. Virology 180(2):849–852

105. Chong H, Starkey W, Vile RG (1998) A replication-competent retrovirus arising from a split-function packaging cell line was generated by recombination events between the vector, one of the packaging constructs, and endogenous retroviral sequences. J Virol 72(4):2663–2670
106. Martinez I, Dornburg R (1996) Partial reconstitution of a replication-competent retrovirus in helper cells with partial overlaps between vector and helper cell genomes. Hum Gene Ther 7(6):705–712
107. Kemler I, Barraza R, Poeschla EM (2002) Mapping of the encapsidation determinants of feline immunodeficiency virus. J Virol 76(23):11889–11903
108. Browning MT, Mustafa F, Schmidt RD, Lew KA, Rizvi TA (2003) Delineation of sequences important for efficient packaging of feline immunodeficiency virus RNA. J Gen Virol 84(Pt 3):621–627
109. Kemler I, Azmi I, Poeschla EM (2004) The critical role of proximal gag sequences in feline immunodeficiency virus genome encapsidation. Virology 327(1):111–120
110. Evans JT, Garcia JV (2000) Lentivirus vector mobilization and spread by human immunodeficiency virus. Hum Gene Ther 11(17):2331–2339
111. Sastry L, Xu Y, Cooper R, Pollok K, Cornetta K (2004) Evaluation of plasmid DNA removal from lentiviral vectors by benzonase treatment. Hum Gene Ther 15(2):221–226
112. Liu ML, Winther BL, Kay MA (1996) Pseudotransduction of hepatocytes by using concentrated pseudotyped vesicular stomatitis virus G glycoprotein (VSV-G)-Moloney murine leukemia virus-derived retrovirus vectors: comparison of VSV-G and amphotropic vectors for hepatic gene transfer. J Virol 70(4):2497–2502
113. Gallardo HF, Tan C, Ory D, Sadelain M (1997) Recombinant retroviruses pseudotyped with the vesicular stomatitis virus G glycoprotein mediate both stable gene transfer and pseudotransduction in human peripheral blood lymphocytes. Blood 90(3):952–957
114. Dai C, McAninch RE, Sutton RE (2004) Identification of synthetic endothelial cell-specific promoters by use of a high-throughput screen. J Virol 78(12):6209–6221
115. Lotti F, Menguzzato E, Rossi C, Naldini L, Ailles L, Mavilio F, Ferrari G (2002) Transcriptional targeting of lentiviral vectors by long terminal repeat enhancer replacement. J Virol 76(8):3996–4007
116. Jager U, Zhao Y, Porter CD (1999) Endothelial cell-specific transcriptional targeting from a hybrid long terminal repeat retrovirus vector containing human prepro-endothelin-1 promoter sequences. J Virol 73(12):9702–9709
117. Grande A, Piovani B, Aiuti A, Ottolenghi S, Mavilio F, Ferrari G (1999) Transcriptional targeting of retroviral vectors to the erythroblastic progeny of transduced hematopoietic stem cells. Blood 93(10):3276–3285
118. Fassati A, Bardoni A, Sironi M, Wells DJ, Bresolin N, Scarlato G, Hatanaka M, Yamaoka S, Dickson G (1998) Insertion of two independent enhancers in the long terminal repeat of a self-inactivating vector results in high-titer retroviral vectors with tissue-specific expression. Hum Gene Ther 9(17):2459–2468
119. Vile R, Miller N, Chernajovsky Y, Hart I (1994) A comparison of the properties of different retroviral vectors containing the murine tyrosinase promoter to achieve transcriptionally targeted expression of the HSVtk or IL-2 genes. Gene Ther 1(5):307–316
120. Arbuthnot P, Bralet MP, Thomassin H, Danan JL, Brechot C, Ferry N (1995) Hepatoma cell-specific expression of a retrovirally transferred gene is achieved by alpha-fetoprotein but not insulinlike growth factor II regulatory sequences. Hepatology 22(6):1788–1796

121. Follenzi A, Battaglia M, Lombardo A, Annoni A, Roncarolo MG, Naldini L (2004) Targeting lentiviral vector expression to hepatocytes limits transgene-specific immune response and establishes long-term expression of human antihemophilic factor IX in mice. Blood 103(10)3700–3709
122. De Palma M, Venneri MA, Naldini L (2003) In vivo targeting of tumor endothelial cells by systemic delivery of lentiviral vectors. Hum Gene Ther 14(12):1193–1206
123. Bess JW Jr, Gorelick RJ, Bosche WJ, Henderson LE, Arthur LO (1997) Microvesicles are a source of contaminating cellular proteins found in purified HIV-1 preparations. Virology 230(1):134–144
124. Rolls MM, Webster P, Balba NH, Rose JK (1994) Novel infectious particles generated by expression of the vesicular stomatitis virus glycoprotein from a self-replicating RNA. Cell 79(3):497–506
125. Baekelandt V, Claeys A, Eggermont K, Lauwers E, De Strooper B, Nuttin B, Debyser Z (2002) Characterization of lentiviral vector-mediated gene transfer in adult mouse brain. Hum Gene Ther 13(7):841–853
126. Scherr M, Battmer K, Eder M, Schule S, Hohenberg H, Ganser A, Grez M, Blomer U (2002) Efficient gene transfer into the CNS by lentiviral vectors purified by anion exchange chromatography. Gene Ther 9(24):1708–1714
127. Yamada K, McCarty DM, Madden VJ, Walsh CE (2003) Lentivirus vector purification using anion exchange HPLC leads to improved gene transfer. Biotechniques 34(5):1074–1078, 1080
128. Baekelandt V, Eggermont K, Michiels M, Nuttin B, Debyser Z (2003) Optimized lentiviral vector production and purification procedure prevents immune response after transduction of mouse brain. Gene Ther 10(23):1933–1940
129. Moller-Larsen A, Christensen T (1998) Isolation of a retrovirus from multiple sclerosis patients in self-generated Iodixanol gradients. J Virol Methods 73(2):151–161
130. Christensen T, Sorensen PD, Hansen HJ, Moller-Larsen A (2003) Antibodies against a human endogenous retrovirus and the preponderance of env splice variants in multiple sclerosis patients. Mult Scler 9(1):6–15
131. Fujisawa R, McAtee FJ, Favara C, Hayes SF, Portis JL (2001) N-terminal cleavage fragment of glycosylated Gag is incorporated into murine oncornavirus particles. J Virol 75(22):11239–11243
132. Coleman JE, Huentelman MJ, Kasparov S, Metcalfe BL, Paton JF, Katovich MJ, Semple-Rowland SL, Raizada MK (2003) Efficient large-scale production and concentration of HIV-1-based lentiviral vectors for use in vivo. Physiol Genomics 12(3):221–228
133. Kafri T, van Praag H, Ouyang L, Gage FH, Verma IM (1999) A packaging cell line for lentivirus vectors. J Virol 73(1):576–584
134. Klages N, Zufferey R, Trono D (2000) A stable system for the high-titer production of multiply attenuated lentiviral vectors. Mol Ther 2(2):170–176
135. Farson D, Witt R, McGuinness R, Dull T, Kelly M, Song J, Radeke R, Bukovsky A, Consiglio A, Naldini L (2001) A new-generation stable inducible packaging cell line for lentiviral vectors. Hum Gene Ther 12(8):981–997
136. Xu K, Ma H, McCown TJ, Verma IM, Kafri T (2001) Generation of a stable cell line producing high-titer self-inactivating lentiviral vectors. Mol Ther 3(1):97–104
137. Page KA, Landau NR, Littman DR (1990) Construction and use of a human immunodeficiency virus vector for analysis of virus infectivity. J Virol 64(11):5270–5276
138. Mochizuki H, Schwartz JP, Tanaka K, Brady RO, Reiser J (1998) High-titer human immunodeficiency virus type 1-based vector systems for gene delivery into nondividing cells. J Virol 72(11):8873–8883

139. Kobinger GP, Weiner DJ, Yu QC, Wilson JM (2001) Filovirus-pseudotyped lentiviral vector can efficiently and stably transduce airway epithelia in vivo. Nat Biotechnol 19(3):225–230
140. Landau NR, Page KA, Littman DR (1991) Pseudotyping with human T-cell leukemia virus type I broadens the human immunodeficiency virus host range. J Virol 65(1):162–169
141. Chan SY, Speck RF, Ma MC, Goldsmith MA (2000) Distinct mechanisms of entry by envelope glycoproteins of Marburg, Ebola (Zaire) viruses. J Virol 65(1):4933–4937
142. Hanawa H, Kelly PF, Nathwani AC, Persons DA, Vandergriff JA, Hargrove P, Vanin EF, Nienhuis AW (2002) Comparison of various envelope proteins for their ability to pseudotype lentiviral vectors and transduce primitive hematopoietic cells from human blood. Mol Ther 5(3):242–251
143. Sandrin V, Boson B, Salmon P, Gay W, Negre D, Le Grand R, Trono D, Cosset FL (2002) Lentiviral vectors pseudotyped with a modified RD114 envelope glycoprotein show increased stability in sera and augmented transduction of primary lymphocytes, CD34+ cells derived from human and nonhuman primates. Blood 100(3):823–832
144. Takeuchi Y, Cosset FL, Lachmann PJ, Okada H, Weiss RA, Collins MK (1994) Type C retrovirus inactivation by human complement is determined by both the viral genome and the producer cell. J Virol 68(12):8001–8007
145. Kang Y, Stein CS, Heth JA, Sinn PL, Penisten AK, Staber PD, Ratliff KL, Shen H, Barker CK, Martins I, Sharkey CM, Sanders DA, McCray PB Jr, Davidson BL (2002) In vivo gene transfer using a nonprimate lentiviral vector pseudotyped with Ross River Virus glycoproteins. J Virol 76(18):9378–9388
146. Kumar M, Bradow BP, Zimmerberg J (2003) Large-scale production of pseudotyped lentiviral vectors using baculovirus GP64. Hum Gene Ther 14(1):67–77
147. Schauber CA, Tuerk MJ, Pacheco CD, Escarpe PA, Veres G (2004) Lentiviral vectors pseudotyped with baculovirus gp64 efficiently transduce mouse cells in vivo and show tropism restriction against hematopoietic cell types in vitro. Gene Ther 11(3):266–275
148. Ikeda Y, Takeuchi Y, Martin F, Cosset FL, Mitrophanous K, Collins M (2003) Continuous high-titer HIV-1 vector production. Nat Biotechnol 21(5):569–572
149. Keckesova Z, Ylinen LM, Towers GJ (2004) The human and African green monkey TRIM5alpha genes encode Ref1 and Lv1 retroviral restriction factor activities. P Natl Acad Sci USA 101(29):10780–10785
150. Hatziioannou T, Cowan S, Von Schwedler UK, Sundquist WI, Bieniasz PD (2004) Species-specific tropism determinants in the human immunodeficiency virus type 1 capsid. J Virol 78(11):6005–6012
151. Hatziioannou T, Cowan S, Goff SP, Bieniasz PD, Towers GJ (2003) Restriction of multiple divergent retroviruses by Lv1 and Ref1. EMBO J 22(3):385–394
152. Cowan S, Hatziioannou T, Cunningham T, Muesing MA, Gottlinger HG, Bieniasz PD (2002) Cellular inhibitors with Fv1-like activity restrict human and simian immunodeficiency virus tropism. P Natl Acad Sci USA 99(18):11914–11919
153. Towers G, Bock M, Martin S, Takeuchi Y, Stoye JP, Danos O (2000) A conserved mechanism of retrovirus restriction in mammals. P Natl Acad Sci USA 97(22):12295–12299
154. Aagaard L, Mikkelsen JG, Warming S, Duch M, Pedersen FS (2002) Fv1-like restriction of N-tropic replication-competent murine leukaemia viruses in mCAT-1-expressing human cells. J Gen Virol 83(Pt 2):439–442
155. Kozak CA, Chakraborti A (1996) Single amino acid changes in the murine leukemia virus capsid protein gene define the target of Fv1 resistance. Virology 225(2):300–305

156. Benit L, De Parseval N, Casella JF, Callebaut I, Cordonnier A, Heidmann T (1997) Cloning of a new murine endogenous retrovirus, MuERV-L, with strong similarity to the human HERV-L element and with a gag coding sequence closely related to the Fv1 restriction gene. J Virol 71(7):5652–5657
157. Best S, Le Tissier P, Towers G, Stoye JP (1996) Positional cloning of the mouse retrovirus restriction gene Fv1. Nature 382(6594):826–829
158. Bock M, Bishop KN, Towers G, Stoye JP (2000) Use of a transient assay for studying the genetic determinants of Fv1 restriction. J Virol 74(16):7422–7430
159. Besnier C, Ylinen L, Strange B, Lister A, Takeuchi Y, Goff SP, Towers GJ (2003) Characterization of murine leukemia virus restriction in mammals. J Virol 77(24):13403–13406
160. Hughes JF, Coffin JM (2001) Evidence for genomic rearrangements mediated by human endogenous retroviruses during primate evolution. Nat Genet 29(4):487–489
161. Lander ES et al. (2001) Initial sequencing and analysis of the human genome. Nature 409(6822):860–921
162. Venter JC et al. (2001) The sequence of the human genome. Science 291(5507):1304–1351
163. Li WH, Gu Z, Wang H, Nekrutenko A (2001) Evolutionary analyses of the human genome. Nature 409(6822):847–849
164. Lee K, KewalRamani VN (2004) In defense of the cell: TRIM5alpha interception of mammalian retroviruses. P Natl Acad Sci USA 101(29):10496–10497
165. Jones PL, Veenstra GJ, Wade PA, Vermaak D, Kass SU, Landsberger N, Strouboulis J, Wolffe AP (1998) Methylated DNA and MeCP2 recruit histone deacetylase to repress transcription. Nat Genet 19(2):187–191
166. Nan X, Ng HH, Johnson CA, Laherty CD, Turner BM, Eisenman RN, Bird A (1998) Transcriptional repression by the methyl-CpG-binding protein MeCP2 involves a histone deacetylase complex. Nature 393(6683):386–389
167. Sen GC (2001) Viruses and interferons. Annu Rev Microbiol 55:255–281
168. Keckesova Z, Ylinen LM, Towers GJ (2004) The human and African green monkey TRIM5alpha genes encode Ref1 and Lv1 retroviral restriction factor activities. P Natl Acad Sci USA 101(29):10780–10785
169. Hatziioannou T, Perez-Caballero D, Yang A, Cowan S, Bieniasz PD (2004) Retrovirus resistance factors Ref1 and Lv1 are species-specific variants of TRIM5alpha. P Natl Acad Sci USA 101(29):10774–10779
170. Perron MJ, Stremlau M, Song B, Ulm W, Mulligan RC, Sodroski J (2004) TRIM5alpha mediates the postentry block to N-tropic murine leukemia viruses in human cells. P Natl Acad Sci USA 101(32):11827–11832
171. Yap MW, Nisole S, Lynch C, Stoye JP (2004) Trim5alpha protein restricts both HIV-1 and murine leukemia virus. P Natl Acad Sci USA 101(29):10786–10791
172. Towers GJ, Hatziioannou T, Cowan S, Goff SP, Luban J, Bieniasz PD (2003) Cyclophilin A modulates the sensitivity of HIV-1 to host restriction factors. Nat Med 9(9):1138–1143
173. Hofmann W, Schubert D, LaBonte J, Munson L, Gibson S, Scammell J, Ferrigno P, Sodroski J (1999) Species-specific, postentry barriers to primate immunodeficiency virus infection. J Virol 73(12):10020–10028
174. Dorfman T, Gottlinger HG (1996) The human immunodeficiency virus type 1 capsid p2 domain confers sensitivity to the cyclophilin-binding drug SDZ NIM 811. J Virol 70(9):5751–5757
175. Owens CM, Yang PC, Gottlinger H, Sodroski J (2003) Human and simian immunodeficiency virus capsid proteins are major viral determinants of early, postentry replication blocks in simian cells. J Virol 77(1):726–731

176. Bieniasz PD (2004) Intrinsic immunity: a front-line defense against viral attack. Nat Immunol 5(11):1109–1115
177. Sheehy AM, Gaddis NC, Choi JD, Malim MH (2002) Isolation of a human gene that inhibits HIV-1 infection and is suppressed by the viral Vif protein. Nature 418(6898):646–650
178. Mangeat B, Turelli P, Caron G, Friedli M, Perrin L, Trono D (2003) Broad antiretroviral defence by human APOBEC3G through lethal editing of nascent reverse transcripts. Nature 424(6944):99–103
179. Zhang H, Yang B, Pomerantz RJ, Zhang C, Arunachalam SC, Gao L (2003) The cytidine deaminase CEM15 induces hypermutation in newly synthesized HIV-1 DNA. Nature 424(6944):94–98
180. Loewen N, Fautsch M, Peretz M, Bahler C, Cameron JD, Johnson DH, Poeschla EM (2001) Genetic modification of human trabecular meshwork with lentiviral vectors. Hum Gene Ther 12:2109–2119
181. Saenz D, Teo I, Olsen JC, Poeschla E (2005) Restriction of Feline Immunodeficiency Virus by Ref1, LV1 and Primate TRIM5a Proteins. J Virol (in press)
182. Wang G, Slepushkin V, Zabner J, Keshavjee S, Johnston JC, Sauter SL, Jolly DJ, Dubensky TW Jr, Davidson BL, McCray PB Jr (1999) Feline immunodeficiency virus vectors persistently transduce nondividing airway epithelia and correct the cystic fibrosis defect [see comments]. J Clin Invest 104(11):R55–R62
183. Yap MW, Nisole S, Stoye JP (2005) A single amino acid change in the SPRY domain of human Trim5-alpha leads to HIV-1 restriction. Curr Biol 15(1):73–78

Adv Biochem Engin/Biotechnol (2005) 99: 193–260
DOI 10.1007/10_008

Published online: 1 November 2005

Production and Formulation of Adenovirus Vectors

Nedim E. Altaras[1] · John G. Aunins[1] · Robert K. Evans[2,6] · Amine Kamen[5] · John O. Konz[3] (✉) · Jayanthi J. Wolf[4]

[1]Fermentation and Cell Culture, Merck Research Laboratories, Sumneytown Pike, West Point, Pennsylvannia 19486-0004, USA

[2]Vaccine Pharmaceutical Research, Merck Research Laboratories, Sumneytown Pike, West Point, Pennsylvannia 19486-0004, USA

[3]Biologics Development and Engineering, Merck Research Laboratories, Sumneytown Pike, West Point, Pennsylvannia 19486-0004, USA
john_konz@merck.com

[4]Live Viral Vectors, Merck Research Laboratories, Sumneytown Pike, West Point, Pennsylvannia 19486-0004, USA

[5]Biotech Research Institute, National Research Council, 6100 Royalmount Avenue, Montreal, Quebec, H4P 2R2, Canada

[6]*Present address:*
VaxInnate, 3 Cedar Brook Dr., Cranbury, New Jersey 08512, USA

Abstract Adenovirus vectors have attracted considerable interest over the past decade, with ongoing clinical development programs for applications ranging from replacement therapy for protein deficiencies to cancer therapeutics to prophylactic vaccines. Consequently, considerable product, process, analytical, and formulation development has been undertaken to support these programs. For example, "gutless" vectors have been developed in order to improve gene transfer capacity and durability of expression; new cell lines have been developed to minimize recombination events; production conditions have been optimized to improve volumetric productivities; analytical techniques and scaleable purification processes have advanced towards the goal of purified adenovirus becoming a "well-characterized biological"; and liquid formulations have been developed which maintain virus infectivity at 2–8 °C for over 18 months. These and other advances in the production of adenovirus vectors are discussed in detail in this review. In addition, the needs for the next decade are highlighted.

Keywords Adenovirus · Analytical characterization · Cell culture · Cell lines · Formulation · Purification

Abbreviations

AdV	adenovirus
Ad2	adenovirus type 2
Ad5	adenovirus type 5
AEX	anion exchange
ARM	adenovirus reference material
CPE	cytopathic effect
DELFIA	dissociation-enhanced flouroimmunoassay
ELISA	enzyme-linked immunosorbent assay
GMP	good manufacturing practice
hdAdV	helper-dependent adenovirus
HDEP	helper-dependent E1-positive particle
hpi	hours post-infection
MOI	multiplicity of infection
P/I	particle-to-infectivity ratio
Pfu	plaque-forming unit
RCA	replication-competent adenovirus
rpHPLC	reversed-phase high performance liquid chromatography
RSD	relative standard deviation
$TCID_{50}$	tissue culture infectious dose at 50%
QPCR	quantitative polymerase chain reaction
vp	viral particles

1 Introduction

Adenoviruses continue to be a leading vector choice for gene transfer, with products in development for applications ranging from cancer therapeutics to prophylactic vaccines to replacement therapies for genetic deficiencies [1–3]. In fact, in the period 1989–2004 over 240 clinical trials were conducted using adenoviruses worldwide (> 180 in the US), primarily adenovirus type 5 (Ad5) [4]. In 2004, a milestone was reached with the first licensed adenovirus gene transfer product for humans; an Ad5 vector expressing p53 was licensed by China's State Food and Drug Administration for treatment of head and neck squamous cell carcinoma using a dose of 1×10^{12} viral particles [5].

Along with interest in the clinical use of adenoviruses, needs have arisen for vectors and cell lines with varying properties, methodologies for product characterization and release, production and purification processes that are scaleable and economical, and formulations enabling sufficient stability to guarantee an acceptable shelf-life. In this review, we highlight the many advances in these areas, and identify requirements for further research. In order to capture a comprehensive and current picture, we emphasize both the published and patent literature.

We begin by discussing the common methods used for analysis of both purified and unpurified adenovirus (AdV) samples. These methods are critical to both process development and the release of GMP (good manufacturing practices) supplies to be used in clinical studies, and are referred to throughout the remainder of the review. Next, we discuss the options and rationale for vector design, and we describe the common complementing cell lines. Cultivation of AdV vectors including media selection, modes of operation (suspension vs. adherent, batch vs. fed-batch, and so on), and critical optimization and control parameters are then discussed, with an emphasis on current industrial practice. Adenovirus purification methodologies are addressed similarly, with a focus on the rational selection of unit operations for a scaleable process. Current industrial practices are compared, and critical issues such as the clearance of empty capsids and host cell DNA are reviewed in detail. Next, the development of remarkably stable liquid formulations is discussed, with an emphasis on excipient selection based on the mechanisms of inactivation. Finally, we briefly recap critical developments and highlight areas in need of additional research to support the next generation of adenovirus-based gene transfer products.

2 Analytical methods for process and product characterization

2.1 Introduction

Adenovirus (AdV) vectors are currently regarded by the FDA as complex biological products with applications for both vaccines and gene therapy, although considerable effort is being made to develop analytical methods that more closely define adenovirus structure and activity. In addition to requirements that are part of the Code of Federal Regulations (CFR), the FDA considers past experiences with gene therapy applications to issue additional safety guidance for using adenovirus vectors. The accuracy of adenovirus dose measurements is an important safety concern as several different philosophies and analytical methods can be used to assign a dose. In order to standardize assays between different groups, an adenovirus reference material (ARM) was produced and is distributed through the American Type Culture Collection (ATCC) [6]. The purpose of the ARM is to provide a reference Ad5 virus stock for use in validating assays and internal standards for adenoviral particle concentration and infectious titer.

Other important safety concerns include the production of replication competent adenoviruses (RCA) and the manufacture of viruses with high viral particle to infectious unit (P/I) ratios. RCA can arise if a recombination event occurs between sequences in an E1-deleted vector and homologous E1 sequences present in the manufacturing cell line, while limitations on P/I ratios are established to minimize the exposure of patients to high concentrations of inactive adenovirus particles. The current US FDA recommendations with respect to these concerns are < 1 RCA per 3×10^{10} vp and < 30 vp pfu^{-1} [7]. Both recommendations were made based on routine observations of Ad5 manufacturing capabilities. Because these recommendations are based on the Ad5 serotype, their applicability to other serotypes will need to be explored.

In response to these concerns by regulatory agencies, assays have been developed to demonstrate the safety, purity and potency of adenovirus vectors throughout the production process as well as the final product.

2.2 Mass assays

Mass assays are intended to measure the total amount of adenovirus in preparations, whether inactive or active. They can attempt to measure intact capsids, or the protein or nucleic acid content of the capsids. The US FDA guidance provided for cell and gene therapy products states that a mass assay should have a limit of detection that extends below 10^8 vp mL^{-1} [8], although greater sensitivity may be required if clinical dosing is below this level.

One of the earliest methods of determining adenovirus particle concentration was direct counting of particles using electron microscopy [9]. Another simple method for determining adenovirus particle concentration is to disrupt viral capsids using detergent (usually sodium dodecyl sulfate) and heat, measure the absorbance at 260 nm, and convert to particle number using a published or determined extinction coefficient [10, 11]. This methodology, which is often referred to as the "Maizel method", relies on similarity of composition between the test article and the standard used to derive the extinction coefficient. The presence of impurities, or heterogeneity in capsid composition (such as full-to-empty particle ratio) can bias the extinction coefficient and render this assay less accurate, so it is limited to purified AdV preparations. The Maizel method has an assay precision of approximately 10% RSD (relative standard deviation) and a sensitivity of 10^{11} vp mL^{-1} for spectrophotometers with a standard 1 cm pathlength.

More precise particle concentration methods have been developed using analytical anion exchange high performance liquid chromatography (AEX-HPLC) with UV absorbance detection at 260 nm [12, 13]. In this method, the peak area is converted to virus concentration by linear regression to a standard curve of calibrated standards previously quantitated using an independent mass assay, usually the Maizel method. This method is particularly valuable for concentration measurements in crude process intermediates, where adenovirus particles can be specifically measured despite representing a small fraction of the total protein mass ($\sim$ 1%) (see Fig. 1). Since these assays rely on an independent calibration of a standard preparation, the accuracy of the assay is no better than the Maizel method. Typically, assay precisions of about 10% RSD are achievable for crude preparations and 5% RSD for purified preparations. Sensitivity is roughly 10^9 vp for crude lysates and 10^8 vp for purified samples [12].

Reverse-phase HPLC (RP-HPLC) has also been used to dissociate intact adenovirus into its structural components such that the concentration of individual structural proteins (such as Protein II/Hexon) in a virus sample can be converted to equivalents of virus particle concentration [14]. This assay also relies on independent calibration of a standard preparation, and can only be applied with accuracy to highly purified samples lacking free viral proteins.

Other mass assays involve viral nucleic acid detection, which can result in 10–100-fold greater sensitivity than the above assays. Nonspecific assays for purified preparations can be based on the fluorescent dye PicoGreen, which selectively binds double-stranded DNA, to obtain a measure of the amount of adenovirus DNA in a sample. A physical particle count of intact virus can then be derived utilizing the stoichiometric relationship between

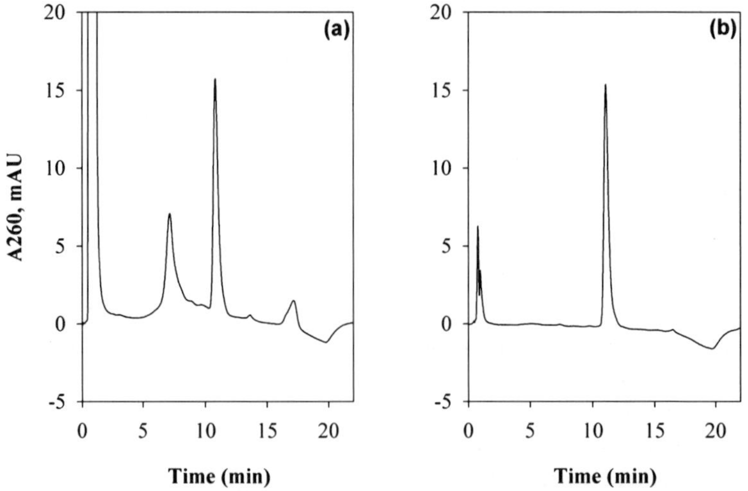

Fig. 1 Anion exchange chromatograms of **a** crude and **b** purified adenovirus on Source 15Q resin. AdV is at 11 min, while the species at 8 min is predominantly hexon protein (Konz, 1999, unpublished data)

adenovirus DNA and intact particles [15]. These assays are useful for samples free from cellular nucleic acids, and unencapsidated AdV genomes. The PicoGreen assay has a reported precision of 6% RSD and a sensitivity of 3×10^8 vp mL^{-1} [15].

Adenovirus-specific genome assays can also be used based on quantitative PCR (QPCR) with primers specific for an adenovirus sequence and an oligonucleotide probe to measure amplified PCR product. Adenovirus genome concentration is calculated by linear regression of the test article cycle threshold values to a standard curve constructed with serial dilutions of either plasmid DNA encoding adenovirus sequences or a virus standard preparation [16, 17]. These assays are useful for either crude or purified samples, and owing to their specificity have the potential for reduced interference from impurities, especially cellular nucleic acids. Because they can be calibrated to a purified DNA standard there is also the potential for improved accuracy in the determination of active product (full virus capsids). If the assay is employed to determine "total" virus, a factor must be applied to account for empty particles in the preparation. These assays are very precise (approximately 5% RSD) with a limit of detection of 10^6–10^7 AdV-genomes mL^{-1}. Similar methods can be run after a viral amplification step to monitor viral shedding in clinical samples [18].

2.3
Infectivity assays

Infectivity assays are useful to gauge the amount of active product in a preparation, and have long been used in the analysis and quality control of live virus vaccine and gene therapy products. Infectivity assays rely on direct or indirect observations of cell infection using cells permissive to replication of the vector, such as the Human Embryonic Kidney 293 cell line [19].

Traditionally, plaque assays and 50% Tissue Culture Infectious Dose ($TCID_{50}$) assays have been used, both of which rely on the detection of a viral cytopathic effect (CPE) [20, 21]. In brief, virus stocks are diluted to the end-point and adsorbed onto a cell monolayer. In plaque assays, medium containing agarose gel is overlaid on a Petri dish or multi-well plate monolayers to prevent viral diffusion and the dishes are observed for plaque formation after 10–12 days. In $TCID_{50}$ assays, 96-well format monolayers are typically used with serial dilutions of virus across rows of the plate. Cytopathic effect caused by viral replication in the individual wells is scored as positive or negative after a 12 day infection and the Spearman-Kärber or other statistical method [22] is used to calculate a 50% endpoint $TCID_{50}$ mL^{-1} value. Owing to Poisson distribution, the actual concentration of virus is $TCID_{50}/0.69$.

These assays are very valuable in that they can return an infectivity measurement for the sample independent of a standard. In practice, however, the titers from both the plaque and $TCID_{50}$ assays depend on the operating conditions and have variable sensitivity and precision (often with high variability) [23], and so the results are often calibrated to the titer of a standard virus preparation assayed alongside the test article. There are multiple reasons for poor accuracy and precision. Virus adsorption is sometimes not quantitative and gives rise to an underestimation of the total virus content [24]. The assays are also susceptible to artifacts from virus aggregation, or failure to induce detectable replication or CPE. Owing to these physical and biological considerations, infectivity assays inherently underestimate the total activity of a virus preparation, so they should always be considered as relative measures of activity that are defined by assay operating conditions. Hence it is difficult to derive a meaningful general guidance or specification for the P/I ratio.

Aside from these limitations, when using CPE for detection of replication, these assays usually require over a week of virus culture for scoring, so they are not suitable for a process monitoring situation where rapid and precise results are needed. As a result, either viral protein (via ELISA) or DNA-based detection methods have been developed for scoring these assays. Based on the experimental observation that newly synthesized viral DNA appearance is very rapid after inoculation of the monolayer, and is proportional to the virus input concentration over a large range of MOI, novel QPCR-based infectivity assays have been developed that are extremely precise and rapid, and that have extraordinary dynamic range. In such an assay, virus dilutions (non-endpoint) are applied to cell monolayers in a multiwell plate (typically 96-well) and infection is allowed to proceed for a short period, around 24 hours. QPCR is conducted on the wells, and the infectivity is calculated by comparison to a standard curve generated by a sample of known infectivity. The assay has a dynamic range of 4–6 orders of magnitude and has been shown to be superior to conventional infectivity assays in throughput, rapidity and precision [25]. Other infectivity assays using flow cytometry or quantitative immunofluorescence have been developed to measure viral protein subcomponents after infection [26, 27]. As with many assays, an independent calibration of the standard used to calculate infectivity is required; this is usually done via plaque or $TCID_{50}$ assay with enough repetitions to ensure good precision.

2.4 Transgene expression (potency) assays

Unlike live virus vaccines where achievement of infection is the goal, the aim of an adenovector is to induce in vivo expression of a heterologous gene product. As a result, verification of the functionality of the expression cassette (and, if active, the heterologous protein) is an important component of AdV characterization packages [7].

Enzyme-linked immunosorbent assays (ELISA) or dissociation-enhanced fluoroimmunoassays (DELFIA) can be used to quantify the level of in vitro gene expression from adenoviral vectors. These assays are conducted using cells permissive to adenovirus infection but not replication, such as the adenocarcinoma line A549. Assays can be formatted to either examine total transgene production in a culture, or can be formatted to determine the number of transgene expressing units in an endpoint-dilution format (either plaque-like or $TCID_{50}$-like). Scoring of "bulk" assays can be done by comparison of total protein production to a standard virus preparation. When operated in an endpoint dilution mode, scoring of transgene-expressing units can be done via immunostaining of the product protein in situ and counting spots or positive wells. In some instances custom reporter cell lines are used that fluoresce upon transgene expression [28].

Operated in bulk mode, transgene expression assays require a standard preparation to generate a calibration curve. The precision of protein quantitation relative to QPCR causes these assays to be relatively less precise compared to a QPCR-based infectivity assay, although robotic ELISA techniques may soon erase this disadvantage. Dynamic range is also constrained relative to QPCR-based infectivity techniques, and the lower limit of sensitivity, approximately 10^8-10^9 vp mL^{-1} is significantly higher than QPCR-based assays. Operated in endpoint dilution mode, transgene expressing unit assays can be comparable in precision and accuracy to infectivity assays (they reduce to an alternative method for infectious virus detection), although they do not have the speed advantage of QPCR-based assays owing to lower sensitivity of protein detection methods, and heterogeneity of protein expression between wells can render adjudication of positivity difficult. Last, transgene expression assays are subject to the same artifacts of operating conditions and sample properties (aggregation) that affect infectivity assays.

After establishing the trangene expression capacity of a vector for representative production lots, infectivity measurements should provide a reasonable correlate of potency in lieu of a transgene expression assay. Defects in capsid assembly or disruptions in protein integrity that affect delivery of the AdV genome to the nucleus, thereby abolishing infectivity, should affect transgene expression proportionately. Since the expression cassette represents a small fraction of the viral genome, probability considerations dictate that random mutations or disruptions to vector DNA integrity are *more* likely to abolish infectivity than transgene expression, with the possible exception of "gutless" vectors with large transgene inserts. Hence, an infectivity estimate should, if anything, underestimate "potency". The role of transgene expression assays in routine product release is not entirely clear, given their tendency for greater variability than the best infectivity and mass assays. Nevertheless, current regulatory posture leans towards a requirement for such assays.

2.5 Purity and residuals

Product purity and contamination with either residual cellular components or reagents used in the manufacturing process are patient safety and regulatory concerns. As noted above, lack of purity can also pose issues for measurement of virus concentration.

In order to determine whether non-AdV proteins are present in the product, protein purity can be visualized using SDS-PAGE followed by Coomassie Blue or Silver staining of separated proteins. The identity of adenovirus proteins can be proposed by the known molecular weights of the adenovirus structural proteins. MALDI-TOF with RP-HPLC can also be used to supplement the gel analysis by confirming that the major bands originate from adenovirus structural proteins, and to determine whether the minor bands are of adenovirus, cell line, or other origin. In addition to electrophoresis methods, a RP-HPLC method can be used to dissociate intact adenovirus and obtain a characteristic pattern of the viral proteome [14, 29, 30]. Residual transgene or adenoviral proteins in the purified adenovirus preparation can be quantitated using ELISA or Western Blot.

Presence of non-AdV DNA is a concern, particularly when the cell line used is tumorigenic. Recombinant E1-deleted Ad5 vectors are most commonly propagated in the 293 and PER.C6 cell lines which enable virus replication through constitutive expression of the E1 gene [19, 31]. E1 expression likely contributes to the tumorigenic phenotype when assayed in the nude mouse model [32]. As a result, the presence of host cell DNA in the product represents a theoretical risk, and demonstration of adequate DNA clearance is required. DNA can be measured using a number of techniques including the use of fluorescent dyes [33], but more specific measures may be preferable. For example, residual genomic DNA from the cell line can be specifically quantitated using Q-PCR with primers and probes against a human gene. The current requirements are less than 10 ng per dose if the cell line is tumorigenic [34]. In addition, if the DNA that is present is degraded to a size of less than 100–200 bp in length, then the need for tumorigenicity assays is less urgent [7]. When significant cellular DNA is present in the product, manufacturers should consider determining whether the DNA is free, promiscuously encapsidated, or extracapsidly associated with particles (such as through entrapment in aggregates) [35].

Adenovirus assembly intermediates might be present in the final product although a good purification process should eliminate the majority of these intermediates. Immature adenoviruses, referred to as empty capsids or light intermediates, contain the protein capsid but lack the DNA core. Immature adenoviruses lack the ability to transfer genes but still contain adenovirus proteins that at high levels could cause toxic effects [36], so the amount of empty capsids in the final product should be quantitated. Empty capsids lack DNA, so they are less dense than full capsids and can be separated using cesium chloride gradient centrifugation. After centrifugation, the empty capsids will be present as an upper band of less dense protein relative to the mature virus band. The contents of the centrifuge tube can be pumped through an in-line A_{280} detector and the ratio of empty capsid to full capsid can be determined by the ratio of the A_{280} peak heights of the upper band to the virus band [37]. Another method, based on the finding that the pVIII (31 kDa) protein is present only in empty capsids, uses RP-HPLC or SDS-PAGE to quantify the level of the 31 kDa protein relative to selected components of the complete virus [37]. These assays need to be further developed and refined in order to be used for product release. In addition, assays to measure the levels of empty capsids in the upstream process need to be developed as the existing assays will not work for samples that contain large amounts of cellular debris. Further identification of the proteins that are unique to adenovirus assembly intermediates might help in developing future assays.

2.6 Characterization

Product characterization should include tests for virus identity and virus aggregation. Tests to confirm the identity of the virus can be performed using restriction analysis or QPCR with primers specific to the adenovirus backbone and the transgene insert.

The aggregation status of the final product can be assayed using photon correlation spectroscopy, also referred to as dynamic light scattering. In this method, virus particles are illuminated with laser light and their vibration is detected by measuring the intensity fluctuations of the scattered light with a photomultiplier. Particle size is then calculated from the recorded scattered-light intensity [38, 39]. Virus aggregation can also be measured indirectly by monitoring the ratio of absorbance at 320 nm to 260 nm in a standard UV spectrophotometer [40]. Disc centrifuge sedimentation can also provide a particle size distribution and identify aggregates [40]. Field flow fractionation coupled with multi-angle light scattering (FFF-MALS) also appears to be attracting interest as a method for aggregate measurement, though we are unaware of any published references pertaining to viruses.

2.7 Safety testing

Testing methods for the demonstration of product safety are applied not only to the final product, but also include master and working cell banks, virus seeds, harvested virus bulks, purified virus, and formulated virus bulks. Detailed guidances for product safety are available from regulatory agencies and have been reviewed elsewhere, but the main considerations are provided below [41–47]. Product sterility must be documented, particularly the lack of mycoplasma, bacteria and fungi. Tests must be performed for specific adventitious viral agents that could have been introduced during the production of the cell substrate or manufacture of the virus. Traditionally, these tests include both in vivo tests in eggs or animals and in vitro tests in cell culture. More recently, highly sensitive virus-specific PCR tests have been included in the panel of safety tests.

In addition to tests for adventitious viruses, it is important to recognize other agents, such as the prions that cause transmissible spongiform encephalopathies. Strategies to address TSE agents involve a determination of the origin of the cell substrate including medical history of the donor and possible exposure of the cell line to bovine-derived materials [32, 48]. In addition, the prion gene could be sequenced to determine whether it contains specific mutations associated with hereditary forms of human prion disease [49, 50]. Evaluating cell substrates at early and late passages by Western Blot for the presence of protease-resistant prion protein is also recommended. When more sensitive in vitro assays are established they could also be used instead of the Western Blot. Some of the strategies mentioned above have been used to examine PER.C6 cells [32] and a derivative of HEK-293 cells [51]. In both cell lines, there was no evidence of the infectious prion protein, PrP^{Sc}.

AdV producer cell lines also need to be evaluated for tumorigenicity [52, 53]. This can be done by injecting nude mice subcutaneously with different amounts of cells and closely observing the animals for tumor appearance [32, 48].

Results from safety tests are combined with the rest of the analytical tests and presented to regulatory authorities for review.

3
Cells and vectors for adenovirus production

3.1
Wild-type AdV vaccines and cell lines

Although licensed adenovirus products for gene therapy and genetic vaccination are scarce, vaccines have been developed against wild-type Ad disease. Historically, killed adenovirus vaccines were developed on monkey kidney cell substrates [54]. Later, live Ad4 and Ad7 vaccines were given to US military recruits to avoid acute respiratory disease [55, 56]; however, at the time of writing the manufacture of this vaccine has ceased and the vaccine is being redeveloped by a new firm. The vaccines consisted of either wild-type serotype 4 or 7, propagated on diploid WI-38 cells [55]. The bulk vaccine was formulated, lyophilized, and placed into orally-administered, enteric-coated tablets or capsules to create the vaccine. Oral administration allowed a limited intestinal infection with no respiratory involvement in most instances. In this manner it was safe to administer wild-type, replication competent adenoviruses (RCA) to a healthy adult population.

WI-38 cells were employed for production as these cells are permissive for Ad replication, and lacked adventitious pathogen issues associated with other cell substrates at the time of vaccine development. These cells are diploid fibroblast human embryonic lung cells; today they enjoy a lengthy safety record, having been used to produce many different human vaccines [57].

Production of these wild-type adenoviruses was straightforward; owing to the low dose ($\sim 10^4$ infectious virus) and limited distribution combined with high productivity per cell, scale of production was small and T-flasks could be used for cultivation. Extensive purification was not necessary since the host cell substrate was diploid and the delivery route oral.

3.2 First- and second-generation AdV vectors and cell lines

3.2.1 First-generation vectors

Initial attempts at construction of adenovirus vectors for gene therapy and genetic vaccination were based on the observation that adenoviruses could be created with approximately 2 extra kilobases (kB) of heterologous genetic information. The recognition that adenovirus deleted in gene functions [58] could be propagated on trans-complementing cell lines dramatically increased the inserted gene size possible and rendered the vectors replication-defective in noncomplementing cells (in vivo). The earliest "first-generation" adenovirus vectors were Ad5 viruses deleted in the E1 region, which was enabled by the E1 complementing human embryonic kidney line HEK 293 [19]. E1 gene deletions typically have a maximum length of $\sim$3 kB. E1 deletion is favorable for gene therapy as it dramatically reduces expression of the remaining AdV sequences [59], and also removes a transforming gene from the product. First-generation vectors deleted in E1 were rapidly followed by vectors also deleted in the E3 region, which affords an extra $\sim$3 kB [60]. The E3 region is not necessary for growth in vitro, but the E3 19-kD gene product is believed to help adenovirus evade immune suppression by down-regulating MHC Class I expression on the cell surface, and could conceivably reduce immune-mediated elimination of transduced cells for those gene therapies requiring persistent transgene expression [61]. Documentation of the effect of E3 presence or deletion, or of vector construction details on vector propagation in culture is relatively lacking. However, it is clear that effects of first-generation vector construction on vector productivity in culture exist [62, 63].

3.2.2
Second-generation vectors

After early trials in animals and humans, it became evident that the prolonged transgene expression desired for gene replacement applications was not possible with first-generation ΔE1, ΔE3 vectors, owing to immune clearance of transduced cells caused by leaky expression of AdV proteins even in the absence of E1 complementation [64]. To reduce expression further, adenoviruses with alterations or deletions in the E2 or E4 regions have been constructed [65].

Changes in the E2 region have been explored which further down-regulate viral DNA replication and gene expression [66–70]. Also, changes in the E4 region have been explored [71–74] as well as both E2 and E4 deletions [75]. However, expression of the E2 and E4 proteins at levels high enough to adequately complement viral replication can be toxic, especially in E4 complementing lines. Production of these vectors has thus been problematic, with productivities in some instances five- to ten-fold lower than first-generation vectors. In vivo data are mixed on whether the additional modifications are beneficial enough to justify the production difficulties for these vectors.

3.2.3
Toxic and conditionally-replicating vectors

Toxic (replicating) vectors, or those expressing cytotoxic transgenes, such as those for cancer therapy, can be difficult to cultivate. Inducible/repressible [76–78] or tissue-specific [79, 80] promoters in the vectors have been devised to allow production of these vectors. In some instances resistant cell lines have been developed to enable production [81, 82]. Vectors which selectively replicate in tumor cells but not in normal cells have been created for cancer therapy; these rely on activated or deleted cellular pathways that the virus can exploit as part of its life cycle [83–88] and require propagation in cell lines that permit their replication [89]. Vectors based on AdV protease deletion have been derived that permit limited replication in cells, and these also require unique complementing lines for production [90].

3.2.4
Nonhuman and non-Group C vectors

Aside from the Ad4 and Ad7 vaccine strains, the majority of development and human clinical trials have been with human group C adenoviruses (Group C comprises serotypes 1, 2, 5, and 6), mainly Ad5, Ad2, and Ad2/5 chimeras. However these serotypes are among the most common adenovirus serotypes in human infection, and substantial fractions of the populace have neutralizing antibodies to these viruses [91, 92]. Also, these viruses are promiscuous in their tissue tropism, when more selective targeting may be desirable for some applications [93]. To address the issue of pre-exposure, desires for repeated dosing, and tissue selectivity, alternative serotypes [94–96], chimeras [97–102], and mutants [103–105] have been explored, sometimes from other species [106–112]. Non-human adenovectors have been explored for veterinary purposes as well [106].

The key to propagating these viruses is provision in the vector or cell line for replication. Targeted vectors with different cellular receptors must have their new receptor present on the production cell line, or be able to use an alternate pathway for infection [85, 113]. Vectors from alternate serogroups must have the appropriate complementing E1 function. As an example, E4orf6 protein complexes with the E1B 55kD protein to perform functions essential to viral mRNA processing, and the two must be matched for proper function and efficient virus propagation [114].

3.2.5 Production cell lines

HEK 293 cells and their derivatives have been commonly used for propagation of the above vectors. These were created by transfection of human embryonic kidney cells with Ad5 DNA fragmented by shearing. As a consequence of the imprecise cleavage method, Ad5 DNA flanking the E1 region was also incorporated, extending into a downstream region coding for protein IX (pIX), a capsid protein essential for virus thermostability [62]. Overlap of cell and vector pIX sequences enables homologous recombination to occur and give rise to replication-competent adenoviruses (RCA) at a frequency of $\sim 10^{-9}$ genome replications [115]. This is undesirable as recombination eliminates the transgene as well as producing a wild-type virus. Nevertheless, HEK 293 cells have been employed widely to produce laboratory as well as human clinical materials for Group C viruses (Ad2 and Ad5), and have supplied the majority of adenoviral human gene therapy and vaccine trials conducted to date.

Various methods have been developed to reduce the frequency of RCA generation for first-generation vectors. Vectors have been designed with degenerate pIX sequences [116], and relocated or deleted pIX genes [117]. These techniques have not found widespread acceptance. Cell lines have also been developed with precisely restricted DNA that avoids significant overlap with vectors [31, 118–121]. These cells have been found useful for vector production. In particular, PER.C6, a human embryonic retinal-derived cell line is finding wide adoption in industry (a commercial cell line, it is unavailable to the academic community) owing to its relatively well-documented history and properties [32]. Although RCA appears to be effectively eliminated in the newer cell lines, a phenomenon known as helper-dependent E1(+) particles (HDEP) can arise over extended passages at high multiplicities of infection [122, 123]. HDEP are particles that have undergone illegitimate recombination to acquire E1 sequences at the expense of other elements in the genome; they are capable of trans-complementing E1-deleted particles when the two coinfect a cell.

There are several issues for cell line selection from the production perspective. Productivity of virus is a major consideration, and the various lines differ in their productivities. As all E1-complementing lines are transformed, there is significant heterogeneity of the population, and subclones can be derived with different virus productivity [124]. A further consideration is the origin of the cell line; lines producing human adenoviruses are invariably based on human tissue, and are often of neuronal origin [125]. Employing human cells for production necessitates special precautions to ensure freedom from adventitious human viruses, and a thorough knowledge of the cell line history, because unlike recombinant protein manufacture any adventitious agents present are not so readily cleared by purification processes (see Sect. 5.3.4).

3.3 "Gutless" or helper-dependent adenovirus vectors

A major drawback of the first generation AdV for many gene replacement applications has been the short-term transgene expression in the absence of host immunosuppression. This is due to cellular immunity against adenoviral antigens but also to the transgene-encoded protein. Helper-dependent adenoviruses (hdAdV) are an improved generation of AdV that do not encode any viral genes [126]. Only two small viral DNA sequences are retained – the inverted terminal repeats (125 bp), and a packaging signal (200 bp) essential for the replication of the viral DNA and its incorporation into the viral capsid. Hence these vectors are sometimes referred to as "gutless" vectors. For propagation, the viral gene functions required for replication are not exclusively provided by the complementing cell, but also by a helper virus. These vectors have an obviously increased capacity for transgene inserts, $\sim$ 36 kb compared to $\sim$ 8 kb for earlier generations of AdV. The transgene insert should be close to the normal genome length (36 kb), or alternately noncoding "stuffer" DNA must be engineered into the vector to satisfy packaging constraints on the genome size; genomes much smaller than the wild-type size are unlikely to be stable [127, 128]. Since hdAdV do not encode any viral genes, they are theoretically safer, less toxic and less immunogenic than previously engineered AdV. Gene transfer experiments with hdAdV in animal models of several human diseases have demonstrated efficient, safe and long-term expression of therapeutic gene products [3, 129–135]; hdAdV could also potentially be used for vaccination.

The first efficient strategy used to produce hdAdV was developed using a 293-derived cell line (293Cre) that encodes the essential E1 region of Ad and the bacteriophage P1 recombinase Cre [136], which is infected with a helper and an hdAdV [129, 130]. The helper vector is a first-generation, ΔE1 AdV that provides all of the necessary enzymes and structural proteins for viral replication and assembly of the viral capsids. It is propagated on a traditional complementing cell line. The packaging signal sequence of the helper virus is flanked on each side by loxP sequences. During infection, recombination mediated by Cre between two loxP sites causes excision of the packaging signal. Without the packaging signal the helper virus DNA can no longer be incorporated into the capsids. Because the packaging signal of the hdAdV is not flanked by the loxP sites, it is not excised. The hdAdV DNA is therefore preferentially incorporated into the capsids. A similar approach using the PER.C6 cell line and the Cre-lox system has been used for generation and propagation of hdAdV in order to further minimize the risk of homologous recombination and generation of RCAs [137]. Subsequently the yeast recombinase FLP enzyme has been used to improve the selective packaging of the hdAdV, and thus reduce the contamination by the helper virus. In this case the packaging sequence of the helper is flanked on each side by frt sites [138].

Despite significant progress in hdAdV vector design, extensive use of hdAdV for gene transfer experiments has been restrained by difficulty in scaling-up the production and controlling the contamination of hdAdV preparations by helper vector. Currently, generation of an hdAdV is initiated by transfecting complementing and recombinase-containing cells with the hdAdV plasmid, followed by helper virus (at $MOI > 1$) a short time later. After cytopathic effect is evident, the crude lysate can be collected and used to infect a large number of plates with or without further helper virus supplementation; oscillations in the ratio of helper to hdAdV on passaging are a natural feature of such systems. Replication at the first step is limited by transfection efficiency, and multiple rounds of amplification are required in order to prepare viral seed lots in quantities sufficient to produce hdAdV at large-scale. Independent research groups report that during serial passages rearrangement and homologous recombination might lead not only to RCA but also to packaging competent helper virus, reducing the efficiency of the recombinase differential packaging strategy. Therefore minimizing the number of passages is a primary condition to large-scale production of hdAdV. As with the first-generation vectors, construction of the hdAdV and helper viruses, along with conditions for the coinfections, affect the outcome of the system in terms of total productivity and final helper : hdAdV ratio [139, 140]. It is obviously desirable to keep the helper : hdAdV ratio as low as possible; with proper design and protocols, the ratio can be less than $1:10^4$. This will also obviously depend on the complementing cell line, its recombinase activity, and hence its thoroughness at processing cleavage sites [141].

4
Cultivation processes for adenovirus vectors

4.1
Introduction

The goals of a commercial adenovirus cultivation process for gene therapy or vaccine applications are 1) to maximize the virus productivity per culture volume per unit time while minimizing the total medium volume used and 2) to provide a consistent, uniform feed for the purification process within a predetermined window of harvest time.

The final choice of cultivation process depends on the total number of virus particles required for the clinical studies or for the market. For the purposes of this section and given the current productivity of adenovirus, 10^{15} virus particles per batch and below is defined as small-scale manufacturing (generally below 100 L working volume) while the cultivation of greater than 10^{15} virus particles per batch is defined as large-scale cultivation. Typically, gene therapy applications, due to the relatively small number of patients involved, will only require the development of a small-scale manufacturing process while some vaccine applications may require the development of a large-scale manufacturing process [2]. For both cases, early development and Phase I clinical supplies can be manufactured using a small scale process.

For small-scale manufacturing processes, it is convenient to use disposable culture-ware such as shake flasks or, more recently, Wave Bioreactors (Wave Biotech, NJ) for suspension cultures and roller bottles, cell factories (Nunclon from Nunc or CellSTACK from Corning) or CellCubes (Corning) for adherent culture. Most disposable culture-ware processes are scaled-up by increasing the number of vessels. The numerous advantages of disposable culture-ware for GMP production, such as avoiding the need for validated cleaning and sterilization, make them the process of choice if a manageable number can be used to produce the desired amount of virus particles. Large-scale manufacturing processes typically utilize stirred-tank bioreactors, with microcarriers for adherent culture, or suspension culture for adapted cells.

The expected virus productivity (cell-specific and volumetric) depends on the overall adenovirus vector design and the cell line used to propagate the virus. In addition to the vector design and cell line, the medium formulation, scale, and culture system result in process optimization leading to unique process parameters. In general, all adenovirus culture processes start by growing the cell line of choice to the desired cell density for infection followed by the introduction of an adenovirus stock to initiate the infection and virus production cycle. In this section, we will review process considerations for the adenovirus cultivation phase, most specifically for first-generation Ad5 vectors.

4.2
Adenovirus cultivation process

4.2.1
Medium for adenovirus cultivation

Almost all major media manufacturers have commercialized specific serum-free formulations designed and optimized for the cultivation of adenovirus vectors in 293 or PER.C6 cells (Hyclone: HyQ SFM4HEK293 and HyQ CDM4Retina; Invitrogen-Gibco: 293 SFMII, CD293 and AEM; Irvine Scientific: IS 293 and IS 293-V; JRH Biosciences Ex-Cell VPRO and Ex-Cell 293). In many cases these formulations have proven to be sufficient for small-scale manufacturing, early phase clinical studies and developmental work. For larger-scale operations or product manufacturing, a customized medium formulation will often be developed to optimize the cultivation of the adenovirus in the cultivation system.

Because of concerns over the introduction of adventitious agents such as BSE and extraneous viruses into the product, medium development efforts avoid the use of primary animal-derived raw materials (such as serum or serum albumin) and, whenever possible, secondary animal-derived raw materials (such as amino acids produced via fermentation processes utilizing animal-derived raw materials).

In general, a medium is first defined and optimized for cell growth. During the infection and virus cultivation phase, the specific consumption rate of several medium components (such as glucose and amino acids) is increased [142], and needs to be taken into account when defining the medium formulation. In addition, there are components that are known to be specifically required for adenovirus cultivation, such as arginine [143].

4.2.2
Cell growth and expansion

The goal of the cell expansion stage is to generate enough host cell mass to produce the adenovirus vector. Generally, the cell expansion process is optimized to ensure consistent and reproducible cell growth while minimizing the number of steps, vessels and manipulations.

As an example, a large-scale operation of a 1000-L suspension culture to be infected at a cell density of 1×10^6 cells mL^{-1} will require approximately a 100 000-fold expansion from a typical cell bank vial containing 1×10^7 cells. With a doubling time in the range of 24–30 hours for typical host cell lines (Sect. 3.2.2), the cell expansion process can be designed in 5–8 stages with split ratios (dilution factor per stage) from 4 to 10 and can take up to 30 days. In principle a solera process can be used for penultimate cell expansion to conserve cell banks and minimize the cell train effort per batch [144].

The final stage of cell expansion typically occurs in the cultivation vessel with the objective of producing the cell mass at the most appropriate physiological state for successful virus infection, replication and packaging.

4.2.3 The infection process

Once the cell-substrate of choice is grown to the predetermined infection cell density, the process parameters are optimized to maximize the efficiency of the virus life cycle. The first steps in the adenovirus life cycle are diffusion, cell-virus interaction/attachment and adsorption of the virus [145]. In this stage, the objective is to define infection process parameters to ensure that the *effective* multiplicity-of-infection (MOI), defined as the number of virus particles that result in a successful virus replication cycle, is reproducible from batch-to-batch, and that the kinetics of that infection are reproducible. In addition, the lowest possible MOI is desired in order to minimize the requirements of the virus stock.

For a static culture of adherent cells, the diffusion of the virus to the cell monolayer dominates the virus transport to the cells [146]. Diffusion of the virus is also the rate limiting step in the case of suspension culture [147]. In suspension cultures, the variable cell aggregation state can become important and may result in a non-uniform infection process. Medium formulations that minimize cell aggregation containing reduced Ca^{2+} levels [148] or anti-clumping agents such as dextran sulfate [149] may allow for a more reproducible infection process.

The optimum MOI is dependent on the virus seed properties and virus construct. The seed properties to consider include: the P/I ratio, the stability of the virus during storage and handling, and the stability of the virus during the infection process. The virus seed preparation process should be well characterized to ensure that the productivity and properties of the virus seed do not change from batch-to-batch. When the properties of virus seeds have large batch-to-batch variability, the cultivation process in turn may not be robust.

Determination of an optimal MOI is considered in order to ensure that the cells are infected with an adequate number of viral particles for maximum volumetric virus productivity. Because AdV infections are fairly rapid and lytic, processes are usually configured as single-round infections with MOI $\gg 1$ to ensure all cells are infected at inoculation. Although MOI > 1 needs to be established, infecting the cells with more virus than is optimally required increases the requirements for a virus seed inventory, and may not improve virus productivity. For Ad5 cultivation processes, a MOI of 10–200 virus particles/cell is typically used (Table 1), resulting in an amplification ratio of greater than 200.

An example of the effects of MOI of an Ad5 vector in PER.C6 cells is shown in Fig. 2. The percentage of cells expressing hexon protein was measured via

Table 1 Adenovirus cultivation examples

Company	Canji	Schering	Introgen	Merck	Berlex	Shenzhen SiBiono GeneTech
References	[150]	[151]	[152, 153]	[154]	[155, 156]	[5]
Serotype	Ad5	Ad5	Ad5	Ad5	Ad5	Ad5
Cell Line	293 GT	293	293	PER.C6	PER.C6	SBN-Cel (293 Subclone)
Culture Method	Microcarrier/ Perfusion	Microcarrier/ Batch & Perfusion	Batch with perfused cell diluted prior to infection	Batch	Batch	Attachment/ CelliGen packed-bed/ Perfusion
Scale	5-L	160-L	100-L	240-L	3- or 10-L	14-L
Infection Cell Density (cells mL^{-1})	5×10^6–1×10^7	1×10^6	1×10^6	$< 1 \times 10^6$	1×10^6	5×10^{10}/reactor
Medium	Serum containing	Serum containing	Serum-free	Serum-free	Serum-free	Serum containing

Table 1 (continued)

Company	Canji	Schering	Introgen	Merck	Berlex	Shenzhen SiBiono GeneTech
MOI (vp cell^{-1})	200	50–100		280	10–50	
Volumetric Productivity (vp L^{-1})		1×10^{13}	1×10^{14}	$5–10 \times 10^{13}$		2×10^{13} (2×10^{15} vp/reactor)
Per cell productivity (vp cell^{-1})	10 000–40 000		126 000	100 000	20 000	40 000
Harvest Time (hpi)	48	72	72	48	58	72

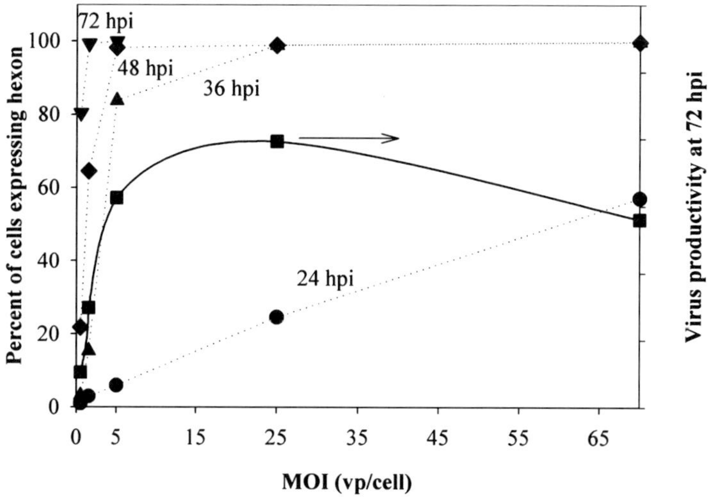

Fig. 2 An example of MOI optimization for adenovirus cultivation in a stirred-tank bioreactor (Altaras et al, unpublished data)

flow cytometry. The percentage of cells expressing hexon protein is used as a marker of successful adenovirus infection and viral protein production. An MOI > 100 vp $cell^{-1}$ (data not shown) was required for approximately all of the cells to express hexon protein within 24 hours post-infection (hpi). An MOI of 25 vp $cell^{-1}$ was sufficient for all of the cells to express hexon protein within 36 hpi. At relatively low MOIs (< 5 vp $cell^{-1}$), not all cells are infected during the first-round of infection, which is also apparent from continued cell growth (data not shown); however, a second round of infection results in all cells expressing hexon protein within 72 hpi [157].

For most Ad5 vectors studied in our laboratory at Merck, the virus productivity versus MOI reaches a plateau around 25 vp $cell^{-1}$ with no difference in virus production at MOI up to 250 vp $cell^{-1}$. However, in the example shown in Fig. 2, the maximum virus production plateau is reached around MOI 25 vp $cell^{-1}$, with a decrease in virus production as MOI is increased to greater than 25 vp $cell^{-1}$. This is an indication of the cytotoxicity of this particular Ad5 vector and emphasizes that the MOI optimization is adenovirus type and vector design specific. In addition, the measurement of virus particles is defined by the specific assay format and the specific standard used, impacting on the direct translation of MOI from one system to another.

4.2.4 The virus cultivation process

Physical parameters such temperature, pH, dissolved oxygen and partial CO_2 pressure should be studied and optimized for each adenovirus cultivation system, although some generalizations can be made. For example, most published adenovirus cultivation processes report low or no virus productivity at pHs below 7 [158, 159].

The optimum virus cultivation temperature can be different to the optimum cell growth temperature (∼ 37 °C) and should be optimized. Jardon and Garnier [158] reported significantly increased virus productivity at 35 °C compared to 37 °C for the cultivation of Ad5 in 293S cells. The temperature was also shown to affect cellular metabolism and the rate of cell death post-infection. Xie et al. [160] have reported that virus productivity in PER.C6 suspension culture could be significantly improved by growing cells at 33 °C prior to virus cultivation at 36.5 °C. Zhang et al. [152] reported an optimum virus cultivation temperature of 35 °C for attached 293 cells and an optimum virus cultivation temperature of 37 °C for suspension 293 cells.

The external pH is an important process parameter that should be optimized for each adenovirus cultivation system. Jardon and Garnier did not report any significant effect of pH (6.7–7.7) during cell growth for 293S cells [158]. However, the virus productivity was significantly impacted at pH 6.7 and 7.7, with significant virus productivity only at pH 7.2. Xie et al. reported no impact of pH on PER.C6 cell growth between pH 7.1–7.6, while a pH of 6.8 resulted in a significant lag phase and reduced cell growth rate [159]. Cell metabolism was very sensitive to pH, with increased glucose consumption and lactate production in more basic conditions. Xie et al. reported an optimum pH of 7.3 for Ad5 cultivation in PER.C6 cells, with a sharp decrease in virus productivity at a pH below 7 and above 7.6.

There are very few published works looking at the effects of other process parameters such as DO and pCO_2 on adenovirus cultivation. In most controlled processes the DO is usually set at greater than 30% of air saturation [154]. In most cases pCO_2 is not controlled and sometimes not even monitored, although Jardon and Garnier [158] report a decrease in virus productivity with increased pCO_2.

4.2.5
The virus cultivation process and harvest timing

Cell growth and adenovirus type 5 production profiles for a typical batch are shown in Fig. 3. In this example, PER.C6 cells are infected with an Ad5 vector during the exponential phase at a cell density below 1×10^6 cells mL^{-1}. After infection, the growth rate decreases and the cells reach a maximum cell density at approximately 24 hours post-infection (hpi). At the same time, an increase in specific oxygen consumption rate, as well as specific glucose consumption [159, 161], becomes apparent, presumably due to increased metabolic activity associated with DNA replication and virus protein production. This increase is observed for up to 24 hpi. Viral DNA replication occurs

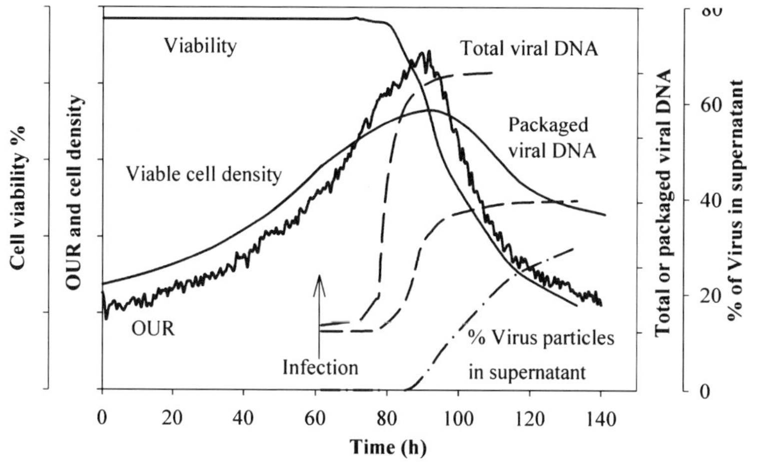

Fig. 3 Representative depiction of adenovirus cultivation in a stirred tank bioreactor

approximately between 10–24 hpi and virus assembly and DNA packaging occurs approximately between 20–48 hpi (Fig. 3).

As virus production progresses, cell viability as measured by trypan blue dye exclusion drops after 24 hpi. We have observed that this drop in viability is adenovirus serotype and transgene dependent. Typically, at 48 hpi, when the volumetric virus productivity is maximum, the cell viability as measured by trypan blue dye exclusion is around 40–80%. At this point, a percentage of the virus (usually less than 50% and as small as 10%) has already been released from lysed cells, but the rest of the virus remains intracellular. The cultivation process can proceed further with no significant increase in virus productivity, but with a significant increase in virus found in the culture medium. Harvest timing and method can be tailored to whether the entire population is desired, or only the intra- or extracellular fraction. The literature is not clear about whether there are quality differences between intra- and extracellular fractions for Ad5. Factors such as desired process reproducibility, extra-cellular virus stability and the requirements of the purification process determine the optimum harvest time, although in most published literature it is defined as approximately 48 hpi. The optimum for a given process will be dependent on AdV serotype, transgene, and also the input MOI.

4.2.6 Improving volumetric productivity beyond the batch process

The cell density at infection is a very important parameter as it impacts on the volumetric adenovirus productivity. However, in a simple batch-mode operation, several reports suggest a relatively narrow range for optimal infection cell density of between $0.5-1.0 \times 10^6$ cells mL^{-1} [142, 154, 158, 159, 162, 163]. In the example presented in Fig. 4 for Ad5 cultivation in a batch stirred tank bioreactor, per cell virus productivity remains constant up to around 0.9×10^6 cells mL^{-1}, but drops abruptly at around 1×10^6 cells mL^{-1}. As a result, the optimum cell density at infection for maximal volumetric virus productivity must be carefully controlled.

The reasons for this drop in per cell productivity at higher cell densities are not currently known, as medium formulations allow maximum cell growth up to $4-5 \times 10^6$ cells mL^{-1}. The decrease in productivity is thought to be due to a limited availability of a nutrient critical for virus growth, due to the accumulation of an inhibitory substance, or a combination of both [162]. More research is warranted to understand and overcome this phenomenon to enable simple batch processes with higher productivity.

A common approach to maintaining cell-specific productivity at increased cell density consists of infecting the cells in fresh medium (via medium exchange) instead of simply adding the virus inoculum to the medium used for cell growth. This mode of operation, which requires a cell separation step, has been frequently reported in the literature for viral vector production [164–166]. In the same experimental system as Fig. 4, the specific productivity can be maintained at a cell density of 2×10^6 cells mL^{-1} with a complete medium exchange prior to infection. In addition, cells grown to 2×10^6 cells mL^{-1} can be diluted down to below 1×10^6 cells mL^{-1} with fresh medium to maintain the specific productivity: this does not result in a significant process improvement as it is equivalent to growing the cells

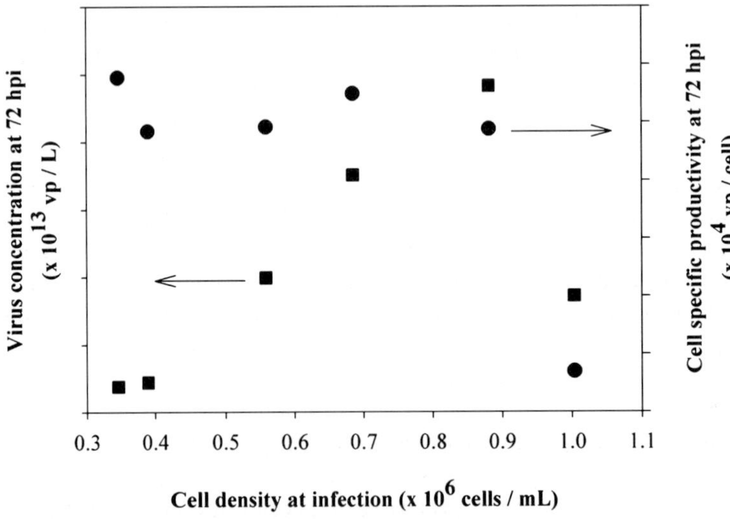

Fig. 4 The effect of cell density on volumetric and specific virus productivity in a stirred tank bioreactor batch culture (Altaras et al, unpublished data)

to 1×10^6 cells mL^{-1} in the final virus cultivation culture volume. Using this approach, Zhang et al. reported [152] a process where cells were grown by perfusion to 1×10^7 cells mL^{-1} in a 10 L working volume Wave Bioreactor and followed by dilution to 1×10^6 cells mL^{-1} for virus cultivation in a 100 L working volume Wave Bioreactor. A more useful approach, complete or partial medium exchange, can be successfully implemented to increase volumetric virus productivity and reduce the cycle time. In another example using 293 cells, Garnier et al. [167] were able to increase the infection cell density from 0.5×10^6 to 1×10^6 cells mL^{-1} with medium exchange, while preserving similar productivity. In the same system, a second medium exchange 24 hpi allowed for infection of a culture at 2×10^6 cells mL^{-1} [167]. It is possible that a further gain in productivity could be achieved with even more frequent media exchanges. The main limitation of such an approach is the implementation of multiple and ever more frequent medium exchanges. While it is relatively easy to perform aseptic batch centrifugation for medium exchange at scales of a few liters, centrifugation is not desirable at scales of hundreds to thousands of liters, and either gravity sedimentation or tangential flow microfiltration is used. At these larger scales it is more practical to develop a fed-batch or perfusion process.

Various fed-batch approaches can be used to maintain specific productivity at higher infection cell densities to increase volumetric virus productivity. For example, the control of glucose and glutamine at low concentrations allowed for an increase in the infection cell density at 3×10^6 cells mL^{-1} [168]. This feeding regime led to a reduction of lactate and ammonia accumulation and lowered increases in osmolality, but the specific virus production was still significantly reduced. Another study reported that a glutamine-controlled fed-batch culture infected at 3×10^6 cells mL^{-1} led to a 10-fold improvement in virus titer compared to a batch culture infected at 1.5×10^6 cells mL^{-1} [169]. In HEK-293 cell cultures infected at 2×10^6 cells mL^{-1}, the addition of glucose, glutamine and amino acids could not maintain the specific productivity of cultures infected at a lower cell density (1×10^6 cells mL^{-1}) [161]. Although easier to implement than sequential-batch, the success of the fed-batch process has been limited.

Perfusion processes provide a means of achieving high cell densities, through continuous feed of fresh media and dilution and removal of inhibiting by-products. Two recent papers [89, 170] demonstrate pilot-scale and industrial feasibility for perfusion in producing Ad vectors.

Adenovirus production for a perfusion culture with an ultrasonic separation device is shown in Fig. 5. In this example, the infection cell density limitation on per-cell productivity encountered in batch and fed-batch operations was overcome [170]. Cells infected at densities up to 2.5×10^6 cells mL^{-1} maintained similar specific productivities because of early feeding in the growth phase, at a rate of 2 reactor volumes per day, improving volumetric virus productivity and cycle time.

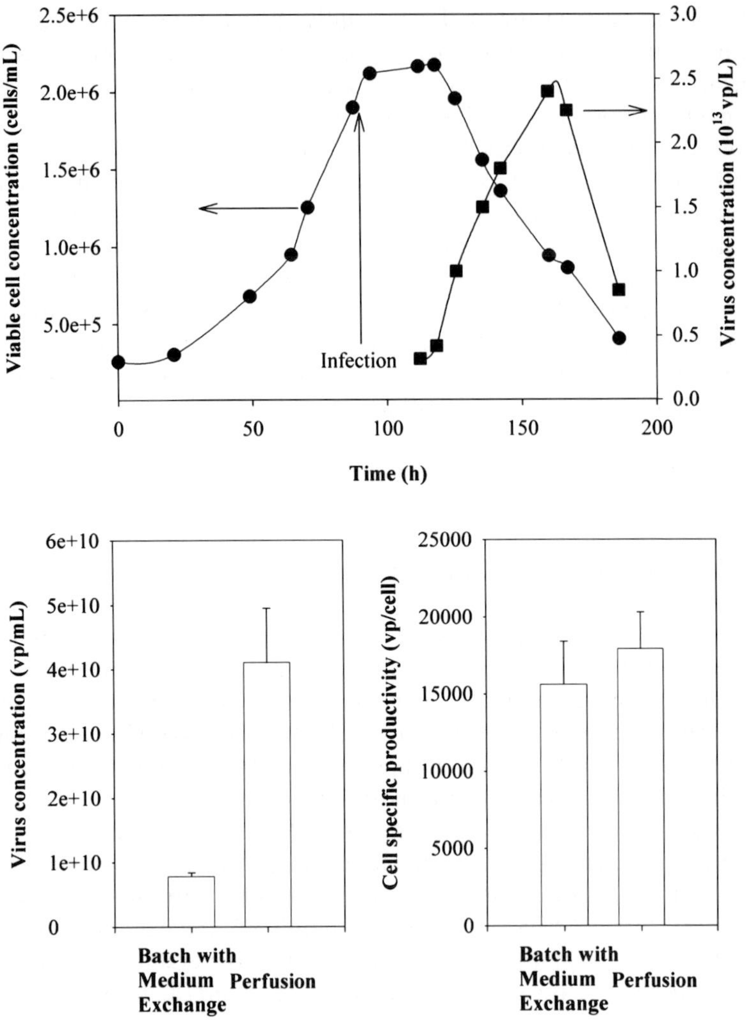

Fig. 5 Improvement in adenovirus volumetric productivity by perfusion. The volumetric and cell-specific productivity between a perfusion process is compared to a batch process with medium exchange at infection (*bottom*). Kinetics of virus concentration and cell viability for the perfusion reactor also are shown (*top*)

4.2.7
Process monitoring

Small-scale adenovirus cultivation processes implemented in disposable culture-ware are typically monitored with off-line sampling. For these processes, process parameters such as pH are not typically controlled, and process definition is required to ensure that the pH remains within a critical range during the cultivation process. To accomplish this, the optimal cell density at infection might be reduced compared to a controlled bioreactor process, or the composition of the gas phase could be changed over time. Buffer composition of the medium may also be adjusted to enable better pH control.

The cultivation of adenovirus can be facilitated by on-line monitoring methods to assure batch-to-batch reproducibility. Cell density at infection is a critical process parameter to ensure maximum and reproducible volumetric virus production. For example, if the doubling time of production cells varies by four hours, the optimum infection window can vary by up to 12 hours. Although off-line methods can be used to monitor cell density, an on-line method to trigger the infection point is very useful for large-scale stirred tank bioreactor processes. Cell density can be estimated by the oxygen uptake rate [156], measured by static or dynamic method, or via capacitance measurement [171]. These methods can also be used to monitor the infection kinetics and to trigger an event-based harvest.

4.3 Helper-dependent adenovirus cultivation

Despite significant progress in hdAdV vector design, extensive use of hdAdV for gene transfer experiments has been restrained by the difficulty of scaling-up production and controlling the contamination of hdAdV preparations by helper virus. Currently, generation of an hdAdV is initiated by transfection of complementing cells which also express recombinase with a linearized DNA plasmid. Then, 12–16 hours post-transfection (hpt), the cells are infected with a helper virus using a ratio of 30–100 total viral particles per cell. After about 40–48 hpt, the culture is harvested. HdAdV replication at this step is limited by transfection efficiency, and yields are typically low compared to infections. Therefore, multiple rounds of hdAdV amplification are required in order to prepare hdAdV viral seed lots in quantities sufficient to produce hdAdV on large scales. The crude lysate from the initial harvest can be used to infect cultures (often using further helper virus supplementation); this process is repeated until the desired scale is achieved.

Independent research groups reported that over several serial passages, rearrangements and/or recombination events lead to RCA and to packaged competent helper virus. These mutations may occur at low frequencies during the transfection stage and only become detectable over several passages. They may also occur during serial passaging when using high MOI. Although the problem can be solved somewhat genetically [140] minimizing the number of passages is a primary condition to large-scale production of hdAdV.

Simultaneous propagation of helper and hdAdV viruses in suspension culture can be performed using a packaging cell line without recombinase expression in order to generate high titer viral stocks in a minimum number of passages [137]. Subsequent hdAdV vector production is completed using

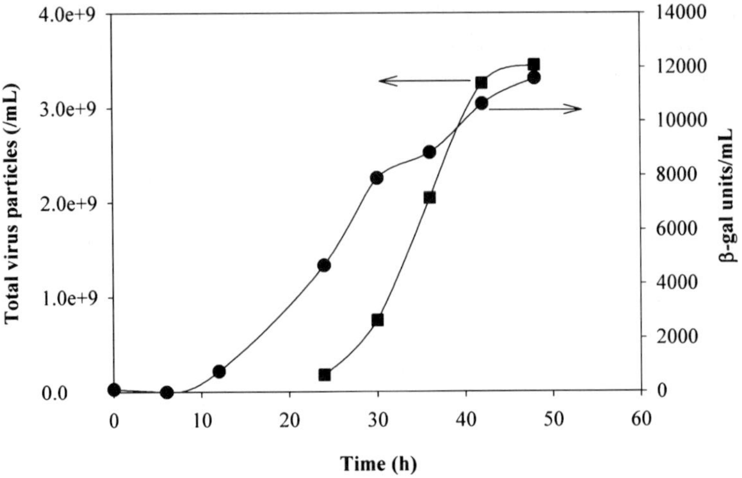

Fig. 6 An example of production of a helper-dependent adenovirus expressing β-galactosidase. Production kinetics in a 3-L bioreactor is monitored after infecting with 100 vp per cell of hdAdV and helper virus at a ratio of 1 to 1. Total viral particles are measured by HPLC. Accumulation of β-gal is an indication of hdAdV activity (Meneses-Acosta et al., unpublished results)

a packaging cell line without the recombinase function in order to selectively propagate hdAdV. Contamination of hdAdV preparation by helper virus is controlled by the recombinase efficiency.

As an example of the hdAV cultivation process, the production of hdAdV in serum-free medium with suspension-adapted 293 cells expressing the FLPe recombinase (293-FLPe) is described. Generation of the hdAdV encoding β-galactosidase [132] was initiated by transfection in 15-cm plates of 293-FLPe. 16 hours post-transfection the cells were infected with a helper virus expressing luciferase at a ratio of 50 vp per cell. Up to four passages were necessary to generate enough material to produce hdAdV-β-gal at the 3 L scale, as shown in Fig. 6. Production in the bioreactor (see Henry et al. [170] for a description of the set-up) was completed after infecting 1×10^6 cells mL^{-1} with 100 vp per cell of hdAdV and helper virus at a ratio of 1 to 1.

5 Purification of adenovirus vectors

5.1 Introduction

Goals for the purification of adenovirus are similar to any other active biological: to maximize yield and specific activity while minimizing impurities and production costs. More specifically, purification goals include the following: (1) clearance of host-cell nucleic acids; (2) clearance of host-cell proteins; (3) clearance of unassembled adenovirus proteins and assembly intermediates; (4) clearance of unpackaged viral DNA; (5) concentration of virus beyond conditions needed for formulation; (6) maintenance (or enhancement) of "specific infectivity"; and (7) clearance of transgene product (if expressed constitutively). Most of these goals will be discussed in detail later.

Early efforts in AdV purification, dating back to the late 1950s, included the use of anion exchange chromatography [172–174] and two-phase extraction [175]; however, these early efforts were hampered by relatively low yields and cumbersome infectivity assays to monitor fractionation. The development of RbCl- and CsCl-density gradient ultracentrifugation of adenovirus in 1962 [176] enabled rapid, parallel purification of adenovirus in quantities sufficient for research purposes based solely on visible band collection. Density-gradient ultracentrifugation separates empty capsids from mature virions, and deferred the technical challenge of reproducibly reducing the empty capsid content by chromatography for almost 30 years.

Scaleable purification methodologies for adenovirus were not seriously revisited until the 1990s, when interest in using adenovirus for gene therapy blossomed. Improvements in chromatographic resins made since 1960 now enable adenovirus particles to be bound and eluted with higher recoveries. In most processes described in the literature, at relevant conferences, in patents, and in patent applications, anion exchange chromatography is the "heart" of the process. This fact is illustrated in Table 2, which summarizes published processes for adenovirus purification, with unit operations listed in the order in which they typically appear in processes. Purification yield data are available in many of the cited references, with overall recoveries typically between 30% and 80%. The application of specific scaleable methodologies, and key issues encountered during AdV purification, are discussed in detail below.

First, a brief description of the classical density gradient methodology is described for those who are interested in purifying small quantities of vectors for research purposes.

5.1.1 Density gradient ultracentrifugation

In this method, infected cells are harvested prior to autolysis to maximize the cell-associated fraction. Cells are resuspended to roughly 2×10^7 cells mL^{-1} and are then generally lysed by freeze/thaw and clarified by batch centrifugation. Other methods of lysis and clarification can be substituted, usually without impact. The virus is normally purified through 2–3 rounds of ultracentrifugation, with careful collection of the virus band. Generally, a smaller centrifuge tube is employed for the second and third rounds since impurities are minimal. The product can be dialyzed or desalted into a stabilizing buffer,

Table 2 Scaleable adenovirus purification methods

Company	Canji	Schering	Bioreliance	Aventis	Introgen	Puresyn	Merck	Berlex
References	[178]	[179, 180]	[181]	[12]	[153]	[182]	[183]	[156]
Serotype	Ad5	Ad5		Ad5		Ad5	Ad5	Ad5
Virus fraction	Cells	Cells (supe unclear)	Cells + supe	Cells + supe	Cells + supe	Cells + supe	Cells + supe	Cells
Lysis method	Free/thaw	Not specified	Microfluidizer	Freeze/thaw	Detergent	Autolysis	Detergent	Shear
Clarification	Batch centrifugation and 0.8/0.2 μm membrane	0.45 μm membrane	Depth Filtration	Batch centrifugation	Depth filtration or MF, and 0.22 μm membrane	2.5/0.3 μm membrane	Depth filtration	Continuous centrifuge (Westfalia CSA-1)
Concentration	Centrifugation of cells prelysis	Not specified	Ultrafiltration	–	Ultrafiltration	Ultrafiltration	Ultrafiltration	Centrifugation of cells prelysis
Nuclease Usage	Yes (Benzonase)	Not specified	Yes	No	Yes (Benzonase)	Yes (Benzonase at end)	Optional	Not specified

Table 2 (continued)

Company	Canji	Schering	Bioreliance	Aventis	Introgen	Puresyn	Merck	Berlex
Primary Purification Step	Anion Exchange (Fractogel DEAE 650M)	Anion Exchange (Fractogel DEAE 650M)	Anion Exchange	Anion Exchange	Anion Exchange (Toyopearl SuperQ 650M)	Anion Exchange (Fractogel DEAE 650M)	Anion Exchange	Anion Exchange (Streamline Q-XL – EBA)
Polishing Step	IMAC	Gel filtration or Hydroxyapatite	Gel filtration	Anion Exchange	None	Reverse Phase (Polyflo)	Not specified	Gel filtration
Other	–	–	–		–	Ultrafiltration to remove Benzonase/ formulate	Precipitation of DNA in lysate	

and this methodology has been used to purify AdV for early-stage clinical trials in some cases. More details describing this methodology are available elsewhere [10, 176, 177]. Equipment is available for the scale-up of ultracentrifugation, with some large-scale continuous-flow ultracentrifuges containing bowls with a capacity of several liters. Although it is theoretically possible to band approximately 10^{15} vp in a single run, we are not aware of any attempts to use this methodology for large-scale manufacture. The cost and toxicity of cesium chloride, as well as concern over the potential to generate infectious aerosols, are likely the main reasons for this technology not being adopted.

5.2 Purification unit operations

5.2.1 Choice of harvest time and virus fraction

Since adenovirus is a lytic virus, and it is impossible to perfectly synchronize the infection and replication process, the virus is split between the cell-associated and supernatant fractions (see Sect. 4.2.5). Therefore, an important process development decision is the selection of the fraction(s) to purify. Although some manufacturers have claimed that the "supernatant virus" is of a lower quality, this has not been substantiated in the literature and may be related to culture conditions. That is, operational specifics (including but not limited to media composition, dissolved oxygen level and pH, and their respective control strategies) may influence the extent of proteolytic clipping, deamidation, oxidative damage, and/or shear-induced damage to virus particles released from the cell. In fact, out of the scaleable processes described in the literature, patents, or patent applications, only two use the cell-associated fraction alone, while six use both cell-associated and supernatant fractions (see Table 2). This suggests that, in most circumstances, virus released to the supernatant has been found to be of comparable quality.

Harvesting only the cell-associated fraction may still be desirable in some circumstances. First, because mammalian cells are cultured at relatively low densities, centrifugation can easily be used to concentrate the batch approximately 20-fold [35, 178]. This concentration reduces the processing volumes for subsequent operations substantially, and can facilitate a freeze-hold if desired. In addition, if the culture medium contains protein hydrolysate or serum, the bulk of these impurities can be eliminated.

5.2.2 Cell lysis

Classical purification procedures for adenovirus use repeated freeze/thaw of the cells. While convenient for small volumes, freeze/thaw is not scaleable for both logistical and technical reasons. Logistically, both freezing and thawing volumes over a liter tends to be slow and cumbersome. For GMP production, both container closure and the reagents used in thaw or freeze baths must be carefully evaluated. Technically, large volume freezing has been shown to lead to cryoconcentration and aggregation for proteins [184]. Similar behavior is likely for adenoviruses, especially in crude solutions, due to their tendency to aggregate and associate with nucleic acids [35].

Other lysis methods, including autolysis, detergent lysis, microfluidization, and shear have all been successfully adopted. Autolysis, the lysis of the cells by the adenovirus, requires no extra equipment and does not introduce additional reagents requiring clearance monitoring. The kinetics of autolysis are likely to be specific to the adenovirus construct and serotype under evaluation, as well as the culture conditions. In one study using an E1/E4-deleted Ad5, 2–3 additional days of culture were required for complete autolysis [182]. Of particular relevance to the kinetics may be the presence or absence of the E3 region in the construct. The "adenovirus death protein" (ADP), an E3-encoded 11.6 kDa protein, accelerates cell lysis substantially. Initiation of lysis occurs within 2–3 days for ADP+ vectors compared to 5–6 days for ADP-mutants [185]. The extension of time in the bioreactor when using autolysis may compromise facility productivity, increase the risk of contamination, and potentially lead to product degradation. These risks, and the potential productivity loss, must be evaluated against the benefits.

Mammalian cells are sensitive to shear, which make shear-based lysis techniques, including microfluidization, relatively simple. Complete virus release in a microfluidizer requires relatively low pressures ($\sim$1000 psi [181]); however, pressures above 1000 psi appear to result in a pressure-dependent reduction in viral concentration by anion exchange HPLC, with less than 20% recovery at 5000 psi [186]. One might speculate that the loss of fiber protein, or the penton (the complex of fiber and penton base proteins) would be caused by shear since the fibers extend outward from the viral capsid; however, this has not been conclusively demonstrated as the cause of particle loss. Shear can also be used as a lysis method in combination with other unit operations. Shear in tangential flow filtration systems is generated by flow through the membranes, valves, and pumps. Consequently, these systems can be used to couple lysis with either concentration (with UF membranes less than about 500 kDa) or clarification (with microfiltration membranes exceeding 0.2 micron) [187]. In addition, Monica et al. [156] described a process in which cells collected by continuous centrifugation are lysed during ejection from the bowl. Challenges with these approaches exist in the need to scale-up shear processes in complex systems.

Detergent lysis is a simple and robust alternative. Nonionic detergents (such as Tween, Triton, Brij) are commonly used to solubilize membranes and, as a result, release adenovirus from cells with varying kinetics and efficiency. Triton X-100 has previously been shown to be effective for releasing hepatitis A virus at a concentration of 0.1% [188]. For adenovirus release from PER.C6, we have obtained release comparable to or exceeding other methods using Triton X-100. Polysorbate-80 (Tween-80) also has been shown to be effective at a concentration of 1% [153].

5.2.3 Clarification

Centrifugation

Batch centrifugation is a convenient laboratory technique for clarifying small (< 1 L) volumes of lysate. It is most useful in basic research studies when parallel purification of multiple constructs is normal. In this case, the differential settling rates between the cell debris and AdV particles are utilized. The sedimentation coefficient for AdV is 4.42×10^{-5} mm g – min^{-1} [23], which means that at 10 000 x g, AdV migrates 0.4 mm min^{-1}. Pelleting of virus, as shown by Huyghe et al. [178], must be considered but is only a concern when the path length for centrifugation is short (such as 500 μL in a microfuge tube). For larger volumes, batch centrifugation is logistically impractical. In that case, continuous centrifugation can be considered. It is particularly important to avoid the generation of infectious aerosols when scaling from closed batch processes to continuous centrifuges. The challenges associated with managing exposure risk often are sufficient to eliminate centrifugation as an option but, as demonstrated by Monica et al. [156], can be overcome.

Depth filtration

For intermediate scales of purification from cell culture processes (1–2000 L of culture), depth filtration represents a simple and cost-effective operation for clarification [189, 190]. Depth filters are available from a number of manufacturers with varying characteristics. In general, they are cellulose fiber beds with a resin used for structural integrity. The resin is often positively charged, improving the adsorption of cellular debris. Filtration aids, such as diatomaceous earth and perlite, are often added to further modify the filter properties and adsorption potential. Excellent clarities and yields can be achieved with very short cycle times [183]. Because depth filters are inexpensive, disposal after a single use is acceptable. At very large scales, depth filtration becomes less preferable as a primary clarification method for logistical reasons.

Microfiltration

Microfiltration is used by itself, or as a polishing clarification method, in half of the processes shown in Table 2. Microfilters can be operated in either dead-end or tangential-flow modes. Tangential flow filtration is preferred when microfiltration is to be used as the primary clarification method. This methodology may be particularly well-suited to clarification of viral seeds because it can be run aseptically [187]. However, because adenovirus is quite large (approximately 0.1 micron), moderately large pore sizes are necessary. In addition, cell debris may foul the microfilter, resulting in large yield losses. Finally, microfilters are significantly more expensive than depth filters.

5.2.4 Concentration and buffer exchange

The production of adenovirus in cell culture results in a dilute stream in terms of both product concentration and cell mass. In order to minimize facility size, volume reduction early in purification is desirable. Volume reduction can be achieved by several means, including the following: (1) centrifuging and resuspending infected cells; (2) ultrafiltration; (3) precipitation or extraction of the virus; or (4) virus adsorption on a "capture" chromatography step.

Centrifugation and resuspension of cells was discussed earlier. Mammalian cells pellet easily, and centrifugation is quite convenient at laboratory-scale. The two main drawbacks of this approach are the loss of virus which is not cell-associated and, as mentioned in the clarification section, biosafety considerations upon scale-up.

For capture chromatography, adsorption onto an anion exchange resin is used. Either a packed or expanded bed can be used. This approach requires a trade-off between long processing times for column loading and low binding capacity due to resin particle size (described in Chromatography section below). The use of this option seems to require at least one additional chromatography step to achieve acceptable purity.

Adenovirus precipitation is not included in any of the processes summarized in Table 1, but has been described in the literature using organic flocculation, polyethylene glycol, and ammonium sulfate. Organic flocculation, which is frequently used to concentrate viruses from environmental samples, utilizes a shift to low pH to cause association between virus and denatured protein in flocs which then can be centrifuged or filtered [191, 192]. Though this technique has been applied to Ad40 [193], the aggressive conditions and the need to reverse the flocculation for further purification represent significant challenges.

Polyethylene glycol (PEG) has also been used to precipitate viruses, including hepatitis A [194], Ad40 [193], and Ad5 [195, 196]. Precipitation is believed to occur by increasing the concentration of virus in the volume excluded by polymer. In the Ad40 study, 40% virus recovery was achieved from the pellet following addition of 7% PEG-8000 in the presence of 0.5 M NaCl. The low recoveries in this study were likely due to the dilute conditions (10^5 $TCID_{50}$ mL^{-1}) and nonoptimized precipitation conditions. A more recent study achieved 84% recovery of Ad5 infectivity from the culture supernatant using 10% PEG-6000 and 1.25 M NaCl [196]. PEG precipitation from a cell lysate has not been demonstrated, but would represent an interesting process option if shown to yield a moderate purification factor in addition to concentrating the product.

Ammonium sulfate precipitation utilizes hydrophobic interactions between proteins following partial dehydration. Nearly quantitative recoveries of adenovirus infectivity from culture supernatant have been achieved with incubation at 40% $(NH_4)_2SO_4$ saturation followed by precipitate recovery with centrifugation [196]. Again, successful implementation of this approach in cell lysates and demonstration of moderate purification factor could make such a step appealing for large-scale purification.

Aqueous two-phase extraction has also been used to concentrate viruses. These systems are generated using specific polymer mixtures and the properties of the phases can be modified by the addition of salts. Ad2 has been shown to partition into the bottom phase of a 0.40% dextran sulfate, 0.48% methylcellulose, 0.15 M NaCl system with a concentration ratio of nearly 1000; less encouraging results were obtained with dextran sulfate:PEG systems [174]. There are no recent reports of the use of this technology for adenovirus.

The most common method for concentrating adenovirus and/or exchanging buffers at moderate to large scales is ultrafiltration. Ultrafiltration membranes are anisotropic, with a thin membrane with a controlled pore size distribution generally layered on top of a more rigid support with wide pores. Nominal pore sizes range from about 1 kDa to about 1000 kDa, with the largest sizes overlapping with small microfiltration membranes. Selection of the largest possible membrane that fully retains Ad is important when maximizing flux and impurity clearance. For most membranes and streams, the optimal size will be between 100 and 1000 kDa [153, 182]. At this size, selective retention of virus and permeation of impurity protein is possible. In fact, the large difference in size between Ad and typical host cell proteins makes ultrafiltration unusually powerful. For example, a single ultrafiltration can clear over two logs of impurity protein [183].

Ultrafiltration at moderate to large scales is usually operated in tangential-flow mode. In this mode, the flow across the membrane reduces the concentration polarization that occurs as a result of permeation through the membrane. Particular attention must be paid to minimizing this concentration layer, since adenovirus is prone to aggregation (Sect. 5.3.3). Buffer exchange is accomplished by simultaneously feeding buffer into the retentate tank while permeating through the ultrafiltration membrane. Greater than 99% exchange of small molecules is obtained by diafiltering with five batch volumes of buffer; this is significantly less buffer than required using desalting (size exclusion) chromatography or dialysis.

5.2.5 Chromatographic methods

Anion exchange chromatography

As mentioned earlier, the hexon protein comprises about 50% of the total virion protein and almost all of the exposed capsid. As a result, the hexon properties are believed to be critical for ion exchange behavior. At pH 7, the Ad5 hexon has a theoretical net charge of – 23.8 per chain, or – 17 136 per virion (calculated using Vector NTI from Ad5 hexon sequence, NCBI accession number AAO24084). Since macromolecules with multiple interaction sites tend to exhibit enhanced binding strength, adenovirus affinity for positively charged anion exchange resins exceeds most host cell proteins as well as individual hexon capsomeres (Fig. 1). These properties make anion exchange the primary purification step in all scaleable AdV purification processes. In fact, some processes rely almost entirely on the AEX step to achieve acceptable purities [12, 153].

The work of Shabram et al. [13] showed that, for Source 15Q resin, purified penton and fiber proteins do not bind to the resin when equilibrated with 0.3 M NaCl. Hexon capsomers do bind, and elute at approximately 0.33 M NaCl, about 0.12 M less than the full viral particles. Undigested DNA has a higher charge density than adenovirus, and consequently binds more tightly to anion exchange resins. RNA coelutes on most resins (Q Sepharose XL being the sole published exception [12]) and therefore must be cleared by nuclease treatment or an alternative separation method.

Shabram et al. also analyzed the "light" band from an initial CsCl density gradient separation by anion exchange chromatography. The "light" band, which has a density of about 1.31 g mL^{-1}, is generally agreed to be comprised primarily of empty capsids. A large nonbound peak was seen, as well as peaks which coelute with the purified hexon and AdV particles [13]. The authors suggest that the presence of an AdV particle peak is consistent with imperfect fractionation during the density gradient separation, and that empty capsids likely coelute with hexon protein. The former point was supported by $TCID_{50}$ analysis of the sample, which showed infectivity in the sample, albeit at a high P/I ratio (> 200 vp IU^{-1}). Others have shown that empty capsids coelute with the full particles during anion exchange separations. For example, Vellekamp et al. [37] demonstrated the presence of empty capsids in a preparation purified by anion exchange chromatography (DEAE-Fractogel) followed by gel-filtration chromatography (Superdex-200). An assay that utilized the presence of pVIII (the precursor to Ad protein VIII, that is not present in mature virions) was used to demonstrate that the resolution between empty capsids and full particles on preparative anion exchange was minimal. Slightly better separation was obtained in another preparation using Source 15Q resin [29]. As the capsid of empty particles still contains hexon, penton base, IIIa, and probably fiber [37, 197], it is logical that chromatographic separation by anion exchange would be challenging.

Though limited data is available, published loadings for anion exchange resins range from about 0.5–5 × 10^{12} viral particles (vp) per mL of resin [12, 35, 178, 182, 183]. Corresponding to 0.14–1.4 mg of virus per mL of resin, these loadings are extremely low when compared to typical proteins but are similar to plasmid DNA [198]. Inverse size exclusion chromatography using dextran standards can be used to explain these results. Sepharose resins have no pore volume accessible to solutes with radii greater than 32 nm. Less than 6% of TosoHaas 650 M pore area is accessible to solute with 32 nm radius; less than 1% is accessible to a solute with 48 nm radius [199]. Therefore, Ad particles with a capsid radius of about 40 nm are likely to be nearly fully excluded from these resins and bind primarily on to the outer surface of the particles. Since the surface area-to-volume ratio of a sphere is inversely proportional to the diameter, the highest capacity resins of this type are likely to be those with the smallest particle size. Capacity must be balanced with pressure drop, however, which scales with the square of the diameter. This explains why the smallest resin used is Source 15Q (Amersham), with a diameter of 15 micron [35, 153, 183].

Gel filtration

Separation of molecules by size is possible by gel filtration, which utilizes the fact that varying accessibility to pores results in a difference in effective column volume and therefore a difference in retention. Because the product is not actually bound to the resin, it is extremely gentle and therefore often attempted for viral purifications. The fact that neither the product nor impurities are bound, however, limits the practical loading to about 5% of the column volume.

Gel filtration can be used for polishing purification of adenovirus, with varying degrees of success reported in the literature. Huyghe et al. reported only 15–20% recovery for Toyopearl HW-75F, and reported that the elution of adenovirus was delayed and broad [178]. This observation is consistent with adenovirus being trapped in pores. The manufacturer reports this resin to have pores in excess of 100 nm and a dextran exclusion limit of > 10 000 kDa, exceeding the hydrodynamic radius of adenovirus. Operation in "size exclusion" mode, where the adenovirus is excluded by the pores, is feasible and has been successfully utilized at large scales. Vellekamp et al. [37] report the successful use of Superdex-200 (Amersham Biosciences), which has a dextran exclusion limit of 100 kDa. In addition, buffer exchange of adenovirus using PD-10 desalting columns (which utilize Sephadex G-25 resin with a dextran exclusion limit of 5 kDa) results in quantitative recovery (Konz, unpublished data).

Immobilized metal affinity chromatography

Protein binding to IMAC columns is predominantly through the interaction of histidine residues with chelated metal ions, allowing for selective retention of proteins with histidine clusters. Huyghe et al. have reported the use of immobilized zinc affinity chromatography as a polishing step for Ad5 purification [178]. Adenovirus was bound to the column, washed, and eluted with a gradient from 0 to 500 mM glycine. Recoveries of 49–65% were reported and it improved the protein purity as demonstrated by SDS-PAGE.

Reversed phase chromatography

For adenovirus, reversed phase chromatography is most often associated with disruptive proteome analysis aimed at identifying the quantity of structural proteins and impurities (see Sect. 2.4). Since adenovirus is not held together by covalent interactions, it is unlikely that a bind/elute step using reversed phase chromatography could be developed that has a high yield and does not detrimentally impact product infectivity. Green et al., however, have demonstrated that reversed phase chromatography with an ion pairing agent can be used in flowthrough mode following anion exchange chromatography [182]. Their work examined impurity adsorption onto a polymeric resin (Polyflo from Puresyn, Malvern, PA). In this study, 10% isopropanol and 2 mM tetrabutylammonium phosphate were used in the loading buffer to facilitate selective binding of free viral and host cell proteins without retaining a significant fraction of the virus. At the published virus loading (5×10^{12} vp per mL of resin), the cost of resin, solvent, and solvent disposal should be considered if the ultimate goal is an economic large-scale process. The utility of a reversed phase adsorption will be particularly high in circumstances where adequate purity cannot be achieved by other means.

Other chromatography

The use of hydrophobic interaction chromatography for adenovirus purification also has been described [178]. Reported yields were between 5 and 30% for steps that utilize the bind-and-elute mode. As seen for reversed phase chromatography, it may be possible to obtain higher yields in flow-through mode. Calcium phosphate resins, including Brushite and hydroxyapatite, have also been used historically for the purification of viruses including adenovirus [200]. These "mixed-mode" resins containing positive and negative charge often display unique selectivities. Hydroxyapatite can be used to bind and elute Ad5 with yields in excess of 70% [35].

5.3 Other purification considerations

5.3.1 DNA Clearance

Clearance of host cell DNA is of particular interest to those interested in manufacturing adenovirus for clinical use, since host cell DNA contributes to the theoretical potential for product tumorigenicity (see Sect. 2.4). As a result, some manufacturers have chosen to set rigorous targets (such as 100 pg per dose) based on earlier recommendations [43] and a desire to minimize risk [35]. Assuming a production of 10^4–10^5 Ad5 particles per cell [162], a genome of 10–15 pg [201], and a dose of 10^{11} vp, 5–6 logs of DNA clearance are required to achieve the target of 100 pg per dose. As a result, multiple orthogonal DNA clearance methods are needed. Methods described earlier, including anion exchange, can be effective at reducing DNA levels. Additional operations specifically aimed at reducing nucleic acids are often incorporated. Two options, nuclease treatment and DNA precipitation, are described below.

Nuclease treatment

Treatment with nucleases is the most common orthogonal DNA clearance method employed (see Table 2). Several commercially available nucleases are now made using recombinant techniques minimizing contamination concerns. For AdV purification, Benzonase (EM Science) is often chosen since it digests both DNA and RNA and is stable and active in a variety of solutions. The simplicity and rapidity of nuclease treatment make it appealing for AdV purification for both research and clinical uses. Manufacturers planning for eventual licensure need to be aware of the cost of these nucleases, as their cost can become a substantial fraction of the variable costs when used at high concentrations. This cost may be more significant when the product is intended for prophylactic vaccination (such as an HIV vaccine).

DNA Precipitation

An alternative method of DNA removal is precipitation. Selective precipitation of host cell DNA from Ad5 has been demonstrated using cationic detergents [183]. Cationic detergents bind to the phosphate backbone of the DNA, which results in charge neutralization. The hydrophobic detergent tails then interact to form precipitates. Remarkably, despite the fact that Ad5 is highly negatively charged, it is possible to precipitate most of the DNA without significant Ad5 precipitation. This is demonstrated in Fig. 7. When cell lysate is precipitated, the clarification can remove both the precipitated DNA and cell debris. When coupled with depth filtration, up to three logs of DNA

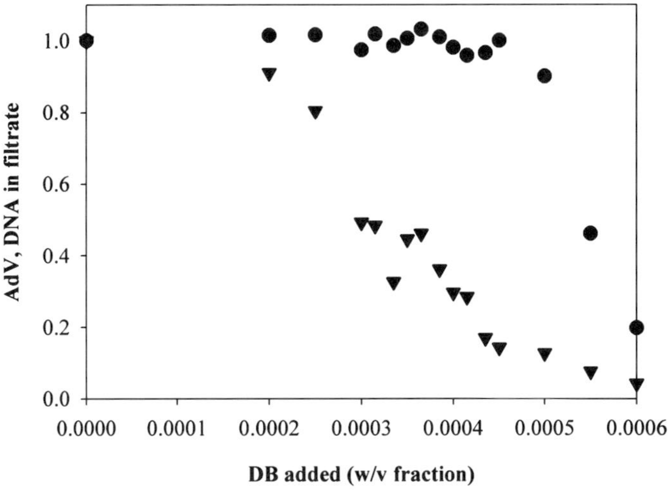

Fig. 7 Selective precipitation of DNA from Ad5 using Domiphen Bromide. Normalized adenovirus concentration (*circles*) and DNA concentration (*triangles*) following a 0.45 micron filtration are shown. Adapted from [183]

clearance can be achieved. In addition to improving the DNA clearance profiles, the performance of subsequent steps is improved and nuclease usage can be reduced or eliminated [183]. A moderate effort, however, may be required to develop appropriate precipitation/clarification conditions for a particular culture condition and scale. As a result, incorporating DNA precipitation into the process is most prudent when process economics and/or extremely high purity are priorities.

5.3.2
Empty capsid clearance

Selective empty capsid removal is one of the most significant AdV purification challenges. As described earlier, neither the size nor protein composition of the capsid is significantly different for empty and full particles. As a result, the development of methodologies to clear empties below a reproducible upper limit is a critical aspect of process development. Unfortunately, there is little data demonstrating these efforts. Clearance in anion exchange chromatography, which was discussed earlier, is possible but challenging. Vellekamp et al. [37] state in their discussion that immobilized metal affinity chromatography and hydrophobic interaction chromatography are better methods for empties clearance, but still only reduce empty capsid levels by 50–80%. The limited purification factors derived from these steps suggest that minimizing the input into the purification process through careful control of culture is critical.

5.3.3
Aggregation

Despite their aggregate negativity, Ad5 capsids are capable of self-association, forming paracrystalline arrays in the nucleus during replication and assembly [202], and forming aggregates in solution. Self-association is presumably mediated by hydrophobic interactions. For purified adenovirus, aggregation is a function of virus concentration, temperature, pH, excipient concentration, and storage container [39, 40, 203, 204]. A recent report demonstrates that Ad5 also is susceptible to aggregation throughout the purification process, and demonstrated that association of host cell nucleic acids with aggregates limits the DNA clearance potential of the process [35]. Viral aggregation is most likely to be an issue when the virus is at high concentrations, for example following density-gradient ultracentrifugation or during elution from a chromatography column. Various approaches can be employed to address this concern, including operating at low Ad concentrations or including excipients aimed at minimizing aggregation.

5.3.4
Viral clearance

If use in clinical studies and eventual product licensure is intended, the potential for contamination with adventitious agents, including viruses and prions, must be carefully addressed. Depending on the cell line, its history and safety testing, and the clinical application, evaluation of the clearance of model or specific agents in the purification process may be required [47]. Given the properties of AdV and the need to maintain infectivity, common viral clearance options may not be available. For example, nanofiltration is now commonly used for both virus and prion clearance for monoclonal antibodies and plasma proteins [205, 206] but it obviously cannot be used during adenovirus purification since the product is approximately four times larger than the parvoviruses and picornaviruses these steps are designed to remove. Therefore, clearance studies must focus on the use of selective inactivation or fractionation (for example with chromatography). Because these methodologies may be less robust than alternatives such as nanofiltration, purification process design and clearance studies must be carefully integrated with a long-term plan for release testing of the cell banks, virus seeds, and production lots in order to ensure that the overall risk assessment for the product is favorable given the intended application.

5.3.5
Helper-dependent systems

The propagation of "gutless" or helper-dependent adenoviruses (hdAdV) requires coinfection with helper adenovirus during the propagation stage (see

Sects. 3.3 and 4.3). As a result, helper virus is an impurity in the feed to the purification process. The separation of helper virus from hdAdV is possible by batch density gradient ultracentrifugation, due to differences in the density caused by differing genome sizes. We are unaware of any scaleable methodologies to achieve this separation, other than continuous ultracentrifugation (see Sect. 5.1.1) which has not been demonstrated. Further research in this area is warranted.

5.3.6 Serotype-specific considerations

Although the vast majority of clinical applications for adenovirus vectors have utilized Ad2 and Ad5, there is increasing interest in the use of alternative serotypes, especially Ad35 [96, 207]. This interest in new serotypes is driven by a desire for vectors with alternative tropism [208–210] as well as a desire to circumvent pre-existing anti-adenovirus immunity [211]. An optimistic view of the potential of adenovirus vectors to act as vaccine vectors and therapeutic agents would require the industry to develop multiple serotypes to minimize overlap in cases where a patient may receive AdV vectors for multiple purposes (for example in a vaccine and, later in life, a cancer therapeutic).

Fortunately, purification of alternative adenovirus serotypes for research purposes is readily accomplished using density-gradient ultracentrifugation. Scaleable processes, like those described in Table 2, must be developed as these new vectors proceed into clinical development. Serotype-specific impacts on purification will arise from a limited number of factors including the assembly efficiency in cell culture, the physical properties of the virion (capsid charge, capsid hydrophobicity, fiber length, fiber knob composition), and the stability to dissociation and/or aggregation.

The physical properties of the virion are most easily analyzed and applied to our understanding of purification behavior. Perhaps the most straightforward example is the charge on the capsid, which is dominated by the hexon protein. A survey of available hexon sequences suggests that the net charge at pH 7 can vary almost three-fold from –10 (Ad40, SwissProt accession number P11819) to –28 (Ad2, NCBI accession number AAP31203). Differences in hexon charge appear to impact ion exchange behavior, with a positive correlation between the net charge and the sodium chloride content required to elute from anion exchange resin [12]. Charge may also affect the selectivity of precipitation reactions and the susceptibility of the virus to aggregation.

In addition, the length of the fiber protein is also highly variable amongst different adenovirus serotypes. Known human serotypes have fibers ranging from 9 nm to 37 nm, with the difference being attributed to a varying number of repeat units in the shaft domain [212–214]. With a capsid diameter of about 80 nm, the diameter of adenovirus particles from fiber knob to fiber knob varies almost two-fold. As a result, it is likely that properties that are impacted by molecular size will vary by serotype. These include sensitivity to shear, capacity on chromatographic resins, and yield through filtration processes.

In addition, Ad5 binding to immobilized metal affinity resins is believed to be mediated by histidine residues in the fiber protein [186]. Therefore, the conservation of these residues and their three-dimensional proximity may significantly impact the utility of that method for other serotypes.

Other properties, such as the hydrophobicity of the capsids, will also impact on purification, but are harder to predict a priori in the absence of serotype-specific three-dimensional structures of key capsid proteins.

6 Development of stable liquid adenovirus formulations

6.1 Introduction

To fully exploit the value of adenoviruses as a platform for gene therapy and vaccine applications it is essential that stable formulations be developed. Historically, lyophilization has been the most successful approach used for the development of stable live virus formulations. But when compared to refrigerated liquids they are at a clear disadvantage in terms of cost, complexity and ease of administration. Despite the drawbacks of lyophilization it is usually required for long-term storage of live viruses. Frozen storage of liquid formulations is one alternative to lyophilization; however, this approach significantly increases the difficulty and cost of maintaining an effective cold chain during distribution. For example, the campaign to eradicate polio is currently being impeded by lack of adequate cold storage for the oral polio vaccine [215]. Although liquid formulations intended for 2–8 °C storage are clearly superior to these alternatives, until recently it was assumed that adenovirus formulations would require lyophilization to achieve long-term 2–8 °C stability [216, 217]. However, much progress has been made toward the development of stable liquid formulations for Ad5 in the last few years. In fact, liquid formulations now exist that are stable during long-term storage at 2–8 °C. A better understanding of inactivation mechanisms and how various formulation parameters affect the stability of Ad5 has been obtained in the course of devising these formulations.

To capitalize on the potential of recombinant adenovirus vectors for international use, formulations should preferably meet several practical acceptance criteria: (1) they should be liquid formulations with an 18–36 month shelf-life when stored at 2–8 °C or above, (2) they should have an acceptable stability profile at ambient temperatures to cover short-term storage and handling in the field, without refrigeration, (3) they should not be adversely affected by freeze-thaw cycling, and (4) they should be compatible with parenteral administration in terms of tonicity and pharmaceutical acceptability of excipients. Although several groups have reported the development of liquid formulations of Ad5, most of them do not meet these criteria.

The development of liquid Ad5 formulations suitable for long-term storage at 2–8 °C is reviewed in detail below, focusing on formulations with ≥ 6 months of published 2–8 °C stability data. The approaches used and the effects of formulation parameters on Ad5 stability are also discussed. Finally, the suitability of the published Ad5 formulations, with regard to the above acceptance criteria, is considered.

6.2 Pharmaceutical approaches for identifying stable adenovirus formulations

Early approaches to the stabilization of live viruses involved the use of high temperature thermal inactivation/denaturation studies [218–220]. The objective of these studies was to better understand conditions for heat disinfection and persistence of virus in the environment, rather than the development of stable formulations [221]. However, an underlying assumption was that excipients and formulation parameters such as pH that increased the thermal melting temperature, or extended the time required for inactivation, would also increase the stability of the virus at lower temperatures. In fact, this assumption was validated for live poliovirus with the identification of $MgCl_2$ as an effective stabilizer [222, 223]. Adenovirus stability studies conducted from 1962–1992 were based on this approach. However, there are no reported correlations between the thermal inactivation time or temperature and Ad5 storage stability. It has been observed that the thermal melting temperature and inactivation rate at the onset of capsid melting (47 °C) do not correlate with either long-term or short-term accelerated storage stability data for Ad5 (Evans et al., unpublished data). These data suggest that the lack of correlation is caused by different inactivation mechanisms predominating during storage and thermal inactivation conditions. Therefore, inhibition of the major inactivation mechanism under high temperature thermal stress has little effect on stability at lower temperatures. The hypothesis that different virus inactivation pathways can dominate at different temperatures, in the same formulation, has been verified by studies conducted with poliovirus [224].

From a historical perspective, the most successful approach for the development of stable live virus formulations has been the use of empirically based studies designed to compare the stability of many candidate formulations in long-term stability studies [225]. The focus on empirical strategies is due to the fact that live viruses, as complex macromolecular assemblies, are susceptible to many different inactivation mechanisms. The existence of multiple inactivation pathways usually makes a formulation development approach based on rational design too complex to have a reasonable probability of success. Moreover, it limits the usefulness of an Arrhenius plot of accelerated stability data to predict stability at lower temperatures. Therefore, screening of candidate formulations must usually be based on real-time stability data at 2–8 °C instead of accelerated data.

Systematic testing and evaluation of Ad5 formulations to determine their storage stability at 2–8 °C began only about six years ago [226]. The results of earlier work with other live virus vaccines probably explain the lack of effort applied to develop stable liquid adenovirus formulations prior to 1998. Since no marketed live virus vaccine is stable in a liquid formulation at 2–8 °C it has been assumed that Ad5 formulations would require lyophilization or the use of high concentrations of protein stabilizers to achieve long-term stability at 2–8 °C [217, 227]. Since 1998 several groups have worked to develop stable liquid formulations of Ad5 and have determined Ad5 stability at multiple temperatures (Table 3). However, until very recently there were no reports on whether stability data obeyed the Arrhenius relationship. It is now clear that in some formulations Ad5 storage stability data are consistent with the Arrhenius equation (from 4 °C to 37 °C), so short-term accelerated stability data are useful for predicting the rate of infectivity loss during long-term storage at 2–8 °C [228].

A more rational approach to live virus formulation development is best exemplified by studies with poliovirus. In a classic series of studies low molecular weight compounds were identified by their ability to bind to and stabilize the poliovirus capsid, based on its three-dimensional structure as determined by X-ray crystallography [229]. Although these compounds enhance the thermal stability of the virus through direct binding to capsid proteins, their ability to stabilize the capsid did not correlate with loss of poliovirus infectivity [230]. Later studies confirmed that the major inactivation pathway at high temperatures involved capsid protein denaturation, while the predominant inactivation pathway at lower temperatures was degradation of the RNA genome [224, 231]. Therefore, notwithstanding the sophistication of these studies to characterize the virus inactivation pathways at high temperatures and then rationally design compounds to inhibit the thermal inactivation mechanism, a more pragmatic strategy based on screening of excipients was ultimately successful at stabilizing poliovirus infectivity.

An alternative approach for developing live virus formulations is based on the combination of rational selection and empirical screening. Using this approach, the rationale for selection of candidate formulations for subsequent screening is based on the use of excipients commonly used to stabilize proteins, DNA and viral capsids (such as cryoprotectants, nonionic surfactants, divalent cations, inhibitors of free radical oxidation). The approach involves the identification of the major inactivation pathways occurring during storage coupled with screening of candidate formulations to identify specific excipients to inhibit these pathways. Recent results using this approach suggest that the major mechanisms of Ad5 inactivation are oxidation by free radicals, adsorption and freeze/thaw damage [228]. Screening of candidate formulations was used to identify the most effective type and concentration of oxidation inhibitors and to optimize the formulation composition in terms of pH. The results clearly indicate the involvement of multiple inactivation mechanisms which require the combination of several different excipients to achieve the desired level of storage stability. Fortunately, there is apparently only one predominant inactivation pathway (oxidation) for Ad5 in formulations containing polysorbate-80 and sucrose to control adsorption and freeze/thaw damage, respectively. Therefore, Arrhenius plots of accelerated stability data were useful for predicting long-term storage stability at 2–8 °C and for guiding formulation development.

6.3
Development of stable adenovirus formulations for long-term storage at 2–8°C

One of the earliest evaluations of storage stability for a human adenovirus was reported in an article by Green and Piña [176] describing the isolation and purification of adenovirus type 2. The formulation was a dilute solution containing only 10 mM Tris, pH 8.1. Ad2 was found to lose about 30% of its infectivity per month at 4 °C in this formulation. The effects of some excipients, including cysteine, glutathione, Ca^{2+} and Mg^{2+} ions, and 1 M NaCl were evaluated, but none of them were found to enhance stability. However, they did report that 0.1% BSA enhanced Ad2 stability. During the period from 1979 to 1998 there were numerous reports of liquid adenovirus formulations for frozen storage but no evaluations of 2–8 °C storage stability. Typically, either glycerol or sucrose was included in the formulation, presumably because of their historical use as cryoprotectants for proteins and live viruses [13, 178, 217, 232].

A summary of the Ad5 formulations reported in the scientific and patent literature that include at least six months of 2–8 °C stability data is shown in Table 3. The first systematic evaluation of the long-term 2–8 °C storage stability of Ad5 in liquid formulations was conducted by Sene et al. [226]. The inadequate stability of Ad5 in glycerol containing formulations up to that time was noted as one of the factors underlying their choice of sucrose as a stabilizer. Their sucrose-based formulations were reported to be stable for one year when stored at 4 °C; however, the high sucrose concentrations used (0.75 M to 1.5 M) result in very hypertonic formulations, making them less desirable than an isotonic formulation for parenteral administration. An analysis of their short-term stability data indicates that Ad5 has a $t_{1/2}$ of $\sim$ 26 hrs at 37 °C for the most stable formulation containing 1 M sucrose.

Shih et al. have developed liquid Ad5 formulations using both human serum albumin (HSA) and sucrose as stabilizers [227]. Ad5 was found to be stable for 8.5 months at 4 °C, with an estimated $t_{1/2}$ of 3.6 weeks at 20 °C (Table 3). Although this formulation appears promising in terms of its 4 °C stability, additional data are needed to establish the suitability of this formulation for long-term storage. Moreover, recombinant HSA should be used to avoid the use of animal-derived excipients in humans.

Blanche et al. have developed liquid adenovirus formulations based on the use of glycerol as a stabilizer (Table 3) [233, 234]. Ad5 in their Tris-glycerol formulation is reported to be stable for 18 months of 4 °C storage. Although this represents a significant advance in the development of stable adenovirus formulations, there are significant issues associated with the use of high glycerol concentrations in parenteral formulations.

The underlying cause of one concern is the capacity of glycerol to pass freely through cell membranes. This capacity results in the complete hemolysis of human erythrocytes in glycerol solutions of the same osmolality as blood [239, 240]. Therefore, glycerol-based formulations will not be isotonic with blood unless additional excipients, such as salts or sugars, are added to increase the osmolality. Without the use of other excipients to increase the osmolality, glycerol-based formulations will behave as if they are hypotonic and their parenteral use could potentially cause local muscle damage and pain upon IM injection.

Frei et al. and Kovesdi et al. have also reported on the stability of Ad5 compositions after one year of storage at 4 °C [236, 237]. The formulations developed by these groups each contain Tris, $MgCl_2$, and PS-80. Moreover, the pH values of the formulations are similar. However, the Frei et al. formulations contain sucrose and glycerol but do not contain NaCl. By comparison, the Kovesdi et al. formulations contain trehalose and NaCl but no sucrose or glycerol. The stability of Ad5 in these formulations appears to be similar at 4 °C, but the Frei et al. data were generated with a mutant of adenovirus type 5 that does not express protein IX. Because protein IX has an important role in stabilizing the viral capsid, Ad5 may have better stability in this formulation than the Ad5 mutant.

Wu et al. have disclosed a series of liquid formulations for Ad5 [238]. They have reported 4 °C stability data indicating that one of their formulations (AQF4-2) maintains the stability of Ad5 for at least one year. The AQF4-2 formulation includes sucrose and mannitol as well as PS-80 and NaCl, but no glycerol.

Evans et al. have recently reported the development of liquid Ad5 formulations containing inhibitors of free radical oxidation, and have demonstrated that these formulations are suitable for long-term storage of Ad5 at 2–8 °C [228]. It is interesting to note that the base formulation for this development effort, a formulation referred to as A105 (Table 3), provides about the same level of storage stability as the other formulations described in Table 3. However, an evaluation of the stability of Ad5 in the Frei et al. [236] and A105 formulations at both 4 °C and 25 °C indicates that Ad5 is significantly more

Table 3 Ad5 formulation compositions with ≥ 6 months of 2–8 °C storage stability data

References	Representative formulation	2–8 °C stability	Other stability data/comments
[226]	10 mM Tris, 1 M sucrose, 150 mM NaCl, 1 mM $MgCl_2$, 0.005% PS-80, pH 8.5	≥ 1 year	$t_{1/2} \sim 26$ hrs at 37 °C in 10 mM Tris, 1 mM $MgCl_2$, 1 M sucrose, pH 8.5
[227]	10 mM Tris, 5% HSA, 5% sucrose 150 mM NaCl, 2 mM $MgCl_2$, pH 8.4	≥ 8.5 months	$t_{1/2} \sim 3.6$ wks at 20 °C
[233, 234]	20 mM Tris, 20% glycerol, pH 8.4	> 18 months	$t_{1/2} \sim 2.6$ wks at 20 °C
[235]	0.4% sucrose, 0.4% mannitol 0.001% Span 20	$t_{1/2} \sim 8.6$ months	
[216]	GTS formulation 20 mM Tris, 2.5% glycerol, 25 mM NaCl, pH 8.0	≥ 6 months	stable through 5 freeze/thaw cycles
[236]	14 mM Tris, 12 mM Na-PO4 2 mM $MgCl_2$, 2% Sucrose 10% Glycerol, 0.015% PS-80 pH 7.4 to 7.8	≥ 1 year	> 1 week at 25 °C Ad5 mutant minus protein IX

Table 3 (continued)

References	Representative formulation	2–8 °C stability	Other stability data/comments
[237]	10 mM Tris, 5% trehalose 75 mM NaCl, 0.08 mM $MgCl_2$ 0.0025% PS-80, pH 7.8	33% loss in 1 year	$t_{1/2} \sim 6$ wks at 25 °C
[238]	AQF4-2 formulation 10 mM Tris, 5% sucrose, 5% mannitol 150 mM NaCl, 1 mM $MgCl_2$ 0.02% PS-80, pH 8.2	≥ 1 year	$t_{1/2} \sim 2.6$ wks at room temperature
[228]	A105 formulation 10 mM Tris, 5% sucrose, 75 mM NaCl 1 mM $MgCl_2$, 0.005% PS-80, pH 8.0	42% loss in year $t_{1/2} \sim 15$ months	$t_{1/2} \sim 2.5$ wks at 25 °C stable through 12 freeze/thaw cycles
[228]	A195 formulation 10 mM Tris, 10 mM histidine 5% sucrose, 75 mM NaCl 1 mMMgCl$_2$, 0.02% PS-80 0.1 mM EDTA, 0.5% ethanol, pH 7.4	≤ 21% loss in 2 years $t_{1/2} \sim 7$ years	$t_{1/2} \sim 11$ months at 15 °C $t_{1/2} \sim 5.7$ months at 20 °C $t_{1/2} \sim 3.5$ months at 25 °C $t_{1/2} \sim 1$ month at 30 °C $t_{1/2} \sim 13$ days at 37 °C stable through 12 freeze/thaw cycles

stable in A105 at each temperature [228]. The dramatically better stability of Ad5 in the presence of EDTA and ethanol (Table 3) clearly implicates free radical oxidation as an important mechanism of virus inactivation during storage. Furthermore, the results indicate that Ad5 formulated in A195 has a projected half-life of ~ 7 years at 4 °C. Under accelerated stability conditions the half-life of Ad5 in A195 is ~ 3.5 months at room temperature (25 °C), one month at 30 °C and 13 days at 37 °C. This corresponds to < 5% loss of infectivity after one week at room temperature or one day at 37 °C. Therefore, losses of infectivity during manufacturing, short-term storage and handling in the field should be minimal in A195, even without refrigeration. A195 does not contain animal-derived excipients, is isotonic with blood and is stable through at least 12 freeze/thaw cycles over a 1000-fold range of Ad5 concentrations (3×10^8 vp mL^{-1} to 3×10^{11} vp mL^{-1}).

6.4 Impact of formulation variables on Ad5 stability

6.4.1 Adenovirus concentration

Hoganson et al. have examined the concentration dependence of Ad5 stability in the GTS formulation (2.5% glycerol (w/v) and 25 mM NaCl in 20 mM Tris, pH 8.0) during 2–8 °C storage, reporting that Ad5 was stable up to 1.7×10^{11} vp mL^{-1} but was not stable after six months of 2–8 °C storage at 1.7×10^{12} vp mL^{-1} [216]. Ad5 stability in the A195 formulation has been reported to be independent of virus concentration over a range of 3×10^8 vp mL^{-1} to 3×10^{11} vp mL^{-1} when stored for one year at 2–8 °C [228]. There are no published studies on the effect of Ad5 concentration on stability at room temperature. However, Fig. 8 shows the effect of Ad5 concentration on stability after one month of storage at 30 °C in the A105 and A195 formulations. The results indicate that stability is not significantly affected by virus concentration from 1×10^8 vp mL^{-1} to 3×10^{11} vp mL^{-1}.

Although Ad5 concentration has little if any effect on stability in A105 and A195 over a 1000-fold range of virus concentration, generally formulation parameters such as pH and detergent concentration would be expected to have significant effects on the tendency of Ad5 to aggregate or adsorb to surfaces. Therefore, it seems likely that the effect of Ad5 concentration on stability will differ from formulation to formulation, particularly at the extremes of Ad5 concentration. Looking at this in another way, the major degradation mechanism for an Ad5 formulation could be very different at 10^8 vp mL^{-1} and 10^{12} vp mL^{-1}. For example, at low virus concentrations adsorption could be a major problem while aggregation is unlikely to be a significant issue. At high virus concentrations the opposite is more likely. Therefore, a formulation composition that is optimized for stability at

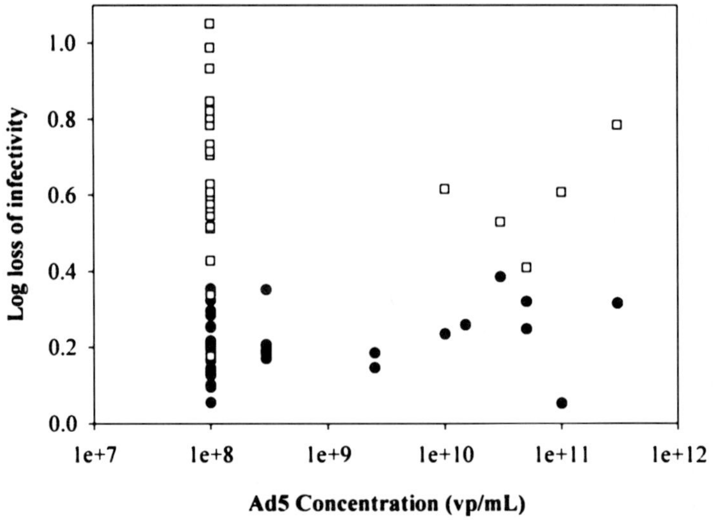

Fig. 8 Effect of Ad5 concentration on storage stability after one month of storage at 30 °C in the A105 (□) and A195 formulations (•)

10^{12} vp mL^{-1} (where aggregation may be the main issue) may not be optimal in terms of stability at 10^8 vp mL^{-1} (where adsorption may be a significant problem), and vice versa. This situation makes comparisons of stability data between formulations difficult to interpret, unless the studies were conducted at nearly the same virus concentration or the authors have shown that virus concentration doesn't affect stability over the virus concentration range of interest.

6.4.2
Adenovirus purity

Many characteristics of an adenovirus formulation can potentially affect stability. One is the purity of the adenovirus itself. Although no systematic study has been made of the effect of Ad5 purity on storage stability, it remains a theoretical possibility that differences in stability noted in the literature may be partially due to differences in the purity of the Ad5 preparations used, particularly for studies conducted at high virus concentrations. Based on reports showing that human serum albumin (HSA) and low concentrations of nonionic surfactants enhance the stability of adenovirus [226, 228] it seems conceivable that residual proteins or detergent in a purified Ad5 preparation might tend to enhance storage stability. However, direct comparisons of laboratory-scale preparations of Ad5 purified by CsCl density gradient ultracentrifugation with Ad5 purified by anion exchange chromatography, at $\sim 10^8$ vp mL^{-1}, have not shown significant differences in stability, based on short-term accelerated stability studies or long-term studies conducted at 2–8 °C (Evans et al., unpublished data).

6.4.3 Inhibitors and enhancers of free radical oxidation

It is known that both proteins and nucleic acids are susceptible to damage by oxygen free radicals, such as the hydroxyl radical [241, 242]. Oxidation by free radicals is also widely recognized as one of the leading causes of degradation of pharmaceuticals, including pharmaceutical formulations of plasmid DNA [243]. The stability of Ad5 is significantly enhanced by several different inhibitors of free radical oxidation, including histidine, EDTA/ethanol, triethanolamine and citrate [228]. In contrast, trace metal ions capable of enhancing the generation of hydroxyl radicals, like Fe^{2+}, are extremely detrimental to Ad5 stability at concentrations below 1 ppm. Reducing agents that can enhance the Fenton cycle of hydroxyl radical generation, like ascorbic acid, are also potent inactivators of Ad5. These results clearly indicate the need to consider all potential sources of trace metal ions and to reduce their levels to the lowest possible concentration. Common sources of trace metal ions include formulation excipients and possibly contact with metal surfaces during manufacturing. Formulations that do not contain inhibitors of free radical oxidation are more susceptible to trace metal ion contamination. This susceptibility would likely manifest itself as greater variability in the lot-to-lot stability of Ad5, since the level of trace metal ions will vary from lot to lot of excipients. Fortunately, formulations have been developed that effectively control free radical oxidation, based on the combination of a metal ion chelator (EDTA) and an inhibitor of free radical oxidation (ethanol) [228]. Not only do these formulations enhance the stability of Ad5 but they also appear to reduce the variability of Ad5 stability from lot to lot of formulation buffer (Evans et al., unpublished data).

6.4.4 Magnesium chloride

A number of enteroviruses are protected against thermal inactivation by 1–2 M concentrations of certain divalent cations [218, 222]. Poliovirus, for example, is stabilized by the presence of $MgCl_2$ at concentrations ranging from as low as 45 mM to as high as 3 M [244]. In fact, $MgCl_2$ is used commercially as a stabilizer of live polio vaccines [245]. Simian and canine adenoviruses are also stabilized by high concentrations of $MgCl_2$ [218, 220]. The rationale for inclusion of $MgCl_2$ in Ad5 formulations appears to be based on these early studies. One of the earliest references to a "virus preservation buffer" for Ad5 describes a formulation containing 1 mM $MgCl_2$ [246]. However, there are no data showing that low mM concentrations of $MgCl_2$ enhance the storage stability of Ad5. In fact, the stability of Ad5 is reported to be unaffected by $MgCl_2$ in formulations with and without inhibitors of free radical oxidation [227, 228]. However, 1 mM $MgCl_2$ does slightly increase the thermal melting temperature of Ad5 (by $\sim 2\,^\circ C$), indicating that Mg^{2+} ions interact with and stabilize the viral capsid under high-temperature thermal stress conditions [228].

6.4.5 Detergent type and concentration

Nonionic detergents have been shown to enhance adenovirus stability [226, 235, 237, 247, 248], with polysorbates being the most commonly used. However, there are reports of formulations without detergents that appear to stabilize Ad5 for a year or more at 4 °C [216, 227, 233]. The reason for this difference is not clear but it is interesting to note that two of the latter formulations contain glycerol, suggesting that glycerol may be able to substitute for nonionic detergents in some formulations and within some range of Ad5 concentration. There are claims in the patent literature that Brij, Pluronic and Triton surfactants enhance adenovirus stability, but the reported data do not address these claims. In formulations without glycerol the presence of a nonionic detergent appears critical for Ad5 stability [228]. Polysorbate-80 (PS-80) not only inhibits adsorption of Ad5 to glass surfaces but also appears to reduce the rate of virus inactivation, probably by inhibiting aggregation [228]. PS-80 is an effective stabilizer of Ad5 over a wide range of concentrations but virus stability is independent of PS-80 concentration from 0.002% to 0.15% (w/v) [228]. PS-20, PS-40 and PS-80 are interchangeable in terms of their effects on Ad5 storage stability at virus concentrations of $\sim 10^8$ vp mL^{-1} (Evans et al., unpublished data). An examination of a number of different Pluronic surfactants as potential replacements for PS-80 provided disappointing results (Evans et al., unpublished data).

6.4.6 Effect of pH

Some of the earliest published reports on adenovirus stability were studies designed to measure the effect of pH on the thermal inactivation temperature of simian and canine adenoviruses [218, 220]. The results show that these adenoviruses are more resistant to thermal inactivation at pH 5–6 than at pH 7–8. Early thermal inactivation studies of Ad5 show virus capsid disassembly at temperatures above 56 °C, but the effect of pH on thermal stability was not investigated [219]. Rexroad et al. have conducted thermal perturbation studies of Ad5 and have shown loss of capsid quaternary structure at 45 °C, pH 7.4, but the effect of pH was not examined [39]. Using a method similar to that reported by Rexroad et al., the effect of pH on the thermal melting temperature of Ad5 was determined in a formulation containing 50 mM sodium phosphate, 5% sucrose, 75 mM NaCl, 1 mM $MgCl_2$ and 0.005% PS-80, from pH 5.5 to pH 8.5. The results shown in Fig. 9 below indicate that the thermal stability of the Ad5 capsid increases with decreasing pH, suggesting that Ad5 might also have greater storage stability in formulations of lower pH.

The effect of pH on the storage stability of Ad5 has been examined in a number of laboratories. Sene et al. reported that, in formulations containing PS-80 and sucrose, Ad5 is more stable at pH 8.5 to 9.0 than at pH 7.4 to 8.0 [226]. But since these studies were very short-term (72 hours) and were conducted at 37 °C, it isn't clear what the data indicate with regard to the effect of pH on Ad5 stability during long-term storage at 2–8 °C. Croyle et al. examined the effect of pH on Ad5 stability at both 37 °C and 4 °C, but the studies were extremely short in duration (24 hours) and no clear trends were identified [217]. Shin et al. reported that Ad5 was most stable in the pH range of 8.0–8.6, based on one week of storage at room temperature and 37 °C [227]. The effect of pH on storage stability at room temperature was also reported by Hoganson et al., but the studies were only 24 hours in duration, and there was no trend for the effect of pH on Ad5 stability from pH 7.0 to pH 9.0 [216].

Evans et al. reported the effect of pH on the long-term (24 month) storage stability of Ad5 at 15 °C [228]. The results reveal a fairly broad pH optimum for Ad5 stability with pH having relatively little effect on stability from ~pH 6.0 to ~pH 7.6 (pH at 15 °C). Above and below this broad pH optimum the stability of Ad5 is significantly lower. It is worth mentioning that these studies were conducted in formulations containing PS-80 and free radical oxidation inhibitors and that the pH for optimum stability may be different in other formulations. It

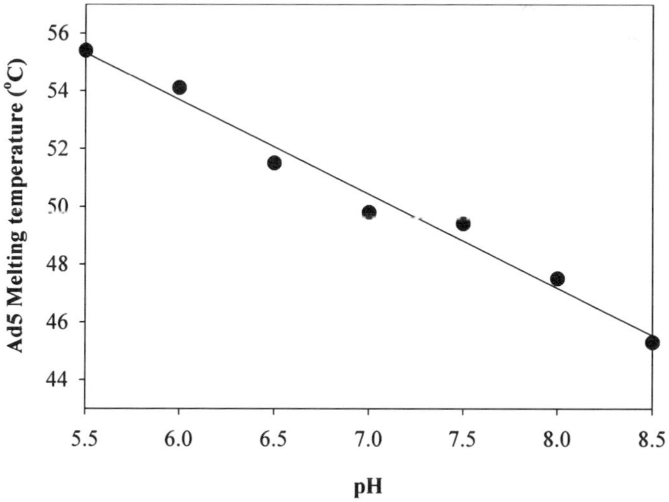

Fig. 9 Effect of pH on thermal melting temperature of Ad5 (Evans et al., unpublished data)

is important to note the dramatically different effects of pH on viral capsid stability and storage stability. Whereas the storage stability of the virus is hardly affected by pH in the range of pH 6.0 to pH 7.6, the stability of viral capsid (as measured by a thermal stress assay designed to measure the temperature at which the fluorescent probe Picogreen gains access to the viral DNA) significantly increases. These results suggest that capsid thermal stability is unrelated to long-term storage stability at 15 °C, in these formulations.

6.4.7 Cryoprotectants

A variety of cryoprotectants have been used to stabilize Ad5 against freeze/thaw damage. Not surprisingly, all of the formulations reported to stabilize Ad5 during long-term storage at 2–8 °C contain a cryoprotectant (Table 3). Although some of the oldest reported adenovirus formulations use glycerol as a stabilizer for frozen storage, some of the more newly developed formulations also contain glycerol [216, 232–234, 236, 246]. Glycerol has been used at concentrations varying from 2.5% to 50%. In one formulation containing a low glycerol concentration (2.5%), the addition of > 10 mM NaCl was necessary to protect against freeze/thaw damage [216]. Sucrose, trehalose and mannitol are also effective cryoprotectants for Ad5. Sucrose-based formulations typically contain 5% sucrose but there are reports of stable Ad5 formulations containing as little as 0.4% sucrose and 0.4% mannitol [235] and as much as 34% sucrose [226]. The combination of 5% sucrose and 75 mM NaCl is not only isotonic with blood but has been found to provide excellent resistance to freeze/thaw-induced inactivation of Ad5 in a wide range of formulation compositions [228]. Recent biophysical studies have shown that raising the sucrose concentration from 2% to 10% increased the thermal stability of the secondary and tertiary structures of Ad5 [39], suggesting that sucrose may also function to stabilize the capsid during storage, particularly during short-term thermal stress.

6.5 Serotype-specific considerations

Although most of the formulation development work on AdV vectors has been focused on group C serotypes, and mainly Ad5, there is a growing interest in developing the next generation of AdV vectors using serotypes outside of group C. Compared to Ad5, much less is known about the stability of other AdV serotypes, but the approaches outlined above for identifying stable formulations should apply to other serotypes. However, it seems likely (based simply on differences in the amino acid sequences of hexon and fiber between Ad5 and non-group C vectors) that formulations will need to be optimized for each serotype (serotype-specific formulation) or perhaps for each AdV group (group-specific).

7
Conclusions and perspectives

Over the past decade, intense interest in the use of adenovirus for gene transfer and vaccination has led to substantial efforts in defining scaleable cultivation and purification processes, focused primarily on "first-generation" Ad5 vectors. Batch culture processes with high specific productivities and reasonable volumetric productivities have been developed. Several manufacturers have also developed scaleable purification processes, primarily relying on anion exchange chromatography. Many of these production processes may be suitable for scale-up to 10^{17} viral particles (25 g) per batch. While detailed process economic studies have not been reported, we believe that some of the processes reported herein are suitable for even low-cost prophylactic vaccines. Despite this conclusion, there are several areas where further advances could benefit the manufacturers of these products. For virus production, the development of a fed-batch process to significantly increase volumetric productivity could improve the process economics. Achieving more productive processes will clearly require a better understanding of the "cell density effect" on virus productivity. More highly productive cell lines could also be developed, and may benefit from such knowledge.

For purification, the development of an economic, robust, and scaleable step for the clearance of empty capsids would decrease the reliance on culture reproducibility and chromatography steps with limited effectiveness. Developments which limit the use of chromatography and membrane filtration (microfiltration and/or ultrafiltration) may also have a measurable impact on cost. Demonstration of selective methods to remove or inactivate potential adventitious agents will also be valuable.

The likelihood of developing 2–8 °C stable liquid formulations of Ad5 appeared small just ten years ago. Yet, we now have reports of liquid Ad5 formulations in which Ad5 is stable for 18 months or more at 2–8 °C. The significance of these results is illustrated by the fact that none of the currently marketed live virus vaccines are liquid formulations suitable for long-term storage at 2–8 °C. These developments have been possible in part due to efforts to understand the inactivation pathways, including free radical oxidation, adsorbation, and freeze/thaw-induced damage. These new formulations have notably enhanced the value of Ad5 for both gene therapy and vaccines, and further improvements will only be highly beneficial if they facilitate storage at ambient temperatures.

With most critical development activities for first-generation Ad5 vectors solved, attention must be redirected to the array of next-generation AdV vectors in development. Included amongst these are an assortment of serotypes of adenoviruses, both human and nonhuman in origin, and chimeras thereof. Intrinsically linked to the vector construction is the design of complementing cell lines, and the ideal systems would be designed with both vector quality (appropriate tropism, low P/I ratio, limited pre-existing immunity in the target population) and vector productivity in mind. There is little doubt that the properties of these vectors will vary from first-generation Ad5's, and will require specific optimization of production. Purification methodologies will be dependent on the surface properties of these vectors, and a detailed understanding of the serotype-specificity of unit operations as well as process sensitivities to shear, aggregation, and association with cellular DNA will be valuable additions to the literature. Formulations may need to be optimized for other serotypes for similar reasons. Finally, movement away from the well-studied human group C serotypes may also require substantial efforts in analytical characterization as well as basic virology, since many serotypes have not been studied to any appreciable extent.

Also included among the next-generation vectors are helper-dependent (gutless) adenoviruses, which may be required to provide the long-term expression needed to satisfy the original premise of gene therapy (for example, replacement therapies). Current production systems for gutless vectors require the use of helper virus, and minimizing helper contamination while maximizing productivity in cell culture is a challenging task, particularly for large-scale manufacturing. A productive system for generating gutless vectors without the need for helper coinfection would be a significant development. If such a system cannot be developed, applications requiring large quantities of vector will be dependent on the development of a scaleable method for helper virus removal. For low-volume/high-value applications, however, density gradient ultracentrifugation may suffice.

A general challenge for all adenoviral products is the development of more sophisticated analytical techniques which could bring the characterization of these products to the level where they are considered more as well-characterized biotech products rather than live-virus vaccines. Enumeration of the "parts" of the virus and demonstrating consistency of composition will be required. This alone will not suffice, however; techniques which can discern the proper assembly of the parts into functional viruses will also be required. Analytical efforts will need to proceed hand-in-hand with cultivation and purification advances, for without a relatively homogeneous starting population with high specific activity (low P/I ratio) the task of defining product consistency and structure-activity relationships will be extremely difficult.

Acknowledgements We thank coworkers who contributed to the generation of unpublished data presented in this work (Fig. 2: Marie-Pierre Gentile and Gina Cremona; Fig. 4: Matthew Kalscheur, Marie-Pierre Gentile, and Benjamin Hughes; Figs. 8 and 9: Denise Nawrocki, Lynne Isopi, and Donna Williams) and Drs. Angelica Meneses-Acosta and Olivier Henry for providing data and assistance with Figs. 5 and 6. We also thank Dr. John Lewis and Dr. Katey Einterz Owen for reviewing portions of this manuscript and providing helpful suggestions.

References

1. St George JA (2003) Gene Ther 10:1135
2. Shiver JW, Emini EA (2004) Annu Rev Med 55:355
3. Morsy MA, Gu M, Motzel S, Zhao J, Lin J, Su Q, Allen H, Franlin L, Parks RJ, Graham FL, Kochanek S, Bett AJ, Caskey CT (1998) Proc Natl Acad Sci USA 95:7866
4. Edelstein ML, Abedi MR, Wixon J, Edelstein RM (2004) J Gene Med 6:597
5. Anonymous (2004) BioPharm Int 17:42
6. RAC (2002) Hum Gene Ther 13:3
7. Bauer SR, Pilaro AM, Weiss KD (2002) In: Curiel DT, Douglas JT (eds) Adenoviral vectors for gene therapy. Academic, New York, p 615
8. CBER (2000) Pharmacopeial Forum 26:56
9. Pinteric L, Taylor J (1962) Virology 18:359
10. Maizel JVJ, White DO, Scharff MD (1968) Virology 36:115
11. Sweeney JA, Hennessey JP Jr (2002) Virology 295:284
12. Blanche F, Cameron B, Barbot A, Ferrero L, Guillemin T, Guyot S, Somarriba S, Bisch D (2000) Gene Ther 7:1055
13. Shabram PW, Giroux DD, Goudreau AM, Gregory RJ, Horn MT, Huyghe BG, Liu XD, Nunnally MH, Sugarman BJ, Sutjipto S (1997) Hum Gene Ther 8:453
14. Lehmberg E, Traina JA, Chakel JA, Chang RJ, Parkman M, McCaman MT, Murakami PK, Lahidji V, Nelson JW, Hancock WS, Nestaas E, Pungor E Jr (1999) J Chromatogr B 732:411
15. Murakami P, McCaman MT (1999) Anal Biochem 274:283
16. Ma L, Bluyssen HA, De Raeymaeker M, Laurysens V, van der Beek N, Pavliska H, van Zonneveld AJ, Tomme P, van Es HH (2001) J Virol Methods 93:181
17. Saulnier P, Vidaud M, Gautier E, Motte N, Bellet D, Escudier B, Wilson D, Yver A (2003) J Virol Methods 114:55
18. Wang F, Patel DK, Antonello JM, Washabaugh MW, Kaslow DC, Shiver JW, Chirmule N (2003) Hum Gene Ther 14:25
19. Graham FL, Smiley J, Russell WC, Nairn R (1977) J Gen Virol 36:59
20. Graham FL, van der Eb AJ (1973) Virology 52:456
21. Wigand R, Kumel G (1977) Arch Virol 54:177
22. Darling AJ, Boose JA, Spaltro J (1998) Biologicals 26:105
23. Nyberg-Hoffman C, Shabram P, Li W, Giroux D, Aguilar Cordova E (1997) Nat Med 3:808
24. Allison AC, Valentine RC (1960) Biochim Biophys Acta 40:400
25. Wang F, Puddy AC, Mathis WC, Hager AG, Louis AA, McMackin JL, Xu J, Zhang Y, Tan CY, Schofield TS, Wolf JJ, Lewis JA (2005) Vaccine 23:4500
26. Weaver LS, Kadan MJ (2000) Methods 21:297
27. Lusky M, Christ M, Rittner K, Dieterle A, Dreyer D, Mourot B, Schultz H, Stoeckel F, Pavirani A, Mehtali M (1998) J Virol 72:2022
28. Ohtani S, Kagawa S, Tango Y, Umeoka T, Tokunaga N, Tsunemitsu Y, Roth JA, Taya Y, Tanaka N, Fujiwara T (2004) Mol Cancer Ther 3:93
29. Blanche F, Monegier B, Faucher D, Duchesne M, Audhuy F, Barbot A, Bouvier S, Daude G, Dubois H, Guillemin T, Maton L (2001) J Chromatogr A 921:39
30. Roitsch C, Achstetter T, Benchaibi M, Bonfils E, Cauet G, Gloeckler R, L'H te H, Keppi E, Nguyen M, Spehner D, Van Dorsselaer A, Malarme D (2001) J Chromatogr B 752:263

31. Fallaux FJ, Bout A, van der Velde I, van den Wollenberg DJ, Hehir KM, Keegan J, Auger C, Cramer SJ, van Ormondt H, van der Eb AJ, Valerio D, Hoeben RC (1998) Hum Gene Ther 9:1909
32. Nichols WW, Lardenoije R, Ledwith BJ, Brouwer K, Manam S, Vogels R, Kaslow D, Zuidgeest D, Bett AJ, Chen L, van der Kaaden M, Galloway SM, Hill RB, Machotka SV, Anderson CA, Lewis J, Martinez D, Lebron J, Russo C, Valerio D, Bout A (2002) In: Curiel DT, Douglas JT (eds) Adenoviral vectors for gene therapy. Academic, New York, p 129
33. Bondoc LL, Tang JCT (1997) Anal Biochem 247:443
34. Peden K (2001) VRBRAC Meeting on "Designer" Cells as Substrates for the Manufacture of Viral Vaccines, 16th May 2001, Gaithersburg, MD
35. Konz JO, Lee AL, Lewis JA, Sagar SL (2005) Biotechnol Prog 21:466
36. Seth P, Fitzgerald D, Ginsberg H, Willingham M, Pastan I (1984) Mol Cell Biol 4:1528
37. Vellekamp G, Porter FW, Sutjipto S, Cutler C, Bondoc L, Liu YH, Wylie D, Cannon Carlson S, Tang JT, Frei A, Voloch M, Zhuang S (2001) Hum Gene Ther 12:1923
38. Oliver CJ, Shortridge KF, Belyavin G (1976) Biochim Biophys Acta 437:589
39. Rexroad J, Wiethoff CM, Green AP, Kierstead TD, Scott MO, Middaugh CR (2003) J Pharm Sci 92:665
40. Bondoc LL, Fitzpatrick S (1998) J Ind Microbiol Biot 20:317
41. Horaud F (1994) Dev Biol Stand 82:113
42. Brown F, Griffiths E, Horaud F, Petricciani JC (eds) (1998) Dev Biol Std 93
43. CBER (1998) Guidance for industry: Guidance for human somatic cell therapy and gene therapy
44. Grachev V, Magrath D, Griffiths E, Petricciani JC, Chiu YY, Dobbelaer R, Gust I, Hardegree MC, Hayakawa T, Horaud F, Lubiniecki AS, Minor P, Montagnon B, Peetermans J, Ridgeway A, Robertson J, Schild G, Seamon KB (1998) Biologicals 26:175
45. USP (2000) US Pharmacopeia 2:138
46. FDA (1998) International Conference on Harmonization; Q5D: Guidance on quality of biotechnological/biological products: Derivation and characterization of cell substrates used for production of biotechnological/biological products (Notice). Food and Drug Administration, Rockville, MD
47. FDA (1998) International Conference on Harmonization; Q5A: Guidance on viral safety evaluation of biotechnology products derived from cell lines of human or animal origin (Notice). Food and Drug Administration, Rockville, MD
48. Tatalick LM, Gerard CJ, Takeya R, Price DN, Thorne BA, Wyatt LM, Anklesaria P (2005) Vaccine 23(20):2628–38
49. Prusiner SB (1998) P Natl Acad Sci USA 95:13363
50. Ironside J (1998) J Pathol 186:227
51. Butman BT (2004) In: 11th Annual Viral Vectors and Vaccines Conf, 8–11 November 2004, Williamsburg, VA
52. Cook J (2001) VRBRAC Meeting on "Designer" Cells as Substrates for the Manufacture of Viral Vaccines, 16th May 2001, Gaithersburg, MD
53. WHO (1987) Technical Report Series: Acceptability of cell substrates for the production of biologicals. World Health Organization, Geneva
54. Anderson SA, Hilleman MR, Warfield MS, Werner JH (1957) J Am Med Assoc 163:4
55. Hayflick L, Moorhead PS (1961) Exp Cell Res 25:585
56. Top FH Jr, Grossman RA, Bartelloni PJ, Segal HE, Dudding BA, Russell PK, Buescher EL (1971) J Infect Dis 124:148
57. Fletcher MA, Hessel L, Plotkin SA (1998) Dev Biol Stand 93:97

58. Berkner KL, Sharp PA (1983) Nucleic Acids Res 11:6003
59. Rich DP, Couture LA, Cardoza LM, Guiggio VM, Armentano D, Espino PC, Hehir K, Welsh MJ, Smith AE, Gregory RJ (1993) Hum Gene Ther 4:461
60. Haj-Ahmad Y, Graham FL (1986) J Virol 57:267
61. Lee MG, Abina MA, Haddada H, Perricaudet M (1995) Gene Ther 2:256
62. Ghosh-Choudhury G, Haj-Ahmad Y, Graham FL (1987) EMBO J 6:1733
63. Youil R, Toner TJ, Su Q, Casimiro D, Shiver JW, Chen L, Bett AJ, Rogers BM, Burden EC, Tang A, Chen M, Emini EA, Kaslow DC, Aunins JG, Altaras NE (2003) Hum Gene Ther 14:1017
64. Wivel N, Gao G, Wilson J (1999) The development of human gene therapy. Cold Spring Harbor Laboratory Press, Cold Spring Harbor, NY, p 87
65. Yeh P, Perricaudet M (1997) FASEB J 11:615
66. Fang B, Koch P, Roth JA (1997) J Virol 71:4798
67. Zhou H, O'Neal W, Morral N, Beaudet AL (1996) J Virol 70:7030
68. Gorziglia MI, Kadan MJ, Yei S, Lim J, Lee GM, Luthra R, Trapnell BC (1996) J Virol 70:4173
69. Amalfitano A, Chamberlain JS (1997) Gene Ther 4:258
70. Langer SJ, Schaack J (1996) Virology 221:172
71. Armentano D, Zabner J, Sacks C, Sookdeo CC, Smith MP, St George JA, Wadsworth SC, Smith AE, Gregory RJ (1997) J Virol 71:2408
72. Baxi MK, Robertson J, Babiuk LA, Tikoo SK (2001) Virology 290:153
73. Gao GP, Yang Y, Wilson JM (1996) J Virol 70:8934
74. Krougliak V, Graham FL (1995) Hum Gene Ther 6:1575
75. Gorziglia MI, Lapcevich C, Roy S, Kang Q, Kadan M, Wu V, Pechan P, Kaleko M (1999) J Virol 73:6048
76. Massie B, Mosser D, Koutromannis M, Vitte-Mony I, Lamoreaux L, Couture F, Paquet L, Guibault D, Dionne J, Chahla D, Joelicoeur P, Langelier Y (1998) Cytotechnology 28:53
77. Edholm D, Molin M, Bajak E, Akusjarvi G (2001) J Virol 75:9579
78. Tietge UJ, Kozarsky KF, Donahee MH, Rader DJ (2003) J Gene Med 5:567
79. Gotoh A, Ko SC, Shirakawa T, Cheon J, Kao C, Miyamoto T, Gardner TA, Ho LJ, Cleutjens CB, Trapman J, Graham FL, Chung LW (1998) J Urol 160:220
80. Morelli AE, Larregina AT, Smith-Arica J, Dewey RA, Southgate TD, Ambar B, Fontana A, Castro MG, Lowenstein PR (1999) J Gen Virol 80(Pt 3):571
81. Glenn GM, Chatterjee S (2001) Cancer Gene Ther 8:566
82. Rodriguez R, Choudhury W (2003) PCT Patent Application W/O 2003087348
83. Kirn D, Martuza RL, Zwiebel J (2001) Nat Med 7:781
84. Sunamura M, Hamada H, Motoi F, Oonuma M, Abe H, Saitoh Y, Hoshida T, Ottomo S, Omura N, Matsuno S (2004) Pancreas 28:326
85. Geoerger B, Vassal G, Opolon P, Dirven CM, Morizet J, Laudani L, Grill J, Giaccone G, Vandertop WP, Gerritsen WR, van Beusechem VW (2004) Cancer Res 64:5753
86. Maemondo M, Saijo Y, Narumi K, Kikuchi T, Usui K, Tazawa R, Matsumoto K, Nakamura T, Sasaki K, Takahashi M, Niitsu Y, Nukiwa T (2004) Cancer Res 64:4611
87. McCormick F (2001) Nat Rev Cancer 1:130
88. Kanerva A, Hemminka A (2004) Drugs Future 29:359
89. Yuk IH, Olsen MM, Geyer S, Forestell SP (2004) Biotechnol Bioeng 86:637
90. Oualikene W, Lamoureux L, Weber JM, Massie B (2000) Hum Gene Ther 11:1341
91. Sprangers MC, Lakhai W, Koudstaal W, Verhoeven M, Koel BF, Vogels R, Goudsmit J, Havenga MJ, Kostense S (2003) J Clin Microbiol 41:5046

92. Aste-Amezaga M, Bett AJ, Wang F, Casimiro DR, Antonello JM, Patel DK, Dell EC, Franlin LL, Dougherty NM, Bennett PS, Perry HC, Davies ME, Shiver JW, Keller PM, Yeager MD (2004) Hum Gene Ther 15:293
93. Glasgow JN, Bauerschmitz GJ, Curiel DT, Hemminki A (2004) Curr Gene Ther 4:1
94. Alvarez RD, Barnes MN, Gomez-Navarro J, Wang M, Strong TV, Arafat W, Arani RB, Johnson MR, Roberts BL, Siegal GP, Curiel DT (2000) Clin Cancer Res 6:3081
95. Gao W, Robbins PD, Gambotto A (2003) Gene Ther 10:1941
96. Reddy PS, Ganesh S, Limbach MP, Brann T, Pinkstaff A, Kaloss M, Kaleko M, Connelly S (2003) Virology 311:384
97. Yun CO, Cho EA, Song JJ, Kang DB, Kim E, Sohn JH, Kim JH (2003) Hum Gene Ther 14:1643
98. Cashman SM, Morris DJ, Kumar-Singh R (2004) Virology 324:129
99. Ophorst OJ, Kostense S, Goudsmit J, De Swart RL, Verhaagh S, Zakhartchouk A, Van Meijer M, Sprangers M, Van Amerongen G, Yuksel S, Osterhaus AD, Havenga MJ (2004) Vaccine 22:3035
100. Pereboeva L, Komarova S, Mahasreshti PJ, Curiel DT (2004) Virus Res 105:35
101. Nicol CG, Graham D, Miller WH, White SJ, Smith TA, Nicklin SA, Stevenson SC, Baker AH (2004) Mol Ther 10:344
102. Youil R, Toner TJ, Su Q, Chen M, Tang A, Bett AJ, Casimiro D (2002) Hum Gene Ther 13:311
103. Varga CM, Wickham TJ, Lauffenburger DA (2000) Biotechnol Bioeng 70:593
104. Worgall S, Busch A, Rivara M, Bonnyay D, Leopold PL, Merritt R, Hackett NR, Rovelink PW, Bruder JT, Wickham TJ, Kovesdi I, Crystal RG (2004) J Virol 78:2572
105. Nicklin SA, Baker AH (2002) Curr Gene Ther 2:273
106. Ferreira T, Alves P, Aunins J, Carrondo M (2005) Gene Ther 12:S73–S83
107. Wu Q, Tikoo SK (2004) Virus Res 99:9
108. Kremer EJ (2004) J Gene Med 6(Suppl 1):S139
109. Renaut L, Colin M, Leite JP, Benko M, D'Halluin JC (2004) Virology 321:189
110. Glasgow JN, Kremer EJ, Hemminki A, Siegal GP, Douglas JT, Curiel DT (2004) Virology 324:103
111. Roy S, Gao G, Lu Y, Zhou X, Lock M, Calcedo R, Wilson JM (2004) Hum Gene Ther 15:519
112. Farina SF, Gao GP, Xiang ZQ, Rux JJ, Burnett RM, Alvira MR, Marsh J, Ertl HC, Wilson JM (2001) J Virol 75:11603
113. Einfeld DA, Brough DE, Roelvink PW, Kovesdi I, Wickham TJ (1999) J Virol 73:9130
114. Orlando JS, Ornelles DA (2002) J Virol 76:1475
115. Lochmuller H, Jani A, Huard J, Prescott S, Simoneau M, Massie B, Karpati G, Acsadi G (1994) Hum Gene Ther 5:1485
116. Robert JJ, Gauffeny I, Maccario J, Jullien C, Benoit P, Vigne E, Crouzet J, Perricaudet M, Yeh P (2001) Gene Ther 8:1713
117. Hehir KM, Armentano D, Cardoza LM, Choquette TL, Berthelette PB, White GA, Couture LA, Everton MB, Keegan J, Martin JM, Pratt DA, Smith MP, Smith AE, Wadsworth SC (1996) J Virol 70:8459
118. Fallaux F, Van der Velde I, Kranenburg O, Cramer S, Houweling A, van Ormondt H, Bout A, Hoeben R, Valerio D, van der Eb A (1995) Gene Ther 2:S7
119. Schiedner G, Hertel S, Kochanek S (2000) Hum Gene Ther 11:2105
120. Gao GP, Engdahl RK, Wilson JM (2000) Hum Gene Ther 11:213
121. Massie B (1999) U.S. Patent 5,891,690
122. Murakami P, Havenga M, Fawaz F, Vogels R, Marzio G, Pungor E, Files J, Do L, Goudsmit J, McCaman M (2004) J Virol 78:6200

123. Murakami P, Pungor E, Files J, Do L, van Rijnsoever R, Vogels R, Bout A, McCaman M (2002) Hum Gene Ther 13:909
124. Ling WL, Longley RL, Brassard DL, Armstrong L, Schaefer EJ (2002) Gene Ther 9:907
125. Shaw G, Morse S, Ararat M, Graham FL (2002) FASEB J 16:869
126. Kochanek S, Schiedner G, Volpers C (2001) Curr Opin Mol Ther 3:454
127. Bett AJ, Prevec L, Graham FL (1993) J Virol 67:5911
128. Alemany R, Dai Y, Lou YC, Sethi E, Prokopenko E, Josephs SF, Zhang WW (1997) J Virol Methods 68:147
129. Parks RJ, Chen L, Anton M, Sankar U, Rudnicki MA, Graham FL (1996) P Natl Acad Sci USA 93:13565
130. Ng P, Parks RJ, Cummings DT, Evelegh CM, Sankar U, Graham FL (1999) Hum Gene Ther 10:2667
131. Zou L, Zhou H, Pastore L, Yang K (2000) Mol Ther 2:105
132. Gilbert R, Dudley RW, Liu AB, Petrof BJ, Nalbantoglu J, Karpati G (2003) Hum Mol Genet 12:1287
133. Kim IH, Jozkowicz A, Piedra PA, Oka K, Chan L (2001) P Natl Acad Sci USA 98:13282
134. Maione D, Wiznerowicz M, Delmastro P, Cortese R, Ciliberto G, La Monica N, Savino R (2000) Hum Gene Ther 11:859
135. Schiedner G, Morral N, Parks RJ, Wu Y, Koopmans SC, Langston C, Graham FL, Beaudet AL, Kochanek S (1998) Nat Genet 18:180
136. Abremski K, Hoess R (1984) J Biol Chem 259:1509
137. Sakhuja K, Reddy PS, Ganesh S, Cantaniag F, Pattison S, Limbach P, Kayda DB, Kadan MJ, Kaleko M, Connelly S (2003) Hum Gene Ther 14:243
138. Umana P, Gerdes CA, Stone D, Davis JR, Ward D, Castro MG, Lowenstein PR (2001) Nat Biotechnol 19:582
139. Parks RJ, Bramson JL, Wan Y, Addison CL, Graham FL (1999) J Virol 73:8027
140. Sandig V, Youil R, Bett AJ, Franlin LL, Oshima M, Maione D, Wang F, Metzker ML, Savino R, Caskey CT (2000) P Natl Acad Sci USA 97:1002
141. Ng P, Evelegh C, Cummings D, Graham FL (2002) J Virol 76:4181
142. Maranga L, Aunins J, Zhou W (2005) Biotechnol Bioeng 90(5):645–655
143. Heilman CA, Rouse H (1980) Virology 105:159
144. Radlett P, Pay T, Garland A (1985) Dev Biol Stand 60:163
145. Shenk TE (2001) In: Knipe DM, Howley PM (eds) Fields virology. Lippincott Williams & Wilkins, Philadelphia, PA, p 2265
146. Mittereder N, March KL, Trapnell BC (1996) J Virol 70:7498
147. Dee K, Shuler M, Wood A (1997) Biotechnol Bioeng 54:191
148. Peshwa M, Kyung Y-S, McClure D, Hu W-S (1993) Biotechnol Bioeng 41:179
149. Dee K, Shuler M, Wood A (1997) Biotechnol Bioeng 54:206
150. Giroux D, Goudreau AM, Ramachandra M, Shabram P (1999) US Patent 5,994,134
151. Condon RGG, Connelly NV, Frei A, Glowacki E, Yabannavar V, Batandolo S (2004) US Patent 6,783,983
152. Zhang S, Pham H, Senesac P, Clarke P (2004) In: 11th Annual Viral Vectors and Vaccines Conf, 8–11 November 2004, Williamsburg, VA
153. Zhang S, Thwin C, Wu Z, Cho T (2001) US Patent 6,194,191 B1
154. Xie LZ, Metallo C, Warren J, Pilbrough W, Peltier J, Zhong T, Pikus L, Yancy A, Leung J, Aunins JG, Zhou WC (2003) Biotechnol Bioeng 83:45
155. Monica TJ, Montgomery T, Ayala JL, Schoofs GM, Whiteley EM, Roth G, Garbutt JJ, Harvey S, Castillo FJ (2000) Biotechnol Prog 16:866

156. Monica TJ, Whitely E, Castillo F, Tan M, Tse S, Ja G, Rodriguez E, Pungor E, Nastaas E (2000) In: 7th Annual Viral Vectors and Vaccines Conf, 6–9 November 2000, Lake Tahoe, NV
157. Gentile M-P, Altaras NE, Youil R, Toner TJ, Aunins J, Zhou W (2002) In: 224th American Chemical Society National Meeting, 18–22 August 2002, Boston, MA
158. Jardon M, Garnier A (2003) Biotechnol Prog 19:202
159. Xie L, Pilbrough W, Metallo C, Zhong T, Pikus L, Leung J, Aunins JG, Zhou W (2002) Biotechnol Bioeng 80:569
160. Xie L, Pilbrough W, Metallo C, Warren J, Peltier J, Aunins J (2002) In: 8th Cell Culture Engineering Conf, 1–6 April 2002, Snowmass, CO
161. Nadeau I, Gilbert PA, Jacob D, Perrier M, Kamen A (2002) Biotechnol Bioeng 77:91
162. Nadeau I, Kamen A (2003) Biotechnol Adv 20:475
163. Schoofs G, Monica TJ, Ayala J, Horwitz J, Montgomery T, Roth G, Castillo FJ (1998) Cytotechnology 28:81
164. Iyer P, Ostrove JM, Vacante D (1999) Cytotechnology 30:169
165. Merten OW, Manuguerra JC, Hannoun C, van der Werf S (1999) Dev Biol Stand 98:23
166. Frazzati-Gallina NM, Paoli RL, Mourao-Fuches RM, Jorge SA, Pereira CA (2001) J Biotechnol 92:67
167. Garnier A, Cote J, Nadeau I, Kamen A, Massie B (1994) Cytotechnology 15:145
168. Nadeau I, Garnier A, Cote J, Massie B, Chavarie C, Kamen A (1996) Biotechnol Bioeng 51:613
169. Lee YY, Yap MG, Hu WS, Wong KT (2003) Biotechnol Prog 19:501
170. Henry O, Dormond E, Perrier M, Kamen A (2004) Biotechnol Bioeng 86:765
171. Blumentals I, Warren J, Scott R, Otero JM (2004) In: 9th Cell Culture Engineering Conf, 7–13 March 2004, Cancun, Mexico
172. Klemperer HG, Pereira HG (1959) Virology 9:536
173. Haruna I, Yaoi H, Kono R, Watanabe I (1961) Virology 13:264
174. Philipson L, Albertsson PA, Frick G (1960) Virology 11:553
175. Philipson L (1960) Virology 10:459
176. Green M, Pina M (1963) Virology 20:199
177. Kanegae Y, Makimura M, Saito I (1994) Jpn J Med Sci Biol 47:157
178. Huyghe BG, Liu X, Sutjipto S, Sugarman BJ, Horn MT, Shepard HM, Scandella CJ, Shabram P (1995) Hum Gene Ther 6:1403
179. Tang JC, Vellekamp G, Bondoc LL (2001) US Patent 6,261,823 B1
180. Cannon-Carlson SV, Cutler C, Vellekamp GJ, Voloch M (2002) US Patent Application 2002/0064860 A1
181. Murphy C, Nahapetian A, Iyer P, Gore N, Fogle P, Moreland T, Wilhelm E (1998) 5th Annual Viral Vectors and Vaccines Conf, 16–19 November, 1998, Williamsburg, VA
182. Green AP, Huang JJ, Scott MO, Kierstead TD, Beaupre I, Gao GP, Wilson JM (2002) Hum Gene Ther 13:1921
183. Goerke AR, To BCS, Lee AL, Sagar SL, Konz JO (2005) Biotechnol Bioeng 91(1):12–21
184. Eckhardt BM, Oeswein JQ, Bewley TA (1991) Pharm Res 8:1360
185. Tollefson AE, Scaria A, Hermiston TW, Ryerse JS, Wold LJ, Wold WS (1996) J Virol 70:2296
186. Shabram P, Vellekamp G, Scandella C (2002) In: Curiel DT, Douglas JT (eds) Adenoviral vectors for gene therapy. Academic, New York, p 167
187. Hughes BS, Kalscheur M, Chen J, Subramanian S, Zhou W, Altaras NE (2004) 227th American Chemical Society National Meeting, 27 March–1 April 2004, Anaheim, CA
188. Hagen AJ, Aboud RA, DePhillips PA, Oliver CN, Orella CJ, Sitrin RD (1996) Biotechnol Appl Biochem 23:209

189. Singhvi R, Schorr C, O'Hara C, Xie L, Wang DIC (1996) BioPharm 9:35
190. Yavorsky D, Blanck R, Lambalot C, Brunkow R (2003) Pharmaceut Technol March:62
191. Guttman-Bass N, Armon R (1983) Appl Environ Microbiol 45:850
192. Sobsey MD, Oglesbee SE, Wait DA (1985) Appl Environ Microbiol 50:1457
193. Enriquez CE, Gerba CP (1995) Water Res 29:2554
194. Hagen AJ, Oliver CN, Sitrin RD (1996) Biotechnol Prog 12:406
195. Schagen FH, Moor AC, Cheong SC, Cramer SJ, van Ormondt H, van der Eb AJ, Dubbelman TM, Hoeben RC (1999) Gene Ther 6:873
196. Schagen FH, Rademaker HJ, Rabelink MJ, van Ormondt H, Fallaux FJ, van der Eb AJ, Hoeben RC (2000) Gene Ther 7:1570
197. Stilwell JL, McCarty DM, Negishi A, Superfine R, Samulski RJ (2003) J Virol 77:12881
198. Sagar SL, Chau YG, Watson MP, Lee AL (2002) In: Rathore AS, Vella G (eds) Scale-up and optimization in preparative chromatography: Principles and biopharmaceutical applications, vol 88. Marcel Dekker, New York, p 251
199. DePhillips P, Lenhoff AM (2000) J Chromatogr A 883:39
200. Burness ATH (1969) In: Habel K, Salzman NP (eds) Fundamental techniques in virology. Academic, New York, p 94
201. Kraiselbuld EP, Gage I, Weissbach A (1975) J Mol Biol 97:533
202. Puvion-Dutilleul F, Besse S, Pichard E, Cajean-Feroldi C (1998) Biol Cell 90:5
203. Croyle MA, Gerding K, Quick KS (2003) Bioprocess J 2:35
204. Galdiero F (1979) Arch Virol 59:99
205. O'Grady J, Losikoff A, Poiley J, Fickett D, Oliver C (1996) Dev Biol Stand 88:319
206. Hughes B, Bradburne A, Sheppard A, Young D (1996) Dev Biol Stand 88:91
207. Sakurai F, Mizuguchi H, Yamaguchi T, Hayakawa T (2003) Mol Ther 8:813
208. Havenga MJ, Lemckert AA, Ophorst OJ, van Meijer M, Germeraad WT, Grim bergen J, van Den Doel MA, Vogels R, van Deutekom J, Janson AA, de Bruijn JD, Uytdehaag F, Quax PH, Logtenberg T, Mehtali M, Bout A (2002) J Virol 76:4612
209. Roelvink PW, Lizonova A, Lee JG, Li Y, Bergelson JM, Finberg RW, Brough DE, Kovesdi I, Wickham TJ (1998) J Virol 72:7909
210. Mei YF, Lindman K, Wadell G (2002) Virology 295:30
211. Barouch DH, Pau MG, Custers JH, Koudstaal W, Kostense S, Havenga MJ, Truitt DM, Sumida SM, Kishko MG, Arthur JC, Korioth-Schmitz B, Newberg MH, Gorgone DA, Lifton MA, Panicali DL, Nabel GJ, Letvin NL, Goudsmit J (2004) J Immunol 172:6290
212. Green NM, Wrigley NG, Russell WC, Martin SR, McLachlan AD (1983) EMBO J 2:1357
213. Stouten PF, Sander C, Ruigrok RW, Cusack S (1992) J Mol Biol 226:1073
214. Shayakhmetov DM, Lieber A (2000) J Virol 74:10274
215. Minor PD (2000) Virology 268:231
216. Hoganson DK, Ma JC, Asato L, Ong M, Printz MA, Huyghe BG, Sosnowski BA, D'Andrea MJ (2002) Bioprocessing: Tech Dev Opp 1:43
217. Croyle MA, Roessler BJ, Davidson BL, Hilfinger JM, Amidon GL (1998) Pharm Dev Technol 3:373
218. Casto BC, Hammon WM (1966) P Soc Exp Biol Med 122:1216
219. Russell WC, Valentine RC, Pereira HG (1967) J Gen Virol 1:509
220. Yamamoto T (1967) Can J Microbiol 13:1139
221. Wigand R, Bachmann P, Brandner G (1981) Arch Virol 69:61
222. Wallis C, Melnick JL (1962) Virology 16:504
223. Melnick JL, Wallis C (1963) P Soc Exp Biol Med 112:894
224. Rombaut B, Verheyden B, Andries K, Boeye A (1994) J Virol 68:6454
225. Burke CJ, Hsu TA, Volkin DB (1999) Crit Rev Ther Drug 16:1

226. Sene C (1998) PCT Patent Application WO 98/02522
227. Shih S-J, Mcglennon KR, Moody D (2000) PCT Patent Application WO 00/09675
228. Evans RK, Nawrocki DK, Isopi LA, Williams DM, Casimiro DR, Chin S, Chen M, Zhu DM, Shiver JW, Volkin DB (2004) J Pharm Sci 93:2458
229. Grant RA, Hiremath CN, Filman DJ, Syed R, Andries K, Hogle JM (1994) Curr Biol 4:784
230. Andries K, Rombaut B, Dewindt B, Boeye A (1994) J Virol 68:3397
231. Dimmock NJ (1967) Virology 31:338
232. Green M, Wold WS (1979) Methods Enzymol 58:425
233. Blanche F, Shih S-J (2000) PCT Patent Application WO 00/61726
234. Blanche F, Cameron B, Somarriba S, Maton L, Barbot A, Guillemin T (2001) Anal Biochem 297:1
235. Croyle MA, Cheng X, Wilson JM (2001) Gene Ther 8:1281
236. Frei A, Kwan HKH, Sandweiss VE, Vellekamp GJ, Yuen P-H, Bondoc LL, Porter FWP, Tang JC-T, Ihnat P (2003) US Patent Application 0157066 A1
237. Kovesdi I, Ransom SC (2003) US Patent Application 0153065 A1
238. Wu Z, Zhang S (2004) US Patent 6,689,600 B1
239. Jumaa M, Muller BW (1999) Eur J Pharm Sci 9:207
240. Reich I, Schnaare R, Sugita ET (1995) In: Gennaro AR, Chase GD, Marderosian AD, Hanson GR, Hussar DA, Medwick T, Rippie EG, Schwartz JB, White HS, Zink GL (eds) Remmington: the science and practice of pharmacy. Mack Printing Co, Easton, PA, p 613
241. Li S, Schoneich C, Borchardt RT (1995) Biotechnol Bioeng 48:490
242. Middaugh CR, Evans RK, Montgomery DL, Casimiro DR (1998) J Pharm Sci 87:130
243. Evans RK, Xu Z, Bohannon KE, Wang B, Bruner MW, Volkin DB (2000) J Pharm Sci 89:76
244. Fujioka R, Kurtz H, Ackermann WW (1969) P Soc Exp Biol Med 132:825
245. Melnick JL (1994) In: Plotkin SA, Mortimer EA (eds) Vaccines. WB Saunders, Philadelphia, PA
246. Hannan C, Raptis LH, Dery CV, Weber J (1983) Intervirology 19:213
247. Setiawan K, Cameron FH (2003) PCT Patent Application WO 03/049763
248. Lehmberg E, Pungor E (2003) US Patent Application 0232018 A1

Author Index Volumes 51–99

Author Index Volumes 1–50 see Volume 50

Subject Index